In-Situ X-ray Tomographic Study of Materials

In-Situ X-ray Tomographic Study of Materials

Special Issue Editors

Eric Maire
Jérôme Adrien
Philip J. Withers

MDPI • Basel • Beijing • Wuhan • Barcelona • Belgrade • Manchester • Tokyo • Cluj • Tianjin

Special Issue Editors

Eric Maire
National Institute of Applied Sciences of Lyon
France

Jérôme Adrien
National Institute of Applied Sciences of Lyon
France

Philip J. Withers
University of Manchester
UK

Editorial Office
MDPI
St. Alban-Anlage 66
4052 Basel, Switzerland

This is a reprint of articles from the Special Issue published online in the open access journal *Materials* (ISSN 1996-1944) (available at: https://www.mdpi.com/journal/materials/special_issues/in_situ_x-ray_tomographic_study_of_materials).

For citation purposes, cite each article independently as indicated on the article page online and as indicated below:

LastName, A.A.; LastName, B.B.; LastName, C.C. Article Title. *Journal Name* **Year**, *Article Number*, Page Range.

ISBN 978-3-03936-529-6 (Hbk)
ISBN 978-3-03936-530-2 (PDF)

Cover image courtesy of Parmesh Gajjar.

© 2020 by the authors. Articles in this book are Open Access and distributed under the Creative Commons Attribution (CC BY) license, which allows users to download, copy and build upon published articles, as long as the author and publisher are properly credited, which ensures maximum dissemination and a wider impact of our publications.
The book as a whole is distributed by MDPI under the terms and conditions of the Creative Commons license CC BY-NC-ND.

Contents

About the Special Issue Editors vii

Preface to "In-Situ X-Ray Tomographic Study of Materials" ix

Eric Maire, Stanislas Grabon, Jérôme Adrien, Pablo Lorenzino, Yuki Asanuma,
Osamu Takakuwa, Hisao Matsunaga
Role of Hydrogen-Charging on Nucleation and Growth of Ductile Damage in Austenitic
Stainless Steels
Reprinted from: *Materials* **2019**, *12*, 1426, doi:10.3390/ma12091426 1

Ying Wang, Lars P. Mikkelsen, Grzegorz Pyka and Philip J. Withers
Time-Lapse Helical X-ray Computed Tomography (CT) Study of Tensile Fatigue Damage
Formation in Composites for Wind Turbine Blades
Reprinted from: *Materials* **2018**, *11*, 2340, doi:10.3390/ma11112340 17

Anton Du Plessis, Dean-Paul Kouprianoff, Ina Yadroitsava and Igor Yadroitsev
Mechanical Properties and In Situ Deformation Imaging of Microlattices Manufactured by Laser
Based Powder Bed Fusion
Reprinted from: *Materials* **2018**, *11*, 1663, doi:10.3390/ma11091663 29

Yasin Amani, Sylvain Dancette, Eric Maire, Jérôme Adrien and Joël Lachambre
Two-Scale Tomography Based Finite Element Modeling of Plasticity and Damage in
Aluminum Foams
Reprinted from: *Materials* **2018**, *11*, 1984, doi:10.3390/ma11101984 39

Zeng-Nian Yuan, Hua Chen, Jing-Ming Li, Bin Dai and Wei-Bin Zhang
In-Situ X-ray Tomography Observation of Structure Evolution in
1,3,5-Triamino-2,4,6-Trinitrobenzene Based Polymer Bonded Explosive (TATB-PBX)
under Thermo-Mechanical Loading
Reprinted from: *Materials* **2018**, *11*, 732, doi:10.3390/ma11050732 57

Tristan Lowe, Egemen Avcu, Etienne Bousser, William Sellers and Philip J. Withers
3D Imaging of Indentation Damage in Bone
Reprinted from: *Materials* **2018**, *11*, 2533, doi:10.3390/ma11122533 71

Lavan Kumar Eppanapelli, Fredrik Forsberg, Johan Casselgren and Henrik Lycksam
3D Analysis of Deformation and Porosity of Dry Natural Snow during Compaction
Reprinted from: *Materials* **2019**, *12*, 850, doi:10.3390/ma12060850 85

Jingyi Mo, Enyu Guo, D. Graham McCartney, David S. Eastwood, Julian Bent,
Gerard Van Dalen, Peter Schuetz, Peter Rockett and Peter D. Lee
Time-Resolved Tomographic Quantification of the Microstructural Evolution of Ice Cream
Reprinted from: *Materials* **2018**, *11*, 2031, doi:10.3390/ma11102031 101

Abhishek Shastry, Paolo E. Palacio-Mancheno, Karl Braeckman, Sander Vanheule,
Ivan Josipovic, Frederic Van Assche, Eric Robles, Veerle Cnudde, Luc Van Hoorebeke
and Matthieu N. Boone
In-Situ High Resolution Dynamic X-ray Microtomographic Imaging of Olive Oil Removal
in Kitchen Sponges by Squeezing and Rinsing
Reprinted from: *Materials* **2018**, *11*, 1482, doi:10.3390/ma11081482 115

Marta Peña Fernández, Enrico Dall'Ara, Alexander P. Kao, Andrew J. Bodey, Aikaterina Karali, Gordon W. Blunn, Asa H. Barber and Gianluca Tozzi
Preservation of Bone Tissue Integrity with Temperature Control for In Situ SR-MicroCT Experiments
Reprinted from: *Materials* **2018**, *11*, 2155, doi:10.3390/ma11112155 131

Clément Jailin, Stéphane Roux
Dynamic Tomographic Reconstruction of Deforming Volumes
Reprinted from: *Materials* **2018**, *11*, 1395, doi:10.3390/ma11081395 149

Henry Proudhon, Nicolas Guéninchault, Samuel Forest and Wolfgang Ludwig
Incipient Bulk Polycrystal Plasticity Observed by Synchrotron In-Situ Topotomography
Reprinted from: *Materials* **2018**, *11*, 2018, doi:10.3390/ma11102018 167

Shougo Furuta, Masakazu Kobayashi, Kentaro Uesugi, Akihisa Takeuchi, Tomoya Aoba and Hiromi Miura
Observation of Morphology Changes of Fine Eutectic Si Phase in Al-10%Si Cast Alloy during Heat Treatment by Synchrotron Radiation Nanotomography
Reprinted from: *Materials* **2018**, *11*, 1308, doi:10.3390/ma11081308 185

Johanna Maier, Thomas Behnisch, Vinzenz Geske, Matthias Ahlhelm, David Werner, Tassilo Moritz, Alexander Michaelis and Maik Gude
Investigation of the Foam Development Stages by Non-Destructive Testing Technology Using the Freeze Foaming Process
Reprinted from: *Materials* **2018**, *11*, 2478, doi:10.3390/ma11122478 199

Miguel A. Vicente, Jesús Mínguez and Dorys C. González
Variation of the Pore Morphology during the Early Age in Plain and Fiber-Reinforced High-Performance Concrete under Moisture-Saturated Curing
Reprinted from: *Materials* **2019**, *12*, 975, doi:10.3390/ma12060975 211

Chun Tan, Sohrab R. Daemi, Oluwadamilola O. Taiwo, Thomas M. M. Heenan, Daniel J. L. Brett and Paul R. Shearing
Evolution of Electrochemical Cell Designs for In-Situ and Operando 3D Characterization
Reprinted from: *Materials* **2018**, *11*, 2157, doi:10.3390/ma11112157 233

Markus Osenberg, Ingo Manke, André Hilger, Nikolay Kardjilov and John Banhart
An X-ray Tomographic Study of Rechargeable Zn/MnO_2 Batteries
Reprinted from: *Materials* **2018**, *11*, 1486, doi:10.3390/ma11091486 249

Lei Zhang and Shaogang Wang
Correlation of Materials Property and Performance with Internal Structures Evolvement Revealed by Laboratory X-ray Tomography
Reprinted from: *Materials* **2018**, *11*, 1795, doi:10.3390/ma11101795 263

About the Special Issue Editors

Eric Maire is a CNRS research director at the MATEIS laboratory, part of the University of Lyon. In 1995 he received a Ph.D. in Materials Sciences from the National Institute of Applied Sciences of Lyon. Since then, he has worked for CNRS in MatéIS. He focuses on different aspects of materials science, most of these involving in situ experiments in X-ray computed tomography. He pioneered the technique in 1997 doing in situ tensile tests at the ESRF. He is a co-author of over 250 research articles. He is now head of the MatéIS laboratory, gathering 160 researchers in materials science.

Jérôme Adrien is a CNRS research engineer at the MATEIS laboratory, part of the University of Lyon. In 2004, he received a Ph.D. in Materials Sciences from the National Institute of Applied Sciences of Lyon. For 15 years, he has been managing the laboratory tomographs and contributing to the preparation and execution of synchrotron experiments. He also developed a large number of in situ devices used on both laboratory tomographs and synchrotron devices. He contributed to a large number of publications in collaboration with MATEIS laboratory teams working on various types of materials and with many French and foreign universities.

Philip John Withers FRS FREng obtained his PhD in Metallurgy at Cambridge University and took up a lectureship there, before taking up a Chair in Manchester in 1998. He is the first Regius Professor of Materials and Chief Scientist of the Henry Royce Institute. His interests lie in applying advanced techniques to follow the behavior of engineering and natural materials in real time and in 3D, often as they operate under demanding conditions. In 2008, he set up the Henry Moseley X-ray Imaging Facility, which is now one of the most extensive suites of 3D X-ray Imaging facilities in the world.

Preface to "In-Situ X-ray Tomographic Study of Materials"

X-ray computed tomography (CT) is advancing apace both in terms of the spatial resolution that can be achieved and the rate at which the radiographs necessary to reconstruct a 3D image can be collected. These advances combined with the fact that CT is inherently non-destructive mean that X-ray CT is moving simply from the collection of 3D images to the acquisition of 3D movies. Whether it is to collect information of rapidly changing behaviors, such as fracture, where live streaming of many 1000's of radiographs per second are needed to follow the events in situ, exploiting the intensity of a synchrotron X-ray source, or the longer timescales associated with long term oxidation, where time-lapse ex situ observations made by laboratory CT sources, X-ray CT provides unique insights into the behavior of natural and man-made materials that simply cannot be obtained by any other means.

This book is a collection of chapters. It celebrates the possibilities for gaining inside information through time resolved X-ray CT looking both at the technical developments and opportunities for improving the quality and quantification of the image data we collect to the range of questions that X-ray CT can shine light upon. It is clear from this collection that many different environments and external constraints can be applied to the materials in situ whether to form the material or to observe its degradation or healing. We are sure that these chapters only represent the tip of the iceberg and that time-resolved X-ray CT, whether exploiting the intensity of a synchrotron or the accessibility or laboratory CT systems, will continue to develop into an indispensable characterization tool, alongside optical and electron microscopy. Further, we believe that the imaging modes, whether to detect subtle changes in phase, crystallographic structure or to fingerprint the elements contained in phases, will expand to provide an even richer picture of the internal structure of materials and their evolution over time than we can achieve today. We hope that this book shows the future or time resolved imaging is indeed very bright.

Eric Maire, Jérôme Adrien, Philip J. Withers
Special Issue Editors

Article

Role of Hydrogen-Charging on Nucleation and Growth of Ductile Damage in Austenitic Stainless Steels

Eric Maire [1,*], Stanislas Grabon [1], Jérôme Adrien [1], Pablo Lorenzino [1], Yuki Asanuma [2], Osamu Takakuwa [3,4,5] and Hisao Matsunaga [3,4,6]

1. Univ. Lyon, INSA Lyon, CNRS UMR5510, Laboratoire MATEIS, F-69621 Villeurbanne CEDEX, France; stanisla.grabon@insa-lyon.fr (S.G.); jerome.adrien@insa-lyon.fr (J.A.); pablo.lorenzino@insa-lyon.fr (P.L.)
2. Japan Casting & Forging Corporation, 46-59, Sakinohama, Nakabaru, Tobata-ku, Kitakyushu 804-8555, Japan; asanuma.yuki.964@m.kyushu-u.ac.jp
3. Department of Mechanical Engineering, Kyushu University, 744 Motooka, Nishi-ku, Fukuoka 819-0395, Japan; takakuwa.osamu.995@m.kyushu-u.ac.jp (O.T.); matsunaga.hisao.964@m.kyushu-u.ac.jp (H.M.)
4. Research Center for Hydrogen Industrial Use and Storage (HYDROGENIUS), Kyushu University, 744 Motooka, Nishi-ku, Fukuoka 819-0395, Japan
5. AIST-Kyushu University Hydrogen Materials Laboratory (HydroMate), 744 Motooka, Nishi-ku, Fukuoka 819-0395, Japan
6. International Institute for Carbon-Neutral Energy Research (I2CNER), Kyushu University, 744 Motooka, Nishi-ku, Fukuoka 819-0395, Japan
* Correspondence: eric.maire@insa-lyon.fr; Tel.: +33-472-43-88-61

Received: 6 February 2019; Accepted: 19 April 2019; Published: 1 May 2019

Abstract: Hydrogen energy is a possible solution for storage in the future. The resistance of packaging materials such as stainless steels has to be guaranteed for a possible use of these materials as containers for highly pressurized hydrogen. The effect of hydrogen charging on the nucleation and growth of microdamage in two different austenitic stainless steels AISI316 and AISI316L was studied using in situ tensile tests in synchrotron X-ray tomography. Information about damage nucleation, void growth and void shape were obtained. AISI316 was found to be more sensitive to hydrogen compared to AISI316L in terms of ductility loss. It was measured that void nucleation and growth are not affected by hydrogen charging. The effect of hydrogen was however found to change the morphology of nucleated voids from spherical cavities to micro-cracks being oriented perpendicular to the tensile axis.

Keywords: X-ray tomography (X-ray CT); 3D image analysis; damage; hydrogen embrittlement; stainless steel

1. Introduction

Hydrogen energy is strongly expected as a secondary energy, which can be produced from various renewable energy sources and does not result in carbon dioxide (CO_2) emissions when used as energy fuel in a fuel cell. Thus, hydrogen has a good potential for playing a role in the future development of our ever-growing society. Fuel cell vehicles (FCVs) have recently been commercialized in some countries, and constructions of hydrogen refueling stations have also been promoted. In such systems, various components (e.g., vessels, valves, regulators and metering devices) are exposed to high-pressure hydrogen gas environment. For a safe use of such components, it is necessary to properly understand the degradation of strength properties caused by the interaction of hydrogen with the microstructure, since hydrogen can easily penetrate into the material and causes "hydrogen embrittlement", e.g., ductility loss in tensile test [1,2] and acceleration of fatigue crack growth [3–8] in a number of metallic materials. In addition, the degradation mechanism should also be clarified to review existing standards and regulations reasonably based on scientific grounds.

Austenitic stainless steels are successfully used for components installed in high-pressure hydrogen refueling stations and embarked in FCVs. When the stability of austenite phase is relatively low, plasticity-induced phase transformation from austenite to martensite under plastic deformation, i.e., α' martensitic transformation triggers hydrogen-induced degradation of mechanical properties. The transformed martensite phase has the potential to be a crack initiation site [9] and can be a dominant factor of enhancement of crack propagation [10]. Even in austenitic stainless steels with high austenite stability, i.e., AISI316L, the presence of hydrogen also affects void nucleation and its coalescence behavior under intense plastic deformation [1,2]. Thus, fracture behavior of austenitic stainless steels with the presence of hydrogen depends on the phase stability. Although there exist several observations or analyses on fracture surface or a cross section of hydrogen-charged specimens after rupture, in situ analysis should be more effective so as to understand process and mechanism of the crack or void nucleation and/or coalescence behavior at various strain levels.

Analyzing damage experimentally has recently gained new interest thanks to the availability of 3D imaging techniques applicable for the observation of materials. X-Ray Computed Tomography (X-ray CT) is the most versatile of these new techniques [11,12], even at the nanoscale [13]. The results of X-ray CT can be analyzed quantitatively [14] and yield crucial information about damage evolution in ductile materials [15]. X-ray CT experiments have shown the crucial importance of hydrogen pre-existing pores in standard aluminum alloys on ductile fracture [16]. The technique has also been used to study damage process in a non charged standard AISI316L stainless steel in [17]. It has, however, never been used for steels charged with hydrogen. The goal of the present paper is then to use X-ray CT, able to quantify nucleation and growth of cavities in ductile materials, to assess the effect of hydrogen charging on these two mechanisms. For this, in situ tensile tests in X-ray CT were carried out on different steels with and without hydrogen charging. Both qualitative and quantitative results will be presented in the paper.

2. Materials And Methods

2.1. Materials

In this study, two types of austenitic stainless steels were investigated: AISI316 and AISI316L. The choice of these two metals containing different amounts of carbon was motivated by one main consideration. It is well known that carbon strongly influence the stability of austenite [1,2]. The motivation of this study was to perform in–situ observation of the fracture process of hydrogen–charged austenitic stainless steels with different phase stabilities. The AISI316L was provided by NSSC (Nippon Steel & Sumikin Stainless Steel Corporation, Tokyo, Japan), in the form of a plate 50 × 2500 × 6100 (mm) in dimensions. It was solution-treated at 1120 °C for 4 min and then water–quenched. The AISI316 was provided by Yakin as a plate again (30 × 2000 × 4000 mm in dimensions), solution-treated at 1120 °C for 15 min and then also water–quenched. The composition of these two materials given by the provider is shown in Table 1 and their tensile properties (before hydrogen charging, provided by the manufacturer) are summarized in Table 2. The tensile properties after hydrogen charging were not measured with macroscopic tensile tests but will be analyzed later thanks to the in situ tensile tests. The AISI316L contains 0.04% of carbon, whereas the AISI316 has a carbon concentration about 0.18%. The samples were charged with hydrogen by being exposed to 100 MPa hydrogen gas at 270 °C for 200 h. We know from previous experiments [18] that this results in a hydrogen content of 99.7 mass ppm with a uniform distribution over the cross section of the specimen. Hydrogen exists in both Face-Centered Cubic lattice sites and trap sites. Hydrogen outgassing can be negligible since hydrogen diffusivity in the austenite phase at room temperature is extremely low (10–16 m^2/s).

The samples were subsequently stored in a freezer for a few months before the in situ tensile tests could be carried out in the synchrotron. This was necessary because synchrotron access is difficult to schedule precisely in advance. Electron backscatter diffraction (EBSD) maps of the two samples were acquired for grain size characterization. *Post mortem* Scanning Electron Microscope (SEM) observation of the fracture surface of the broken samples was also carried out.

Table 1. Chemical composition (mass %) of the two materials.

	C	Si	Mn	P	S	Cr	Mo	Ni
AISI316	0.04	0.64	0.93	0.032	0.001	16.83	2.05	10.23
AISI316L	0.018	0.50	0.84	0.021	0	17.45	2.05	12.09

Table 2. Tensile properties of the two materials, as provided on the certification sheet by the manufacturer.

Materials	0.2 % Proof Stress $\sigma_{0.2}$ (MPa)	Tensile Strength σ_B (MPa)	Elongation (%) ε_t (%)
AISI316	263	586	61.0
AISI316L	229	528	66.0

In total, our experimental data base was then composed of four different types of samples: AISI316 and AISI316L hydrogen-charged and non-charged.

2.2. Methods

Before charging, smooth samples, with a useful part of 1 mm in diameter and 5 mm in length, were machined for each of the different materials. Each of the two heads of the samples were threaded (M3) which allowed for screwing additional T shaped tabs that were used to connect the sample's head to the tensile grips. The specimen surface was polished with emery paper and then with a diamond paste to obtain a mirror–like surface finish. After the charging, samples were kept at −85 °C. However, it is noted that, even if the samples were kept at ambient temperatures, the outgassing effect could be negligible since hydrogen diffusivity of these austenitic stainless steels are extremely low.

The in situ tensile experiments were conducted at the European Synchrotron Radiation Facility (ESRF), using the tomography setup available at the ID19 beamline [19]. The tensile rig used earlier introduced in [20] was especially designed for X ray tomography in situ experiments. The cross head speed was set to 1 µm/s. Each sample was screwed between two grips. The rig was placed on the rotation stage, between the X-ray source and the detector (the sample to detector distance was about 15 cm). The sample was rotated around the rotating stage axis while a high number (2000) of 2D X-ray absorption radiographs were recorded by the detector, the pixel size of which was set to 0.6 µm. A PCO DIMAX edge® camera (Kelheim, Germany) was used to digitize the attenuation images. Each 2D X-ray radiograph required an exposure time of 0.01 s per frame. Because of the high attenuation of iron, the energy of the monochromatic beam was set to 50 keV. Using a Filtered Backprojection algorithm implemented at the ESRF (PyHST [21]), this series of radiographs were combined to reconstruct a 3D digital image where each voxel (volume element or 3D pixel) represents the X-ray absorption at that point. Because X-ray tomography is a non-destructive technique, many scans of the same sample could be acquired allowing us to observe damage evolution at various values of the applied strain in the different samples. During the scan acquisition, the displacement of the tensile machine was stopped to prevent blurring. This means that we operated in the so-called *interrupted* in situ mode. The force sensor of the tensile rig provided a measurement of F, the force applied to the sample at each time. The total true strain at each step was calculated from the reconstructed images, as explained later.

The raw volumes obtained from the reconstruction of the tomography scans subsequently needed to be processed in order to be used for the characterization of ductile damage. All the image processing and analysis steps were performed using ImageJ (Freeware 1.48q, Rayne Rasband, National Institutes of Health, USA) [22], a specific freeware available to perform image processing of 3D volumes. The images were first processed by removing the ring artifacts. The second processing consisted in median filtering the volumes (isotropic size of the filter = 2 voxels). This decreased the noise induced by the experimental method. The median-filtered volumes were thereafter binarized by simple thresholding to separate the material phase from the void phase. Each pore of the volume was then detected and labeled using a dedicated image processing procedure. During the last step of the process (labeling), the cavities having a volume smaller than 10 voxels (=2.16 µm^3), likely to be confused with noise, were rejected from the analysis.

The central area of each tensile specimen, where damage is mainly concentrated, was cropped from the initial image for damage quantification. The cropped volume has been chosen in the undeformed state to be a cuboid volume of (300)^3 voxels i.e., (180 µm)^3. It can be assumed (and has been verified for instance in [23]) that this central sub-region undergoes the highest stress triaxiality state and the highest strain during the tensile test. The size was chosen to be sufficiently large for the elementary volume to be representative but also sufficiently small for the strain and triaxiality to be spatially constant inside this sub-volume. During the tensile test, the selected initial volume plastically deforms. The shape of the cuboid volume has been chosen in the present study to change and become more elongated. The calculation of the amount of change to apply to the cube was based on the macroscopic plastic deformation of the sample.

3. Results

3.1. EBSD

Figure 1 shows EBSD maps of the two studied alloys shown here to highlight the grain structure of the material. These were obtained by using a Schottky type FE-SEM (JEOL JSM-7001F) at an acceleration voltage of 15 kV. The grains have a similar size (around 100 µm) and a similar amount of twins can be observed in both samples so the microstructure complies with our expectations for these very well known stainless steels. This value of the grain size is rather large compared to the sample diameter, but there are at least ten grains along the diameter so a total number of about 100 grains in a given section of the sample. This is a sufficient number of grains to insure that the mechanical behavior is not strongly influenced by plasticity gradient effects. The assessment of the texture would be interesting but is out of the scope of this paper.

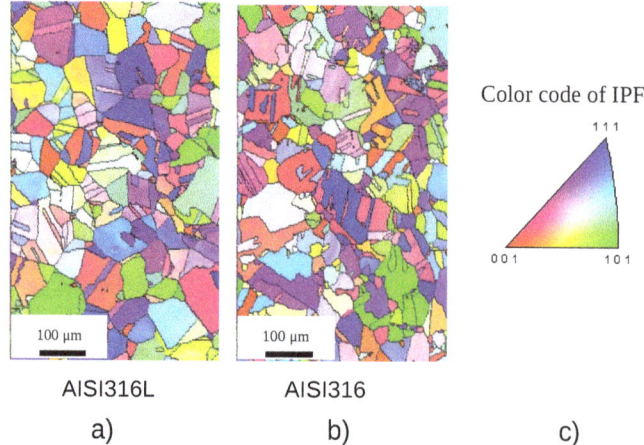

Figure 1. EBSD maps of the two different materials (**a**) non-charged AISI316L steel; (**b**) non-charged AISI316 steel; (**c**) inverse pole figure (IPF) coloring.

3.2. Hardening Curves from In Situ X-Ray Computed Tomography

The sample was mounted vertically in the rig, more or less aligned with the tensile axis, which in turn is more or less parallel to the rotation axis of the rotation stage. This axis is denoted "z" in the following. As already performed in [15,24], from the outer shape of the sample measured using X-ray tomography after segmentation at each deformation step, we could measure the section S of the sample (S being perpendicular to z) as a function of the position of this section along z. From this list of sections $S(z)$, we could determine the coordinates of the location (center of mass) of the minimal section of the sample S_{min}. Because we recorded F at all times, it was then possible to precisely calculate the true stress σ inside this minimal section using the expression:

$$\sigma = \frac{F}{S_{min}}. \tag{1}$$

Assuming no volume dilation of the sample due to damage, it was also possible to precisely calculate the true longitudinal strain, ε, in S_{min} using the following standard expression:

$$\varepsilon = \ln\left(\frac{S^0_{min}}{S_{min}}\right), \tag{2}$$

where S^0_{min} is the value of S_{min} in the initial tensile state. Figure 2 shows the true stress—true strain curves (the hardening curves) recorded during our experiments for all the samples tested. It should in principle start at zero true plastic strain, but, in our case, true strain and true stress were measured at each stop during the interrupted test, with a first step at ε close to 0.2. We have no precise measurement of the yield stress of these samples during the in situ tensile test.

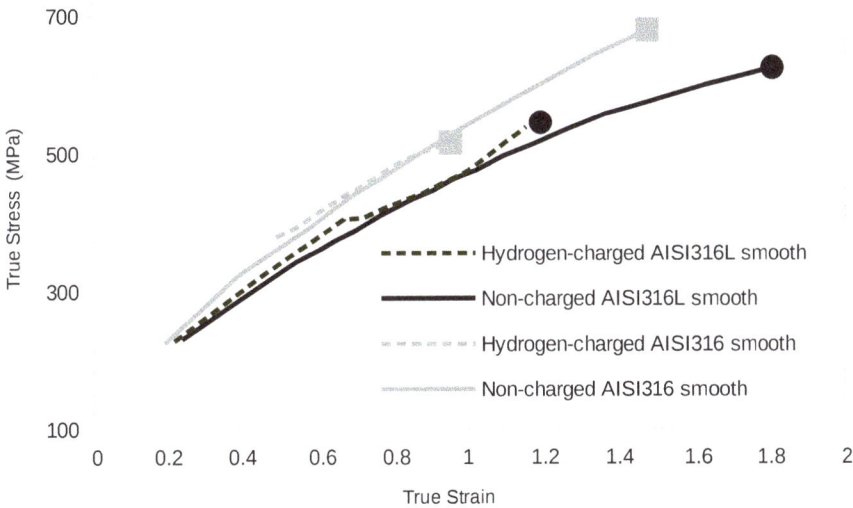

Figure 2. Tensile curve of the different samples (true stress vs. true strain).

The hardening curves are close together with AISI316 hardening a bit more, probably due to its higher C content. We have gathered the ductility values measured from these in situ tensile tests referred to as ε^f in Table 3. We also calculated in this table the decrease in ductility induced by hydrogen charging DDH, calculated as:

$$DDH = \frac{(\varepsilon^f_{Non-charged} - \varepsilon^f_{Hydrogen-charged})}{(\varepsilon^f_{Non-charged})} \times 100. \qquad (3)$$

Table 3. Ductility measured during the in situ tensile tests and calculation of the Decrease in Ductility due to Hydrogen charging (DDH) for the two materials and the two specimen shapes.

Materials	Specimen Shape	Ductility		DDH: Decrease Due to Hydrogen Charging (%)
		Non-Charged	Hydrogen-Charged	
AISI316	Smooth	1.91	0.885	53.7
	Notched	1.66	0.62	62.7
AISI316L	Smooth	1.92	1.35	29.4
	Notched	1.9	1.08	42.9

DDH is clearly higher for the AISI316 than for the AISI316L.

3.3. Qualitative Damage Evolution

Figure 3 shows (as a selected representative example) a volume rendering of the evolution of the cavities in the non-charged AISI316L. It shows similar features compared to what was already observed in [17] on the same type of material. Voids, nucleate grow and coalesce during the severe plastic deformation of the sample, especially in the central region of the notch. This typical evolution is also observed in all the different tested samples and will be quantified further in a subsequent section.

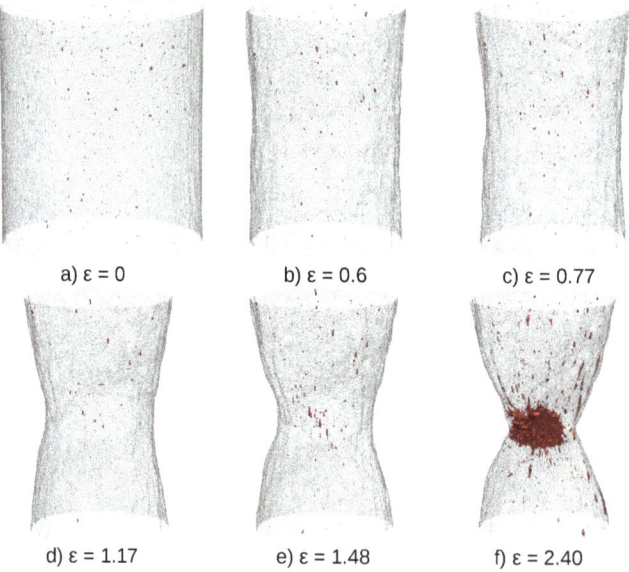

Figure 3. Damage evolution in the non-charged AISI316L sample observed as a volume rendering. The outer surface of the sample is transparent grey and the cavities are the dark red dots. Damage nucleates then grows and finally coalesce in the bottom right image (**a**) initial state; (**b**) true strain ε = 0.6; (**c**) ε = 0.77; (**d**) ε = 1.17; (**e**) ε = 1.48; (**f**) ε = 2.40.

By analyzing carefully volume series like the one shown in Figure 3 for every type of sample, qualitative differences were observed depending on the nature of the material and hydrogen charging. To highlight these differences, Figure 4 compares reconstructed slices of four typical samples in the last step before fracture. It is very clear from these images that the non-charged samples (left column) exhibit a much higher ductility, the section reduction and necking being much higher. Voids have then nucleated, grown and coalesced (see the big coalescence event observed for the non-charged AISI316 sample). This behavior is very typical of ductile metals. The right column shows the effect of hydrogen-charging on the final deformation stage. As already highlighted by Table 3, ductility was clearly reduced (necking is much less pronounced). Micro-cracks perpendicular to the tensile axis are observed in these reconstructions (as can be seen in Figure 5a) in the AISI316 sample but to a lesser extent in the AISI316L.

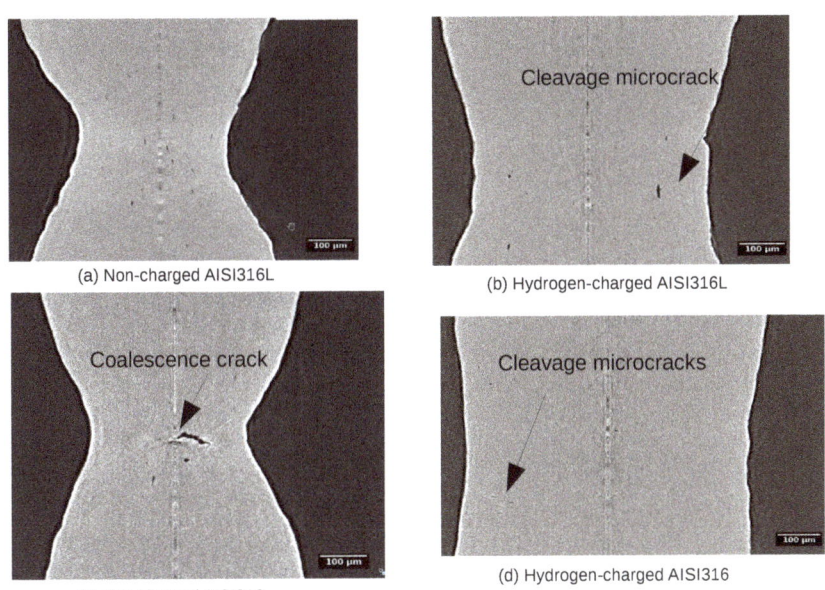

Figure 4. Reconstruction slices extracted parallel to the tensile axis in a central plane for the four samples in the ultimate state before fracture. AISI316 is clearly less deformed at fracture and contains local cleavage microcracks when hydrogen charged. The deformation in the different images are (**a**) $\epsilon = 2.40$ for non-charged AISI316L; (**b**) $\epsilon = 1.36$ for hydrogen-charged AISI316L; (**c**) $\epsilon = 1.92$ for non-charged AISI316; (**d**) $\epsilon = 0.95$ for hydrogen-charged AISI316.

Figure 5 compares 3D renderings of the hydrogen-charged AISI316 and AISI316L in the final stage just before coalescence. The figure clearly show that the AISI316 exhibits many more microcracks than the AISI316L sample in which the voids are elongated along the tensile axis. In the AISI316, cracking from surface can be observed.

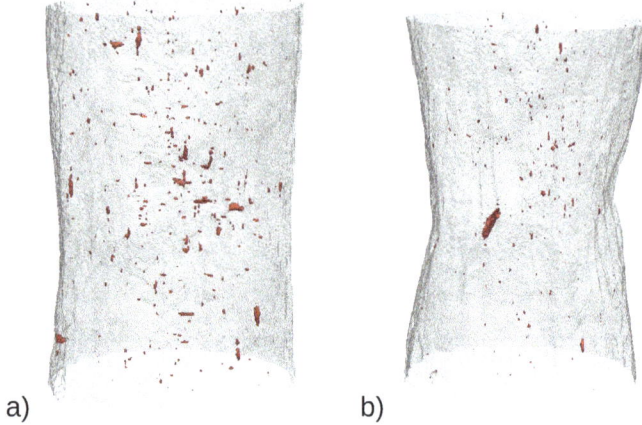

Figure 5. Volume rendering (a) hydrogen-charged AISI316 sample just before coalescence. The amount of penny shaped microcracks is very important. The deformation in the image is $\epsilon = 0.79$; (b) hydrogen-charged AISI316L sample just before coalescence. The morphology of the cavity is elongated along the tensile direction. The deformation in the image is $\epsilon = 1.15$.

3.4. Quantitative Damage Evolution

Void Nucleation

Void nucleation was firstly quantified by calculating the number of cavities n_c in every cropped volume. The mean cavities density N per cubic mm was calculated by dividing n_c by the value of the analysed sub–volume. Figure 6 shows the evolution of the void density N as a function of the true strain for AISI316L and AISI316. Two samples were tested for the non-charged AISI316L; they are both plotted on the figure and they show a similar behavior. Nucleation being quite exponential with strain in steels (already observed in [15]), we have plotted the results in a logarithmic scale. Note first that the number of nucleated cavities is rather small in these samples. This is because of the homogeneous nature of these materials which present very few inclusions where damage can nucleate. It can be seen that, in the non-charged state, N increases very slowly in the AISI316 steel. We have noticed that, surprisingly, for this particular AISI316 sample, nucleation was anomalously small inside the neck and was mainly located at the periphery of the sample, where strain and triaxiality are not at their maximum. This is probably because when nucleation is very scarce, as is the case in these two materials, the random nature of the location of the nucleation site can lead to such surprising observations. We have then decided to reject this sample from the rest of the quantification. From the measurement of the three other materials, where nucleation was more substantial, it appears that hydrogen charging has only a weak effect on the nucleation kinetics, as can be measured by X-ray CT. The dotted curve included in the figure also shows the results that we previously obtained in [17] for a standard non-charged AISI316L sample (of different origin than the one used in this study). For these previous measurements, we were using synchrotron X-ray CT with a pixel size of 1.6 µm so detection capacity was smaller. This explains that the observed number of cavities was smaller in previous study. Despite these differences, we believe that it is comforting to see that the slope of the different curves are similar. The fact that N decreases at high strain for the hydrogen-charged AISI316 can be attributed to coalescence.

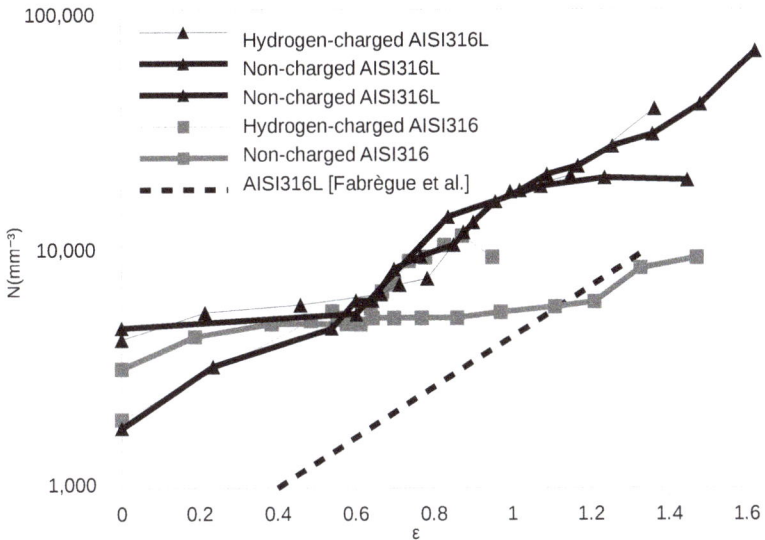

Figure 6. Nucleation quantified as the evolution of the density of cavities with strain. The nucleation in the non-charged AISI316 sample is anomalously small because of a small amount of nucleation in the necked region for this sample. From the behavior of the three other materials, it can be concluded that hydrogen charging has a weak effect on the nucleation kinetics. The measurements by Fabrègue et al. [17] are also shown as a dotted line. These were made using a larger voxel size (1.6 µm pixel size) which explains the lower level of nucleation detected, but the slope of the curve is in line with the measurements of the present paper.

3.5. Void Growth

In the 3D images, each void is composed of a certain amount of voxels and this allowed us to measure the volume of each cavity V_i. From this value, we have then calculated the equivalent diameter of a sphere exhibiting the same volume:

$$D_{eq,i} = (6V_i/p)^{1/3}. \tag{4}$$

It has been shown in previous studies that growth could be easily estimated by quantifying the average value of the largest cavities assumed to remain the same from one strain step to the next [23]. Here, we have chosen to work with the 20 largest cavities. Figure 7 shows the evolution of the average equivalent diameter of the 20 largest cavities in the cropped volume as a function of the true strain for the AISI316L non-charged, and for the two charged materials. For the sake of clarity, we have rationalized the values of D_{eq} by dividing it by its value at 0 strain D_0. Previous measurement [17] is also shown as a dotted curve. In terms of growth, we clearly show again here that hydrogen charging has a weak effect of the growth rate of the cavities.

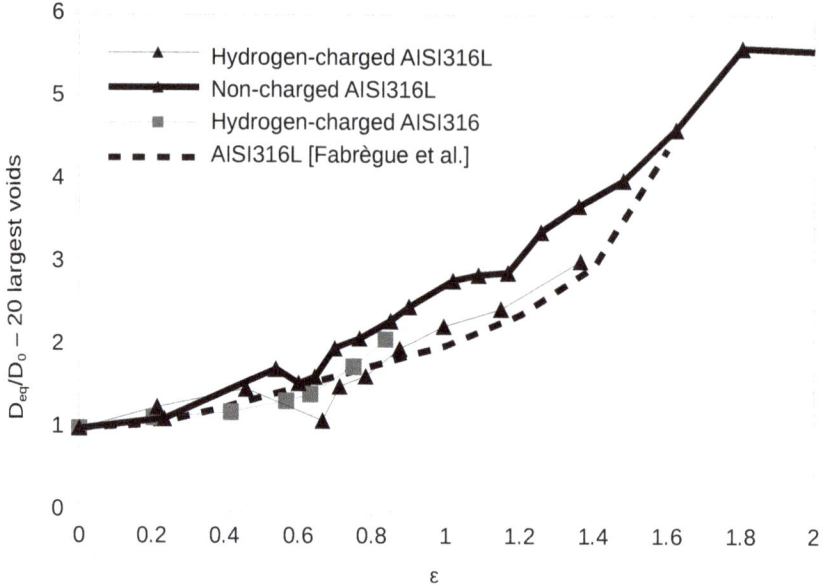

Figure 7. Growth of the average equivalent diameter of the 20 largest cavities with strain.

3.6. Aspect Ratio of the Cavities

The aspect ratio of the cavities has been calculated for the two smooth hydrogen charged samples (AISI316 and AISI316L) just before fracture (same as those shown in Figure 4). For calculating the aspect ratio of the cavities, we used the following simplified formula:

$$Aspect\ ratio = \frac{L_x + L_y}{2L_z}, \tag{5}$$

where L_x, L_y and L_z are the largest dimensions of the cavity along the different directions of the reconstructed volume (remember that z is the tensile axis).

The aspect ratio is smaller than one for cracks and higher when the cavities are elongated along the tensile direction. Figure 8 compares the histogram of the values of the aspect ratio for the 20 largest cavities, for the two charged materials. This comparison is performed for a similar deformation step close to fracture. This histogram confirms the visual impression in Figure 5 i.e., the cavities have a crack–like shape in the case of the AISI316 sample and much less for the AISI316L.

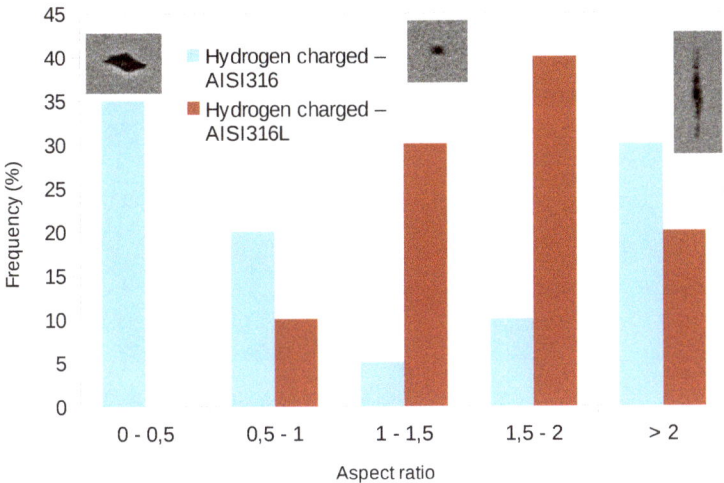

Figure 8. Histogram of the distribution of the aspect ratio of the 20 largest cavities in the two hydrogen charged materials in the state just before fracture.

4. Discussion

It is presumed that the variation in the aspect ratio of voids in hydrogen charged specimens is dominated by void growth in AISI316L and is dominated by the mixture of micro crack propagation and/or void growth in AISI316. As mentioned above, the aspect ratio of 0–0.5 corresponds to micro cracking and that of 0.5–1.0 corresponds to a mixture of micro cracking and/or void growth. For the samples charged with hydrogen, in AISI316, micro cracking dominated the fracture process. By contrast, in AISI316L, the fracture process was dominated by void growth. As a verification of these fracture characteristics, Figure 9 shows the fracture surface of hydrogen charged and non charged specimens as observed using SEM. In non charged specimens, the ordinary void nucleation, growth and subsequent coalescence are the main processes of the fracture in both AISI316 and 316L. As a result, cup and cone fracture occurred. In hydrogen charged AISI316, fracture surface was predominately covered by so called "quasi cleavage" (QC) with some small and elongated dimples. It is noted that QC corresponds to the cavities with the aspect ratio of 0–0.5 in Figure 8. By contrast, in hydrogen charged AISI316L, fracture surface is covered by smaller dimples compared to the non charged case, and QC is rarely observed.

The stability of austenite phase influences the susceptibility of the material to hydrogen-induced cracking. It is well known that nickel is a stabilizer of austenite phase [25], i.e., AISI316L with Ni content of 12.09% has higher stability than AISI316 with that of 10.23%. In AISI316 with lower stability, the austenite phase can easily transform to martensite phase. This transformation occurs above a certain intensity of plastic strain. It is possible that the phase transformation under plastic deformation facilitated the QC in AISI316. It was reported that acceleration of crack growth in austenitic stainless steel corresponds to regions where α' martensite phase is present ahead of the crack tip under load [10]. In this region, hydrogen diffusivity becomes extremely higher compared to that of austenite phase (~ 10–16 m^2/s) [18]. The crack propagates into the transformed martensite phase or interface between austenite phase and α' martensite phase. Koyama et al. investigated crystallographical characteristics of crack propagation in the α' martensite phase and revealed that the crack preferentially propagates along the {100} [9]. Figure 10 shows a proposition for a schematic illustration of the fracture mechanisms. In hydrogen charged AISI316, firstly, the QC is likely to be generated at specimen surface, then to successively propagate concurrently with the small voids nucleation at the central part of the sample. Subsequently, final fracture probably occurs accompanied with elongated voids. It should be noted that, in austenitic stainless steels, hydrogen out-gassing can be assumed to be negligible because hydrogen diffusion in austenitic stainless steel is extremely low at room temperature [18].

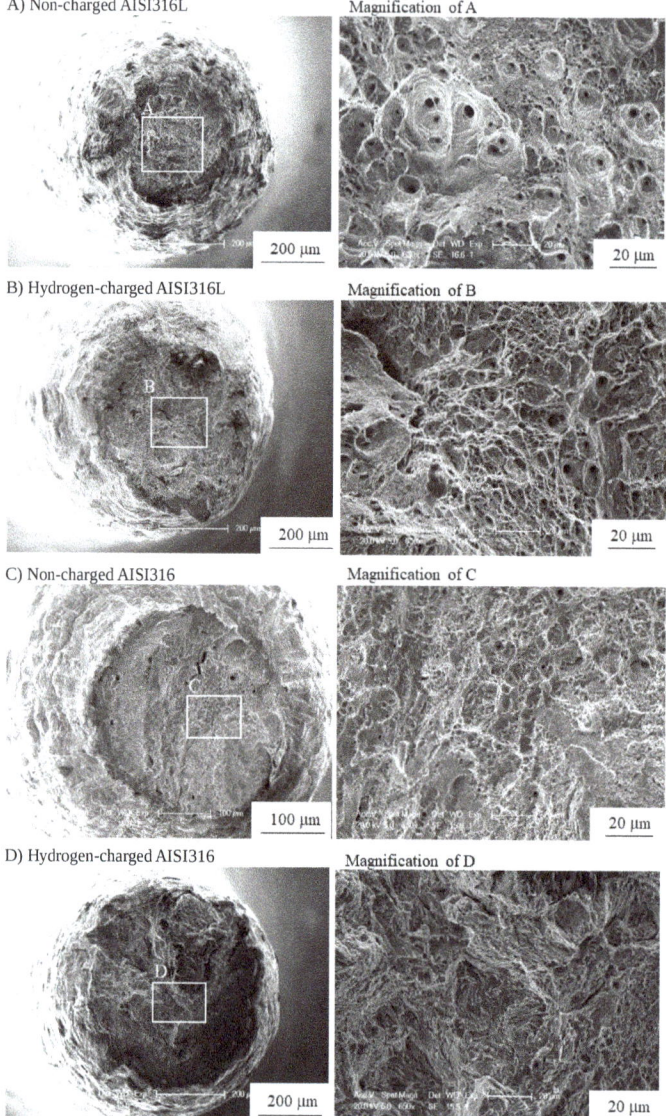

Figure 9. Fractography of non charged and hydrogen-charged samples.

Some of the authors of this paper proposed that a combination of slip localization due to the presence of hydrogen and the phase transformation in the vicinity of the crack tip causes a successive crack propagation [1]. On the other hand, in hydrogen-charged AISI316L, the QC was not generated since the material has a high austenite stability. Therefore, voids nucleated in the central part where stress triaxiality was high, i.e., necked region, with a similar mechanism to non-charged specimen. Hydrogen made the voids easier to coalesce by local shear stress owing to the slip localization, which resulted in void sheet formation [26]. The details in the tensile fracture mechanism of austenitic stainless steels charged with hydrogen have been comprehensively discussed in the literature [2]. The set of results presented above clearly visualize and verify the mechanism of the hydrogen-induced degradation.

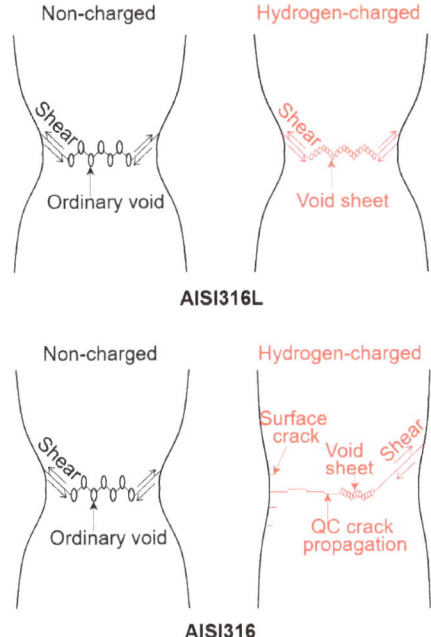

Figure 10. Schematic illustration of fracture mechanism in non-charged and hydrogen-charged AISI316 and 316L.

5. Conclusions

In this paper, we have studied at the microscopic level the effect of charging AISI316 and AISI316L steels with hydrogen. We have used in situ X-ray computed tomography tests to analyze the fracture process of charged and uncharged samples. Our main findings are as follows:

- The ductility is reduced by Hydrogen charging, in a more important way for the AISI316 sample.
- In this material, cavities quickly transform into cracks perpendicular to the tensile axis leading to early fracture.
- By quantifying damage, we have also shown that both nucleation and growth, are not strongly affected by hydrogen charging. This means that the microscopic evolution of damage is not accelerated by the presence of hydrogen.
- The only noticeable difference, and the explanation for the reduction in ductility, is the aspect ratio of the cavities showing again a crack shape in the hydrogen-charged AISI316 leading to earlier macroscopic fracture.

Given these conclusions and regarding potential applications, 316L is better for use in vehicle tanks.

Author Contributions: E.M., J.A. designed and performed experiments, S.G. performed image analysis, P.L., Y.A., O.T., H.M. supervised the study and contributed to drafting of the final version of the manuscript. E.M. wrote the first draft of the manuscript. All the authors discussed the drafts and approved the final manuscript for publication.

Acknowledgments: Elodie Boller is acknowledged for her help as local contact during the ESRF experiment at the ID19 beamline.

Conflicts of Interest: The authors declare no conflict of interest.

References

1. Matsuoka, S.; Yamabe, J.; Matsunaga, H. Criteria for determining hydrogen compatibility and the mechanisms for hydrogen-assisted, surface crack growth in austenitic stainless steels. *Eng. Fract. Mech.* **2016**, *153*, 103–127. [CrossRef]
2. Takakuwa, O.; Yamabe, J.; Matsunaga, H.; Furuya, Y.; Matsuoka, S. Comprehensive understanding of ductility loss mechanisms in various steels with external and internal hydrogen. *Metall. Mater. Trans.* **2017**, *48*, 5717–5732. [CrossRef]
3. Tsay, L.W.; Chen, J.J.; Huang, J.C. Hydrogen-assisted fatigue crack growth of AISI316L stainless steel weld. *Corros. Sci.* **2008**, *50*, 2973–2980. [CrossRef]
4. Ronevich, J.; Somerday, B.P.; San Marchi, C.W. Effects of microstructure banding on hydrogen assisted fatigue crack growth in X65 pipeline steels. *Int. J. Fatigue* **2015**, *82*, 497–504. [CrossRef]
5. Matsunaga, H.; Takakuwa, O.; Yamabe, J.; Matsuoka, S. Hydrogen-enhanced fatigue crack growth in steels and its frequency dependence. *Philos. Trans. R. Soc.* **2017**, *375*. [CrossRef]
6. Matsuoka, S.; Takakuwa, O.; Okazaki, S.; Yoshikawa, M.; Yamabe, J.; Matsunaga, H. Peculiar temperature dependence of hydrogen-enhanced fatigue crack growth of low-carbon steel in gaseous hydrogen. *Scr. Mater.* **2018**, *154*, 101–105. [CrossRef]
7. Wang, S.; Nagao, A.; Sofronis, P.; Robertson, I.M. Hydrogen-modified dislocation structures in a cyclically deformed ferritic-pearlitic low carbon steel. *Acta Mater.* **2018**, *144*, 164–176. [CrossRef]
8. Birenis, D.; Ogawa, Y.; Matsunaga, H.; Takakuwa, O.; Yamabe, J.; Prytz, Ø.; Thøgersen, A. Interpretation of hydrogen-assisted fatigue crack propagation in BCC iron based on dislocation structure evolution around the crack wake. *Acta Mater.* **2018**, *156*, 245–253. [CrossRef]
9. Koyama, M.; Ogawa, T.; Yan, D.; Matsumotoc, Y.; Tasan, C.C.; Takai, K.; Tsuzaki, K. Hydrogen desorption and cracking associated with martensitic transformation in Fe-Cr-Ni-Based austenitic steels with different carbon contents. *Int. J. Hydrogen Energy* **2017**, *42*, 26423–26435. [CrossRef]
10. Ogawa, Y.; Okazaki, S.; Takakuwa, O.; Matsunaga, H. The roles of internal and external hydrogen in the deformation and fracture processes at the fatigue crack tip zone of metastable austenitic stainless steels. *Scr. Mater.* **2018**, *157*, 95–99. [CrossRef]
11. Stock, S. X-ray microtomography of materials. *Int. Mater. Rev.* **1999**, *44*, 141–164. [CrossRef]
12. Stock, S.R. *Microcomputed Tomography: Methodology and Applications*; CRC Press: Boca Raton, FL, USA, 2008.
13. Withers, P.J. X-ray nanotomography. *Mater. Today* **2007**, *10*, 26–34. [CrossRef]
14. Maire, E.; Withers, P.J. Quantitative X-ray tomography. *Int. Mater. Rev.* **2014**, *59*, 1–43. [CrossRef]
15. Landron, C.; Bouaziz, O.; Maire, E.; Adrien, J. Characterization and modeling of void nucleation by interface decohesion in dual phase steels. *Scr. Mater.* **2010**, *63*, 973–976. [CrossRef]
16. Toda, H.; Oogo, H.; Uesugi, K.; Kobayashi, M. Roles of pre-existing hydrogen micropores on ductile fracture. *Mater. Trans.* **2009**, *50*, 2285–2290. [CrossRef]
17. Fabrègue, D.; Landron, C.; Bouaziz, O.; Maire, E. Damage evolution in TWIP and standard austenitic steel by means of 3D X ray tomography. *Mater. Sci. Eng.* **2013**, *579*, 92–98. [CrossRef]
18. Yamabe, J.; Takakuwa, O.; Matsunaga, H.; Itoga, H.; Matsuoka, S. Hydrogen diffusivity and tensile-ductility loss of solution-treated austenitic stainless steels with external and internal hydrogen. *Int. J. Hydrogen Energy* **2017**, *42*, 13289–13299. [CrossRef]
19. Available online: http://www.esrf.eu/home/UsersAndScience/Experiments/StructMaterials/ID19.html (accessed on 4 September 2018).
20. Buffiere, J.Y.; Maire, E.; Adrien, J.; Masse, J.P.; Boller, E. In Situ Experiments with X ray Tomography: An Attractive Tool for Experimental Mechanics. *Exp. Mech.* **2010**, *50*, 289–305. [CrossRef]

21. Available online: http://ftp.esrf.fr/scisoft/PYHST2/ (accessed on 4 September 2018).
22. Schneider, C.A.; Rasband, W.S.; Eliceiri, K.W. NIH Image to ImageJ: 25 years of image analysis. *Nat. Methods* **2012**, *9*, 671. [CrossRef]
23. Landron, C.; Maire, E.; Bouaziz, O.; Adrien, J.; Lecarme, L.; Bareggi, A. Validation of void growth models using X-ray microtomography characterization of damage in dual phase steels. *Acta Mater.* **2011**, *59*, 7564–7573. [CrossRef]
24. Maire, E.; Zhou, S.; Adrien, J.; Dimichiel, M. Damage quantification in aluminium alloys using in situ tensile tests in X-ray tomography. *Eng. Fract. Mech.* **2011**, *78*, 2679–2690. [CrossRef]
25. San Marchi, C.; Michler, T.; Nibur, K.; Somerday, B. On the physical differences between tensile testing of type 304 and 316 austenitic stainless steels with internal hydrogen and in external hydrogen. *Int. J. Hydrogen Energy* **2010**, *35*, 9736–9745. [CrossRef]
26. Cox, T.; Low, J.R. An investigation of the plastic fracture of AISI 4340 and 18 Nickel-200 grade maraging steels. *Metall. Trans.* **1974**, *5*, 1457–1470. [CrossRef]

© 2019 by the authors. Licensee MDPI, Basel, Switzerland. This article is an open access article distributed under the terms and conditions of the Creative Commons Attribution (CC BY) license (http://creativecommons.org/licenses/by/4.0/).

Article

Time-Lapse Helical X-ray Computed Tomography (CT) Study of Tensile Fatigue Damage Formation in Composites for Wind Turbine Blades

Ying Wang [1], Lars P. Mikkelsen [2], Grzegorz Pyka [3] and Philip J. Withers [1,*]

1. Henry Moseley X-ray Imaging Facility, Henry Royce Institute for Advanced Materials, School of Materials, University of Manchester, M13 9PL Manchester, UK; ying.wang-4@manchester.ac.uk
2. Composite Mechanics and Structures, Department of Wind Energy, Technical University of Denmark, DK-4000 Roskilde, Denmark; lapm@dtu.dk
3. Thermo Fisher Scientific Czech Republic, 67200 Brno, Czech Republic; grzegorz.pyka@thermofisher.com
* Correspondence: philip.withers@manchester.ac.uk; Tel.: +44-161-306-4282

Received: 8 October 2018; Accepted: 16 November 2018; Published: 21 November 2018

Abstract: Understanding the fatigue damage mechanisms in composite materials is of great importance in the wind turbine industry because of the very large number of loading cycles rotor blades undergo during their service life. In this paper, the fatigue damage mechanisms of a non-crimp unidirectional (UD) glass fibre reinforced polymer (GFRP) used in wind turbine blades are characterised by time-lapse ex-situ helical X-ray computed tomography (CT) at different stages through its fatigue life. Our observations validate the hypothesis that off-axis cracking in secondary oriented fibre bundles, the so-called backing bundles, are directly related to fibre fractures in the UD bundles. Using helical X-ray CT we are able to follow the fatigue damage evolution in the composite over a length of 20 mm in the UD fibre direction using a voxel size of $(2.75~\mu m)^3$. A staining approach was used to enhance the detectability of the narrow off-axis matrix and interface cracks, partly closed fibre fractures and thin longitudinal splits. Instead of being evenly distributed, fibre fractures in the UD bundles nucleate and propagate locally where backing bundles cross-over, or where stitching threads cross-over. In addition, UD fibre fractures can also be initiated by the presence of extensive debonding and longitudinal splitting, which were found to develop from debonding of the stitching threads near surface. The splits lower the lateral constraint of the originally closely packed UD fibres, which could potentially make the composite susceptible to compressive loads as well as the environment in service. The results here indicate that further research into the better design of the positioning of stitching threads, and backing fibre cross-over regions is required, as well as new approaches to control the positions of UD fibres.

Keywords: helical CT; contrast agent; high cycle fatigue (HCF); fibre break; fibre tows

1. Introduction

Composite wind turbine rotor blades undergo a very large number of fatigue loading cycles during service [1]. As a result, their fatigue performance is a major design factor as high cycle fatigue (HCF) damage can result in unexpected catastrophic failure. Consequently, it is important to understand damage evolution during fatigue, the key damage mechanisms and the interaction between them for the unidirectional (UD) composites employed in wind turbines.

X-ray computed tomography (CT), which has been applied increasingly to materials characterisation [2,3], is superior to most non-destructive techniques in that three-dimensional (3D) information can be obtained non-destructively at a high spatial resolution. Unlike fatigue crack initiation and propagation in homogeneous materials, various damage mechanisms occur cooperatively in composites under cyclic loading, including fibre fracture, matrix cracking, debonding and delamination [4]. Establishing a time evolving 3D map of the complex fatigue damage modes in relation to local microstructure will contribute to the establishment and validation of models of fatigue failure able to better predict the safe life of such composites.

As observed in a number of composite systems, fatigue damage originates from cracks within fibre bundles or individual fibre fractures [5–8], which are on the micron level in size. High-resolution X-ray CT is needed to visualise these features, but high resolution (small voxel size) often means a small field-of-view (FoV), usually much smaller than is sampled by mechanical tests and often shorter than is needed to statistically characterise the failure processes operative in composites unless a significant number of images are stitched together. The use of helical X-ray CT enables a significant length of the sample to be imaged in a single scan. Furthermore the helical method avoids ring artefacts and cone beam artefacts [9], and can lead to high quality images which can be important when imaging low contrast systems or trying to detect events near the spatial resolution limit of the instrument. Consequently, helical CT is well suited to the characterisation of unidirectionally reinforced fibre composites, allowing us to observe the overall damage distribution and to detect localised damage events such as individual fibre fractures. In addition, dye penetrant with high atomic number (e.g., zinc iodide) can be used as stains to improve the detectability of cracks by enhancing the contrast between damage and the bulk material [2,3]. Although the use of staining has limitations in that only cracks connected to the outer surface could be stained and that it could affect the growth of matrix crack/splitting under fatigue [10], Yu et al. [11] assessed the effect of four methods to increase detectability of cracks in X-ray CT imaging and suggested that staining was perhaps the most effective in terms of increasing the sensitivity of cracks to better than 1/10 the spatial resolution.

With regard to reproducibility we have analysed three samples in this project using the strategy of combining helical imaging with staining for damage characterisation. The damage mechanisms observed are similar across all three samples (S1, S2 and S3). In reference [12], the effectiveness of this strategy and the distribution of UD fibre fractures were discussed based on two of these (S1 and S2) fatigued at maximum stress equal to 0.5% initial strain. The damage mechanisms were also found to be similar to those for unstained samples [8].

In this paper, the aim is to track the evolution, distribution and interaction between fatigue damage events and to relate these to the composite microstructure of the glass fibre reinforced polymer (GFRP) using time-lapse helical X-ray CT, and discussion focuses on one sample (S3) fatigued at maximum stress equalling 0.6% initial strain.

2. Materials and Methods

The composite material studied is a glass fibre/polyester composite system typically used in wind-turbine rotor blades. The GFRP sample has a lay-up of [0/b]$_S$, where '0' represents a 0° UD fibre layer and 'b' corresponds to a thin (~100–200 μm) ±80° backing fibre layer. Figure 1 shows the orthogonal virtual CT sections and a 3D rendered CT image of the fibre architecture. The composite panel was manufactured using vacuum assisted resin transfer moulding (VARTM). Specimens having dimensions of 2 mm × 5 mm × 110 mm were cut from the composite panel and GFRP tabs were added to the ends of the test-pieces (see Figure 2a,b). This miniaturised specimen geometry was used in order to obtain a high spatial resolution, given that the full specimen width should ideally lie within the FoV during the scan if the reconstruction algorithm is not to introduce artefacts [9].

Fatigue tests were performed on a hydraulic Instron 8802 (Norwood, MA, USA) mechanical testing machine under load control with a sinusoidal waveform and a stress ratio (R = $\sigma_{min}/\sigma_{max}$, where σ_{min} is the minimum stress and σ_{max} is the maximum stress) of 0.1 at a test frequency of 5 Hz. The maximum stress applied corresponds to 0.6% initial strain on the composite. The fatigue test was interrupted periodically with increasing numbers of fatigue cycles to monitor the damage evolution using ThermoFisher HeliScan (ThermoFisher Scientific, Brno, Czech Republic) helical X-ray CT scanner. The same sample was removed from the loading frame for CT inspection after 0 cycle (N_0), 50,000 cycles (N_1) and 500,000 cycles (N_2) respectively.

After each fatigue increment the sample was removed from the testing frame and stained in zinc iodide solution for 24 h before being imaged on the CT scanner. The zinc iodide solution was prepared following the method used by Nixon-Pearson et al. [13]. It should be noted that the specimen was stained in the absence of any load to open the cracks. Figure 2c shows the imaging set-up. The source voltage was set to 80 kV and filtered by 0.1 mm of stainless steel to remove the low energy X-rays. The exposure time for each projection (radiograph) was 0.52 s with around 20,000 projections acquired in all. The double-helix mode was used to allow reconstruction using filtered-back-projection algorithms. During the scan, the sample stage simultaneously rotates and translates vertically following a helical path with a pitch of ~7.8 mm. The scanned composite volume extends to ~20 mm in height and has a FoV height-to-width ratio of >3 at a pixel size of 2.75 µm, resulting in a total scan time of ~20 h and reconstruction time of ~3 h. High-resolution region-of-interest (RoI) scans at a pixel size of 1 µm were also taken after the time-lapse study to confirm the presence of unstained cracks.

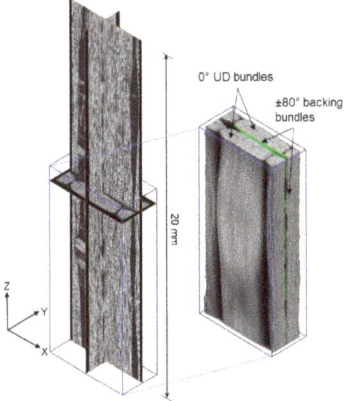

Figure 1. Orthogonal virtual X-ray CT sections (**left**) and a 3D volume rendering (**right**) showing the fibre architecture of the GFRP material where the ±80° backing fibre bundles are rendered in green.

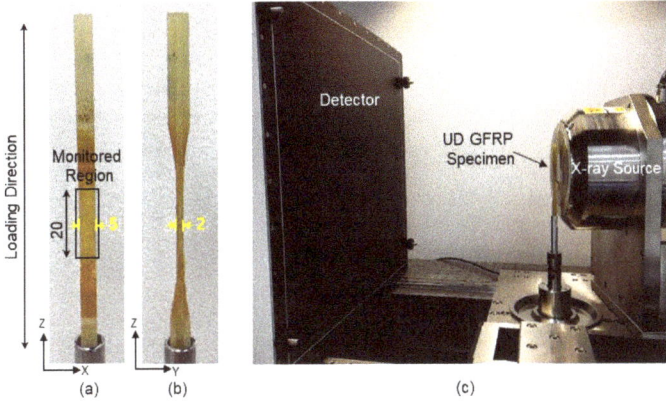

Figure 2. (**a**,**b**) Photographs illustrating the GFRP specimen geometry and the X-ray CT monitored region through interrupted fatigue test (dimensions in mm). (**c**) Photograph showing the helical X-ray CT imaging set-up in the ThermoFisher HeliScan Micro-CT scanner.

3. Results and Discussion

Using traditional circular X-ray CT, it is challenging and time-consuming to locate the RoI to perform time-lapse tracking of the damage evolution due to the limited FoV. Based on previous studies of this material by Jespersen et al. [8,14,15], it is expected that fatigue damage is most likely to initiate from regions where the backing bundles cross-over; however, it is important to confirm whether this is generally true for this material system or whether there are other factors that might localise fatigue damage. Owing to the extended length that can be viewed at high resolution by helical X-ray CT, the overall damage distribution along the composite specimen can be monitored in relation to the composite microstructure (e.g., backing fibre bundle cross-over regions, stitching thread cross-over regions, resin-rich regions, fibre misalignment in UD fibre bundles) at different stages of its fatigue life. In the specimen studied, four main damage modes were observed, namely off-axis cracking in the matrix or backing fibre bundles, fibre fractures, sub-surface debonding and longitudinal splitting in the UD layers. Figure 3 shows the extracted fatigue damage within the specimen after 500,000 cycles, where the fibre fractures and longitudinal splits are visualised with respect to the UD and backing fibre bundles (rendered green). The regions with fibre fractures were manually delineated (see red contours in Figure 3d) to highlight the extent of UD fibre fracture damage in 3D. The zinc iodide dye penetrated most of the highlighted fibre fracture region; however, some fibre fractures in region (e) were not stained as they were not connected to the external surfaces. The increased visibility of thin cracks in the high-resolution RoI images confirms the presence of the unstained cracks as shown in Figure 3e and this damage region was also included in the 3D damage visualisation. The effect of staining on damage detectability in this material was discussed in [12]. In addition, its effect on the observed damage development in this material system has been assessed by comparing the damage status of S1 and S2 after 2 million cycles. For S1, the fatigue test was interrupted at 0.5 million, 1.5 million and 2 million cycles, for repeated staining and imaging at each step; while for S2 the fatigue test was interrupted only after 2 million cycles for staining and imaging. The same damage mechanisms, as observed in S3 here, have been seen in S1 and S2. Moreover, the severity of damage in S1 and S2 is on the same level, with the occurrence of a few small regions of UD fibre fractures near backing fibre bundle cross-over regions. This indicates that there is no obvious influence of staining on the observed damage scenario. The evolution, distribution and interaction of the observed damage mechanisms in S3 will be presented and discussed in the following sections.

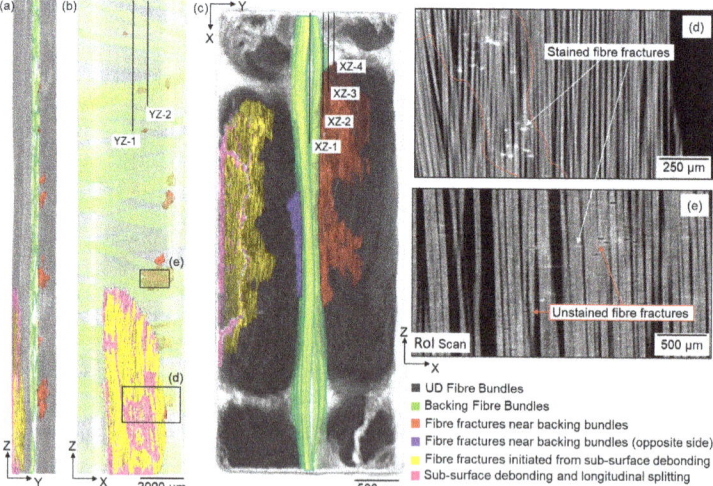

Figure 3. X-ray CT 3D volume rendering of the extracted damage after 500,000 cycles (N_2), where the fibre fractures, longitudinal splits and debonding are visualised with respect to the UD and backing fibre bundles, (**a**) YZ view, (**b**) XZ view, (**c**) XY view (refer to Figure 2a,b for definition of the coordinate system). (**d**) Typical virtual X-ray CT XZ of the region marked in (**b**), illustrating the manual annotation of fibre fracture damage in the CT image. (**e**) High-resolution RoI X-ray CT virtual XZ section of the damage region marked in (**b**) where some fibre fractures were not stained.

3.1. Off-Axis Matrix and Interface Cracks

It has been challenging to observe off-axis cracks in this material by X-ray CT due to the closing of those cracks to below the resolution limit after load removal [15]. With the aid of contrast agent, we are able to observe the development of off-axis matrix cracks from the specimen edges and also off-axis cracks between backing fibres.

Stained off-axis cracks in the matrix were found to have initiated from specimen edges after 50,000 cycles (N_1) as a result of the axial tensile stress (see Figure 4b). After exposure to more fatigue cycles (N_2), the crack density increases dramatically. The observed increase in off-axis cracking has also been reported by Jespersen et al. [15] in a similar material. It is difficult to see the polymer stitching threads as their X-ray absorption capability is similar to that of the matrix material, as can be seen in the magnified views in Figure 4e,f. It has been observed in Figure 4d,f that the off-axis cracks in the matrix tend to be deflected when propagating into the stitching threads. This highlights the effect of stitching threads on delaying the propagation of off-axis matrix cracks from the edges by debonding. Once a crack crosses the column of material containing the stitching threads, it joins the off-axis cracks between backing fibres, as shown in the highlighted ellipse in Figure 4f.

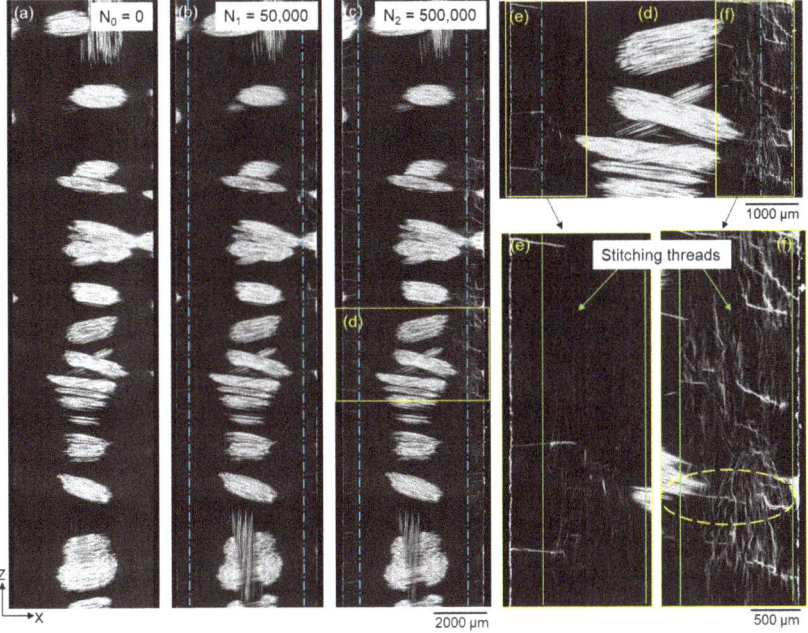

Figure 4. X-ray CT virtual XZ-1 section (position indicated in Figure 3c) of the sample after (**a**) 0, (**b**) 50,000 and (**c**) 500,000 cycles showing the evolution of off-axis matrix cracks initiated from specimen edges. (**d**) Magnified view of the highlighted region in (**c**) showing the off-axis matrix cracks in relation to stitching threads. (**e**,**f**) Magnified views of the highlighted regions in (**d**) showing the polymer stitching threads (light grey) and stained debonding along stitching threads (white). Dashed blue lines indicate the front of off-axis cracks and solid green lines mark the column where the stitching threads are located.

The presence of off-axis cracks in between backing fibres has been reported in a number of fatigue studies on this material [5,15] and is observed here after 500,000 cycles (see Figure 5). Only off-axis cracks that are stained were detectable with the current CT acquisition set-up. These stained off-axis cracks are distributed at different heights in the specimen and multiple ones sometimes occur within a single backing fibre bundle; these tend to be distributed vertically relative to one another due to stress shielding [16]. The regions denoted by the upward arrows in Figure 5a and the highlighted ellipses in the YZ sections shown in Figure 5d,e highlight the joining of stained off-axis matrix cracks in the backing layer and stained fibre fractures in the UD bundles. This confirms the presence of a connecting path between the two damage modes. This observation supports the hypothesis [5] that off-axis cracking in backing fibre bundles triggers fibre fractures in the adjacent UD fibre bundles. It is worth noting that the stained fibre fractures were not necessarily directly connected to the off-axis cracks in some cases, as highlighted by the downward-arrows in Figure 5d,e. However, the fact that these fibre fractures were stained successfully means that they were connected to the free surface, either through cracks running into the specimen surfaces, or tunnel-cracks (through backing bundles) extending to the edges.

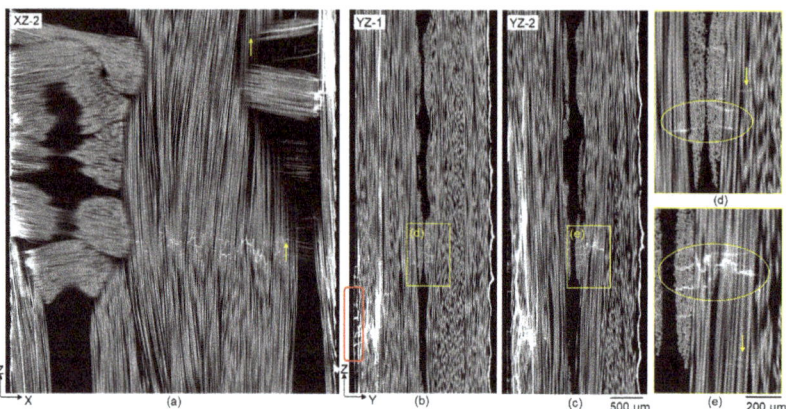

Figure 5. X-ray CT virtual (**a**) XZ-2 and (**b**) YZ-1 and (**c**) YZ-2 sections (positions indicated in Figure 3b,c) recorded after 500,000 cycles showing crack paths connecting off-axis cracks in the backing fibre bundles and fibre fractures in the UD fibre bundles. (**d,e**) Magnfied views of the highlighted regions in (**b,c**) showing the connected off-axis cracks (marked with ellipses) and also isolated UD fibre fractures (marked with downward-arrows).

3.2. Fibre Fractures in UD Fibre Bundles

The nucleation of UD fibre fractures was found to correlate with two micro-structural features; regions where backing fibre bundles cross-over (near mid-thickness of the specimen, see Figure 3a–c) and regions where stitching threads cross-over (near the specimen surface, see the red boxes in Figure 5b and Figure 8c). The development and distribution of UD fibre fractures from these two regions were tracked in the time-lapse CT images, and the results are presented and discussed in this section.

As mentioned above, the regions of UD fibre fractures were manually delineated (see Figure 3d) to help visualise the 3D morphology of this damage mode. Fibre fractures were observed in the UD fibres adjacent to the cross-over regions of backing fibre bundles after 50,000 cycles, which is consistent with the observations of previous studies [5,8]. The onset of UD fibre fracture appears to be localised next to backing bundle cross-over regions (see Figure 6b,d). After 500,000 cycles, seven fibre fracture regions were observed on the right-hand side of the specimen width; these were approximately evenly distributed (~2–3 mm) along the 20 mm length. Out of the seven damage regions, six were next to the backing bundle cross-over regions, while one small region was located next to −80° backing fibres only (see Figure 6c). As can be seen in Figure 3c, it should be noted that most of the UD fibre fractures (rendered red) developed near the UD fibre fracture site (rendered blue) developing on the other side of the backing layer. This site is at the same height (in Z) as the site of damage initiation of fibre fractures after 50,000 cycles (see Figure 6b–e). This is the location where UD fibre fractures propagated most widely within the specimen after 500,000 cycles.

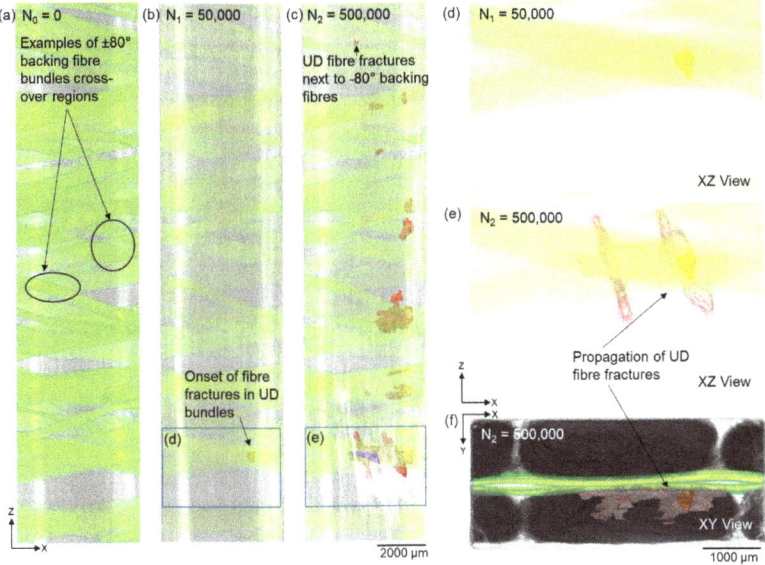

Figure 6. X-ray CT volume rendering of the backing fibre bundles (**a**) after 0 cycles showing the cross-over of the ±80° backing fibre bundles, as well as UD fibre fractures next to backing fibre bundles after (**b**) 50,000 and (**c**) 500,000 cycles, together with magnified views in (**d,e**) and in (**f**) the evolution and distribution of UD fibre fractures initiated from backing bundles after 500,000 cycles (refer to Figure 3 for explanations of colour coding). For 3D visualisation see the supplementary video.

Figure 6e,f shows the propagation of fibre fractures (shown red) between 50,000 and 500,000 cycles. The propagation here was seen to be mostly along the length and width directions rather than through thickness. In the case with backing fibres oriented at ±45°/90° [8] fatigued at maximum strain of 1%, fibre fracture was observed across the full width of the cross-over region and then propagated in the thickness direction. This difference in propagation sequence indicates the necessity of choosing the appropriate arrangement of backing fibres in the composite design for different wind turbine blade requirements. As can be seen in Figure 6e, this damage region has two elongated branches. A local variation in fibre orientation is evident in the 0° fibre bundle (see Figure 7g), which could be the cause of this damage morphology. In other words, 0° fibre alignment could be important in controlling the propagation of UD fibre fractures.

The distribution of UD fibre fractures in relation to the local microstructure within one damage region is revealed by the raw X-ray CT images where individual fibre fractures can be seen (see Figure 7). In the XZ sections near the backing fibre bundles, the distribution of fibre fractures tends to be aligned with the backing fibre bundle in contact with the UD fibres, as shown in the XZ-2 sections. This is also true for the UD fibre fractures at the edge of the UD bundle (see Figure 7a). Further away from the backing bundles, the distribution of fibre fractures is less dependent on backing fibre orientation but more affected by local fibre orientation in the UD bundle, as in the case for the XZ-3 and XZ-4 sections in Figure 7. It is noteworthy that the UD fibre fractures are seldom aligned in a continuous line; instead they tend to cluster in small numbers at different heights.

UD fibre fractures were also observed to occur near one specimen surface after 500,000 cycles (see Figures 3 and 5). Figure 8 shows sub-surface XZ sections in this region through the fatigue cycling. It is evident that sub-surface debonding of stitching threads (see Figure 8b,c) is associated with fracture of the adjacent UD fibres in this region, presumably because of higher local stresses. It is worth noting that a number of longitudinal splits developed, connecting the UD fibre fractures that were relatively more distanced from each other (see Figure 8h). As shown in Figure 3a, this damage site is located on the opposite side of the backing fibre bundles to that where fibre fractures initiate near the cross-overs after 50,000 cycles; this means that the damage sites are in two UD fibre bundles. However, these two damage sites are at a similar height, and the extents of propagation through the bundle thickness are similar. It is worth mentioning that another damage region was found further above the debonded stitching threads, which was caused by attaching the extensometer during the fatigue test.

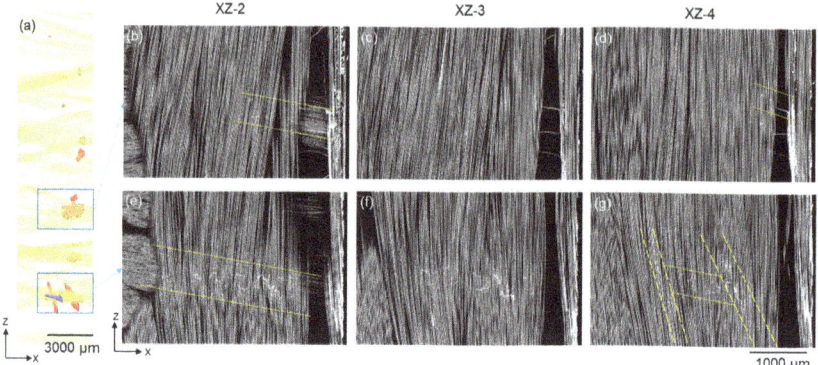

Figure 7. (a) XZ view of the rendered UD fibre fracture damage near the backing bundles after 500,000 cycles. (b–g) X-ray CT virtual XZ sections illustrating the distribution of UD fibre fractures at different positions through thickness in the specimen. The Y-positions of the XZ sections are shown in Figure 3c.

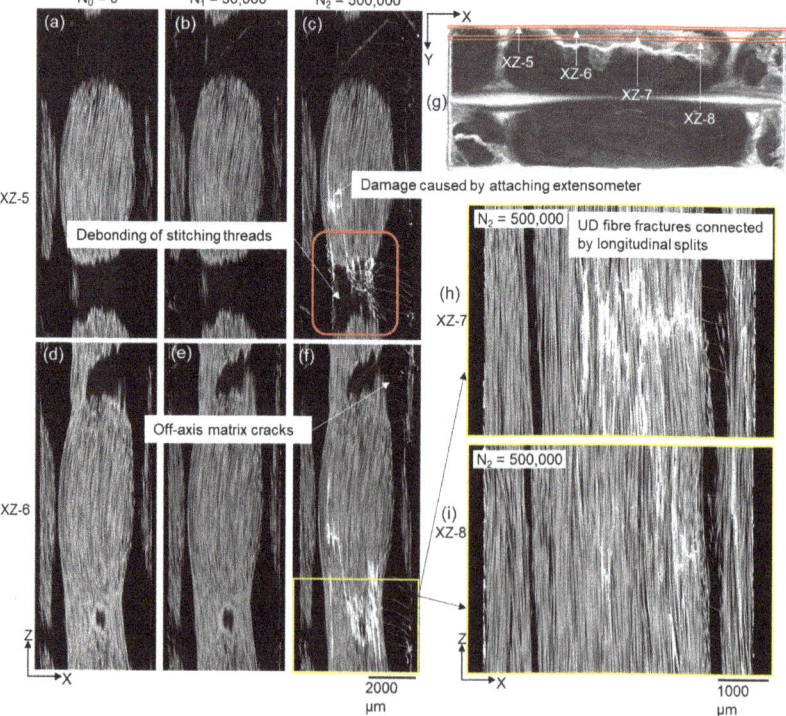

Figure 8. (a–f) and (h,i) Sub-surface X-ray CT virtual XZ sections showing the development of UD fibre fractures from the debonding of stitching threads. The Y-positions of the XZ slices are shown in (g).

3.3. Sub-Surface Debonding of Stitching Threads and Longitudinal Splitting

Extensive longitudinal splitting has been observed in the UD bundle where sub-surface debonding of stitching threads at the cross-over region of the threads has occurred. The longitudinal splits in the UD fibre bundle are extracted and visualised in Figure 9. Longitudinal splitting tends to initially occur in the edge UD bundles due to the edge effect (see Figure 9d). After 500,000 cycles, a longitudinal split originated from a sub-surface debonded region (see Figures 5 and 9). A section of the UD fibre bundle is separated from the full bundle as shown in the XY-1 plane after 500,000 cycles, and also in the 3D view in Figure 9a. Apart from the fact that these splits are closely correlated with the UD fibre fractures, the splits lower the lateral constraint of the originally closely packed UD fibres, which could potentially make the composite susceptible to compressive loads and moisture. It could be inferred that if the wind turbine blades experience bending fatigue, longitudinal splitting could be a detrimental damage mode.

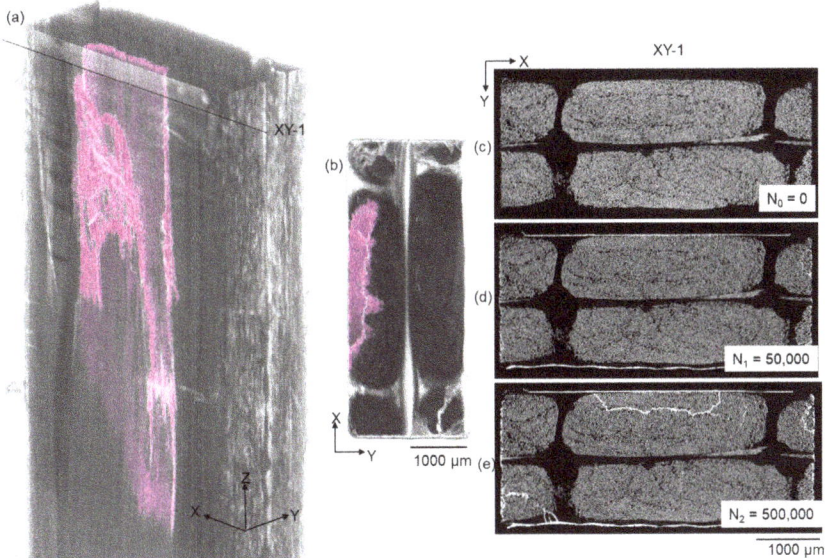

Figure 9. (a) Perspective and (b) plan view of the X-ray CT 3D volume rendering showing the extent of longitudinal splitting (purple), and an X-ray CT virtual XY section (position indicated in (a)) after (c) 0, (d) 50,000 and (e) 500,000 cycles showing the evolution of longitudinal splits.

4. Conclusions

In this paper, a time-lapse ex-situ helical X-ray CT imaging strategy assisted by staining was used to track the development of fatigue damage under tension-tension fatigue in a non-crimp UD GFRP. In essence, the contrast agent could favour the observation of damage where a penetration path exists, but this could be both the weakness and the strength of this method. The weakness is that a damaged region without connection to the outer surface could not be stained; while the strength is that we are able to identify the connection of different damage modes based on the stained path, even for cases where the full crack path is difficult to be ascertained using X-ray CT. This enables us to experimentally prove the hypothesis on the linking of the different fatigue damage mechanisms and also provide insights into their interaction in 3D. Helical X-ray CT makes it experimentally feasible to follow the fatigue damage evolution over a sufficiently long region in the composite along the UD fibre direction. Overall, four main damage modes were identified,

- off-axis matrix and interface cracking,

- UD fibre fracture,
- sub-surface debonding of stitching threads,
- longitudinal splitting.

Off-axis matrix cracks initiating from the specimen edges were sometimes deflected by stitching threads by debonding. Off-axis cracks between backing fibres were found to be associated with UD fibre fractures evidenced by the penetration path of the contrast agent. Moreover, these UD fibre fractures tend to nucleate and propagate locally in the vicinity of cross-over regions of backing bundles instead of being evenly distributed along the UD fibre direction. In addition, UD fibre fractures also tend to be initiated by the presence of extensive debonding and longitudinal splitting, which were found to develop from debonding of the stitching threads near surface. The isolation of the UD fibre bundle caused by longitudinal splitting potentially makes the composite susceptible to compression and bending loads as well as environmental impact in service. It could be inferred from the results here that further research into the better design of the positioning stitching threads, and backing fibre cross-over regions is required in the future, as well as new approaches to fix the positions of UD fibres. The work presented here (all the X-ray CT datasets are available online [17]) could be of significance to the further improvement of analytical and numerical models to predict the fatigue failure of composite materials.

Supplementary Materials: The following are available online at http://www.mdpi.com/1996-1944/11/11/2340/s1, Video S1: A supplementary video accompanies Figure 6.

Author Contributions: Conceptualization, Y.W., L.P.M. and P.J.W.; Investigation, Y.W. and G.P.; Writing-Original Draft Preparation, Y.W.; Writing-Review & Editing, L.P.M. and P.J.W.; Funding Acquisition, L.P.M. and P.J.W.

Funding: This research was funded by the allianCe for ImagiNg of Energy Materials (CINEMA) project under DSF-grant number 1305-00032B.

Acknowledgments: The authors would like to thank LM Wind Power and DTU Wind Energy for sample preparation, and FEI (Thermo Fisher), especially Dirk Laeveren, for technical support and loan of the mk I HeliScan. P.J.W. is grateful to the Engineering and Physical Science Research Council (EPSRC) for funding the Henry Moseley X-ray Imaging Facility (grants EP/F007906, EP/F001452 and EP/I02249X, EP/M010619/1, EP/F028431/1, and EP/M022498/1) and a European Research Council Grant CORREL-CT (No. 695638).

Conflicts of Interest: The authors declare no conflict of interest.

References

1. Nijssen, R.P.L.; Brøndsted, P. Fatigue as a design driver for composite wind turbine blades. In *Advances in Wind Turbine Blade Design and Materials*; Brøndsted, P., Nijssen, R.P.L., Eds.; Woodhead Publishing: Cambridge, UK, 2013; pp. 175–209. ISBN 9780857094261.
2. Wang, Y.; Garcea, S.C.; Withers, P.J. Computed Tomography of Composites. In *Comprehensive Composite Materials II*, 2nd ed.; Beaumont, P.W.R., Zweben, C.H., Eds.; Elsevier: Amsterdam, The Netherlands, 2018; Volume 7, pp. 101–118. ISBN 9780081005347.
3. Garcea, S.C.; Wang, Y.; Withers, P.J. X-ray computed tomography of polymer composites. *Compos. Sci. Technol.* **2018**, *156*, 305–319. [CrossRef]
4. Talreja, R. Fatigue of Composite Materials: Damage Mechanisms and Fatigue-Life Diagrams. *Proc. R. Soc. A Math. Phys. Eng. Sci.* **1981**, *378*, 461–475. [CrossRef]
5. Zangenberg, J.; Brøndsted, P.; Gillespie, J.W. Fatigue damage propagation in unidirectional glass fibre reinforced composites made of a non-crimp fabric. *J. Compos. Mater.* **2014**, *48*, 2711–2727. [CrossRef]
6. Garcea, S.C.; Sinclair, I.; Spearing, S.M. In situ synchrotron tomographic evaluation of the effect of toughening strategies on fatigue micromechanisms in carbon fibre reinforced polymers. *Compos. Sci. Technol.* **2015**, *109*, 32–39. [CrossRef]
7. Yu, B.; Blanc, R.; Soutis, C.; Withers, P.J. Evolution of damage during the fatigue of 3D woven glass-fibre reinforced composites subjected to tension–tension loading observed by time-lapse X-ray tomography. *Compos. Part A Appl. Sci. Manuf.* **2016**, *82*, 279–290. [CrossRef]
8. Jespersen, K.M.; Mikkelsen, L.P. Three dimensional fatigue damage evolution in non-crimp glass fibre fabric based composites used for wind turbine blades. *Compos. Sci. Technol.* **2017**, *153*, 261–272. [CrossRef]

9. Maire, E.; Withers, P.J. Quantitative X-ray tomography. *Int. Mater. Rev.* **2014**, *59*, 1–43. [CrossRef]
10. Spearing, S.M.; Beaumont, P.W.R. Fatigue damage mechanics of composite materials. I: Experimental measurement of damage and post-fatigue properties. *Compos. Sci. Technol.* **1992**, *44*, 159–168. [CrossRef]
11. Yu, B.; Bradley, R.S.; Soutis, C.; Withers, P.J. A comparison of different approaches for imaging cracks in composites by X-ray microtomography. *Philos. Trans. A. Math. Phys. Eng. Sci.* **2016**, *374*, 20160037. [CrossRef] [PubMed]
12. Wang, Y.; Pyka, G.; Jespersen, K.M.; Mikkelsen, L.P.; Withers, P.J. Imaging of composites by helical X-ray computed tomography. In Proceedings of the 21st International Conference on Composite Materials, Xi'an, China, 20–25 August 2017.
13. Nixon-Pearson, O.J.; Hallett, S.R.; Withers, P.J.; Rouse, J. Damage development in open-hole composite specimens in fatigue. Part 1: Experimental investigation. *Compos. Struct.* **2013**, *106*, 882–889. [CrossRef]
14. Jespersen, K.M.; Zangenberg, J.; Lowe, T.; Withers, P.J.; Mikkelsen, L.P. Fatigue damage assessment of uni-directional non-crimp fabric reinforced polyester composite using X-ray computed tomography. *Compos. Sci. Technol.* **2016**, *136*, 94–103. [CrossRef]
15. Jespersen, K.M.; Glud, J.A.; Zangenberg, J.; Hosoi, A.; Kawada, H.; Mikkelsen, L.P. Uncovering the fatigue damage initiation and progression in uni-directional non-crimp fabric reinforced polyester composite. *Compos. Part A Appl. Sci. Manuf.* **2018**, *109*, 481–497. [CrossRef]
16. Quaresimin, M.; Carraro, P.A.; Mikkelsen, L.P.; Lucato, N.; Vivian, L.; Brøndsted, P.; Sørensen, B.F.; Varna, J.; Talreja, R. Damage evolution under cyclic multiaxial stress state: A comparative analysis between glass/epoxy laminates and tubes. *Compos. Part B Eng.* **2014**, *61*, 282–290. [CrossRef]
17. Wang, Y.; Mikkelsen, L.P.; Pyka, G.; Withers, P.J. Time-lapse helical X-ray computed tomography (CT) data of tensile fatigue damage in GFRP. *Zenodo* **2018**. [CrossRef]

© 2018 by the authors. Licensee MDPI, Basel, Switzerland. This article is an open access article distributed under the terms and conditions of the Creative Commons Attribution (CC BY) license (http://creativecommons.org/licenses/by/4.0/).

Article

Mechanical Properties and In Situ Deformation Imaging of Microlattices Manufactured by Laser Based Powder Bed Fusion

Anton Du Plessis [1,*], Dean-Paul Kouprianoff [2], Ina Yadroitsava [2] and Igor Yadroitsev [2]

1. CT Scanner Facility, Stellenbosch University, Stellenbosch 7602, South Africa
2. Department of Mechanical Engineering, Central University of Technology, Free State, Bloemfontein 9300, South Africa; dkouprianoff@cut.ac.za (D.-P.K.); iyadroitsava@cut.ac.za (I.Y.); iyadroitsau@cut.ac.za (I.Y.)
* Correspondence: anton2@sun.ac.za; Tel.: +27-21-808-9389

Received: 30 July 2018; Accepted: 31 August 2018; Published: 9 September 2018

Abstract: This paper reports on the production and mechanical properties of Ti6Al4V microlattice structures with strut thickness nearing the single-track width of the laser-based powder bed fusion (LPBF) system used. Besides providing new information on the mechanical properties and manufacturability of such thin-strut lattices, this paper also reports on the in situ deformation imaging of microlattice structures with six unit cells in every direction. LPBF lattices are of interest for medical implants due to the possibility of creating structures with an elastic modulus close to that of the bones and small pore sizes that allow effective osseointegration. In this work, four different cubes were produced using laser powder bed fusion and subsequently analyzed using microCT, compression testing, and one selected lattice was subjected to in situ microCT imaging during compression. The in situ imaging was performed at four steps during yielding. The results indicate that mechanical performance (elastic modulus and strength) correlate well with actual density and that this performance is remarkably good despite the high roughness and irregularity of the struts at this scale. In situ yielding is visually illustrated.

Keywords: laser powder bed fusion; additive manufacturing; X-ray tomography; in-situ imaging; Ti6Al4V; lattice structures

---

1. Introduction

Additive manufacturing (AM) is an emerging production technique whereby a part with complex geometry can be produced directly from a design file in a layer-by-layer method [1,2]. In the case of laser-based powder bed fusion (LPBF), a single layer of the part is selectively fused using a laser beam that is scanned across a powder bed surface in a series of tracks, new powder is delivered, and the next layer is scanned and fused. Predictably, the part integrity requires that single tracks are stable [3] and well overlapped with one another, as well as layers to prevent unwanted porosity in solid parts. This has been discussed in some detail in a recent review of the use of X-ray microtomography in additive manufacturing [4]. Despite the possibility of irregularities in parts, it is possible to produce parts with excellent mechanical properties when process parameters are optimized (see, for example, Reference [5] for biomedical Ti6Al4V produced by LPBF).

One of the major benefits brought about by additive manufacturing is the ability to produce complex parts, and this is especially true for lattice structures that are regularly spaced and repeating combinations of struts with spaces between them. Lattice structures produced by AM have been the topic of many studies in recent years due to the potential to use these in bone replacement implants [6–8]. In implants, the porous nature of the lattice structure is beneficial to lower the elastic modulus of biocompatible materials to match that of the bone at the implant interface, minimizing the possibility for stress shielding causing loss in bone density in the vicinity. Additionally, the open porous nature allows for bone ingrowth into the lattice, effectively ensuring a good bond with the existing bone.

The investigation of the mechanical properties of lattices produced by AM, and in particular LPBF, is therefore crucial for the adoption of this type of design in implants, along with tailoring its properties for the application of custom shapes that meet local bone density requirements. In general, the mechanical properties of these structures can be predicted by the Ashby–Gibson model for open-cell foams [9,10], with a general relationship for elastic modulus of the lattice (E) as a function of the lattice density (ρ) and elastic modulus of the solid material used (E_s), given as follows:

$$E = \alpha_2 \, E_s \left(\frac{\rho}{\rho_S}\right)^2, \qquad (1)$$

where α_2 is a value between 0.1 and 4 depending on the lattice geometry [9].

In early work by Parthasarathy et al. [11], simple cubic lattices of Ti6Al4V produced by electron beam melting were analyzed by microCT and mechanical testing and it was found that the mechanical properties are weaker than predicted and this was especially so for a model with thinner struts. This might be attributed to manufacturing irregularities such as the rough as-built surface and unexpected porosity inside the struts. Geometric accuracy is often a limitation in additive manufacturing of cellular structures, as is the entrapment of powder in the small pore spaces of these structures [1]. Various LPBF cellular structures in Ti6Al4V have been produced in different unit cell designs and their mechanical properties investigated, for example cubic [12], diamond [13], and combinations of designs including body-centered cubic [14] and minimal surfaces [15]. Besides variations in mechanical performance induced by geometric inaccuracy and manufacturing errors, slight variations also exist in the properties of various lattice designs themselves. This was demonstrated recently by the numerical analysis of various lattice designs, ignoring manufacturing imperfections [16].

It is therefore clear that the only way to fully understand the complex behavior of lattice structures (with many variations in designs and varying amounts of manufacturing errors, which to some extent also depend on the design), is to use high resolution imaging. In prior work, using relatively large lattices with struts more than 1 mm in diameter, compression tests combined with microCT imaging was used to visualize the first yielding crack locations, as shown in Reference [17], with loads up to 140 kN. This was done ex situ by stopping the mechanical test at first yielding and correlating "before" and "after" microCT scans to find cracks/yielding locations. Some work has also previously been done using in situ synchrotron tomography during the loading of small unit cells produced by LPBF [18]. This work showed local strut-scale deformations during yielding and compared experimental results to those predicted by simulation, but was limited to unit cells, which are not necessarily representative of tessellated lattices. Furthermore, the effect of LPBF process parameters on the morphology and mechanical properties of small lattices were investigated using a combination of methods, including microCT, where it was shown that properties may be improved by process optimization and failure occurred at the nodes in that case [19].

In this work, the aim was to investigate the smallest possible lattices that can be produced with a typical LPBF system with a track width of roughly 0.1 mm. In addition to investigating the mechanical properties of such small lattices, the accuracy of these produced microlattices may be useful as a reference for future work. Four different sizes of lattice cube samples were produced, each containing six unit cells in each direction with a diamond unit cell design with porosity of 80% and unit cell sizes 0.6, 0.8, 1.0, and 1.2 mm. Since the geometry and the density is kept constant, the theoretical elastic modulus and yield stress should be identical in all four cases, therefore the aim was to investigate the properties as the struts become thinner with decreasing unit cell size. The mechanical properties of these small lattices is reported and in situ imaging of the lattice deformation using high resolution X-ray tomography is demonstrated.

2. Materials and Methods

Models were designed in Materialize Magics [20] and produced from Ti6Al4V extra low interstitials (ELI) powder by EOSINT M 280 (EOS GmbH—Electro Optical Systems, Krailling, Germany) with a 200 W laser and original parameters Ti64_Performance 1.1.0 (30 µm). Gas atomized Ti6Al4V ELI powder from TLS Technik GmbH & Co. Spezialpulver KG (Bitterfeld-Wolfen, Germany) was used. Particle size distribution was as follows: equivalent diameters (weighted by volume) d_{10} = 12.1 µm, d_{50} = 23.6 µm, and d_{90} = 37.6 µm. The chemical composition fulfilled the requirements of ASTM F136 standard specification for wrought Ti6Al4V ELI alloy for surgical implant applications regarding maximum concentration of impurities (ASTM International, West Conshohocken, PA, USA).

A stress-relief cycle for 3 h at 650 °C [5] was conducted in an argon atmosphere after producing the parts, after which the parts were cut from the build plate using electrical discharge machining. The unit cell design used in this work was the diamond design; the unit cell is shown in Figure 1a. Three samples of each of four designs were produced, the computer aided design (CAD) designs are shown in Figure 1b, with a strut thickness analysis showing that the larger the unit cell, the thicker the strut was, as expected. Strut thickness analysis allowed measurement of the "wall" thickness at every point in the structure. In this case the sphere method was used, which provided the value of the maximal-fitted sphere in every point in the structure. The designs were selected to produce cubes with six unit cells in each direction, with unit cell sizes for the four designs being 0.6, 0.8, 1.0, and 1.2 mm. This ensured that the density was kept constant and was selected to be 20% dense (80% porosity). The physical sample sizes varied from 3.6 to 7.2 mm for the lattice region, and additional solid material was added to the top and bottom to make the total height 8 mm in all cases for simpler loading in the compression cell.

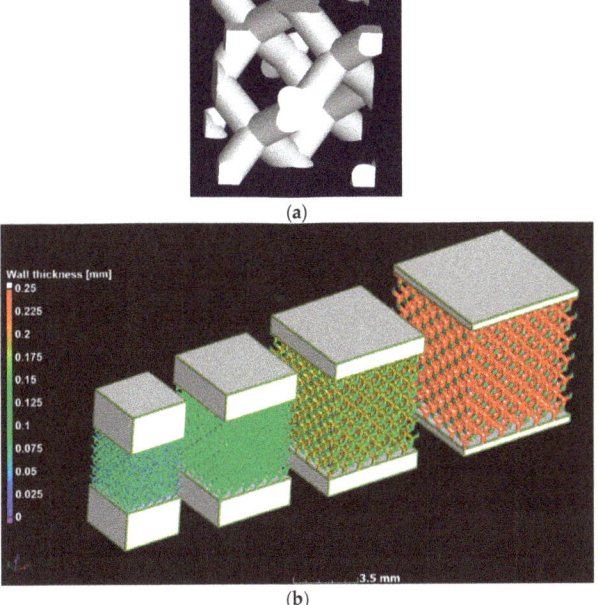

Figure 1. Design of microlattices showing (**a**) a single unit cell design of diamond type, and (**b**) the four full CAD lattice design used with unit cell sizes 0.6, 0.8, 1.0 and 1.2 mm, with strut thickness analysis.

MicroCT scanning was done using laboratory nanoCT as described in Reference [21] using a Deben in-situ loading stage (CT500, Deben UK, London, UK) in a General Electric Nanotom scanner (Nanotom S, General Electric, Wunstorf, Germany). The sample sizes in this work were selected according to the maximum sample size of 10 mm and maximum loading force of 500 N of this loading stage. One sample design that did not fail up to 500 N was additionally subjected to compression tests on a different loading stage to obtain the yield strength. This was the smallest sample with the highest density (Figure 1, sample on the far left).

The microCT voxel size was selected as 4 µm, with 140 kV and 130 µA for the X-ray generation, using a 0.5 mm copper beam filtration and using continuous scanning mode, a total of 3600 images were recorded during a full rotation of the sample. Images were further analyzed in Volume Graphics VGSTUDIO MAX 3.2 (version 3.2, Volume Graphics, Heidelberg, Germany) [22]. Wall thickness analysis used in this work was done with the sphere-method. Due to file sizes and limited computing power, the combined images were resampled in VGSTUDIO MAX to a 10 µm voxel size and 8 bit data depth to reduce file sizes and memory usage to ease the image analysis.

3. Results and Discussion

Samples were manufactured successfully, but microCT analyses showed that the strut thickness across the models did not vary as expected; this is shown in Figure 2 using a strut thickness analysis, analogous to Figure 1. Irregularities were expected at this scale due to various practical limitations that exist when producing small intricate parts during LPBF. The cause of such irregularities can be explained when looking at the minimum size of the designed features with regard to the combined effects of the laser spot size, building direction, layer thickness, and the implemented scanning strategy for core, overhangs, and top surfaces, all having an effect on the amount of detail that could be obtained. Small features were also governed by the single track's width and attached powder particles, which in turn were limited by powder particle size distribution. Accuracy of small overhangs was not only dependent on the layer thickness but also on the loose powder and the inability of the molten pool to penetrate into solid material of the lattice. Therefore, irregular surfaces below the struts were expected [23].

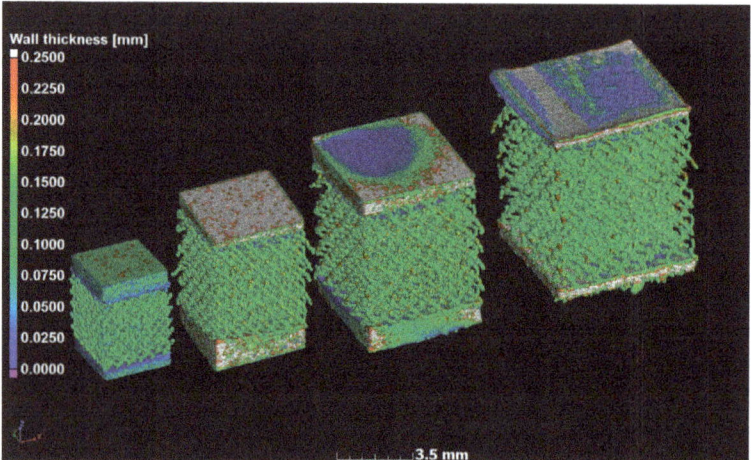

Figure 2. MicroCT scan data 3-D rendering with strut thickness color coding on the microlattices. Models had a 0.6, 0.8, 1.0 and 1.2 mm unit cell size from left to right respectively; here the top and bottom of the samples were slightly cropped as the lattice area was scanned only. Dark blue indicates thin local walls and red indicates thick walls, thickest parts were excluded to highlight the important aspect. All lattice struts have similar thickness (0.1–0.15 mm).

In previous work [24], it was shown that at layer thickness 15–45 μm and similar Ti6Al4V powder and process-parameters, the width of the track was 100–150 μm. The small size of the designed struts, which are close to the single-track width of the laser melting track width, combined with the scanning strategy of the LPBF system used apparently did not allow for variations and these lattices were seemingly all produced with a similar strut thickness deviating from the design thickness as shown in Figures 3a and 4. On the one hand, .stl triangulation of small structures led to an irregular shape of the struts (Figure 4a). Second, analysis of the scanning strategy showed that designed fine structures (less than 300 μm) were scanned by the laser as single lines with process parameters for the skin (contouring). Thus, struts in each of the produced sets of 0.6–1.2 mm units were similar and had thickness of 90–220 μm and they were very rough (Figure 4b). For the total density, the result was that the larger lattice had larger pore spaces, making its actual density lower than designed, as shown in Figure 3b.

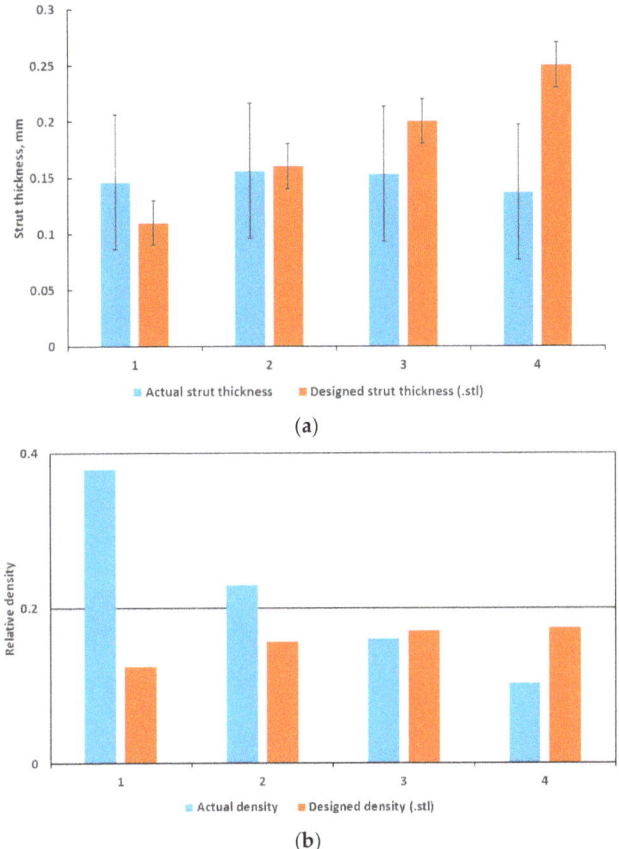

Figure 3. Dimensional assessment of (**a**) actual versus designed strut thickness and (**b**) actual versus designed total density compared to the designed values. Maximum and minimum values are shown as an error margin but indicates the variability within a single strut, as measured manually from 3-D model .stl data and microCT data.

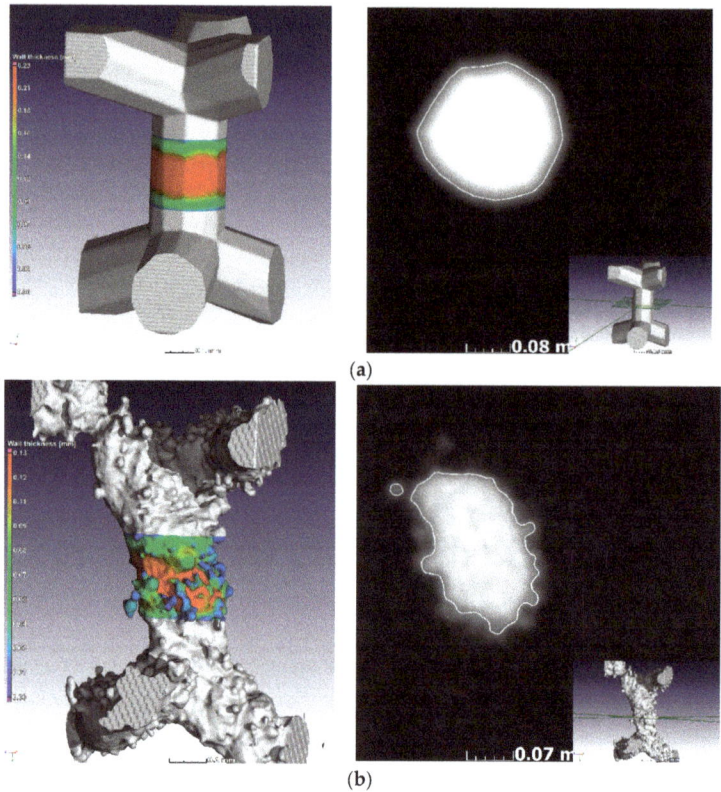

Figure 4. MicroCT voxel view of (**a**) designed and (**b**) manufactured struts and their cross-sections, indicating the actual morphology in each case.

As explained in the previous section, the simplified model of Ashby–Gibson for open cell foams indicates a linear correlation between the elastic modulus and the square of the density. This was found experimentally in this case with a slope of approximately 2.8, as shown in Figure 5.

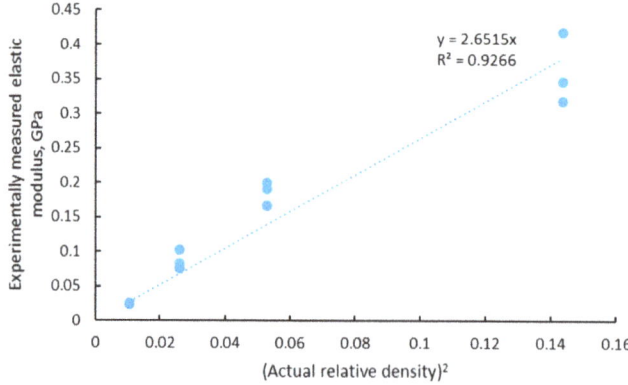

Figure 5. Relationship for experimentally measured elastic modulus and squared relative density of the manufactured LPBF lattices.

The value of 2.8 for the slope was within the expected range of 0.1–4. Previous work with larger lattices built using the same material process parameters showed that the experimental elastic modulus values were 10 and 20 GPa for 50% density lattice structures of two designs, diagonal and rhombic, respectively [17]. This relates to values for alpha (the slope) of 0.35 to 0.7. We can therefore speculate that as the strut thickness reduces, the effect of the rough and irregular surface plays an increasingly important role, increasing the slope and making the structure's mechanical properties more sensitive to changes in density. What is interesting to note here is that since the lattice properties follow the density, the smallest lattice of 3.6 mm (unit cell of 0.6 mm) was the strongest; the yield strength is shown in Table 1 together with the actual relative porosity as measured by microCT. This was due to the similar strut thickness of the four models but shorter strut lengths and hence higher density for the smallest model. This also shows that at this scale, the strength and elastic modulus is strongly correlated with the actual porosity (or density).

Table 1. Experimental data.

Unit Cell Design (mm)	Actual Relative Porosity (%)	Compressive Elastic Modulus (MPa)	Compressive Strength (MPa)	Maximum Load (N)
0.6	63	346	51.1	662
		418	54.8	710
		318	53.4	692
0.8	77	190	9.1	209
		167	9.9	227
		200	10.1	232
1.0	84	83	3.3	117
		77	3.9	139
		102	3.6	130
1.2	90	26	1.0	53
		25	0.9	46
		24	1.1	56

In situ compression allowed imaging of the same lattice prior to full densification, first before loading, then directly after initial yielding, and at a few more representative steps during yielding. This is shown in Figure 6, where red arrows indicate the positions where the loading was stopped and microCT scans were performed. The resulting microCT data is represented for the aligned volumes, with side-by-side slice images through the middle of the lattice, and with 3-D views of the entire lattice. These images indicate that yielding occurred gradually and progressively as struts collapsed in this type of lattice.

The alignment of the scans was simplified by the fact that the sample stayed in the same location in the scan system, and as the load cell works by moving the bottom upwards, the deformation could be imaged more closely on individual struts by making the unloaded scan transparent and visualizing the loaded image. This is done in Figure 7 for a small section (approx. 0.5 mm) to visualize the deformation and collapse of individual struts, in this case taken from the middle (away from the edge of the lattice) near the top of the sample where collapse first occurred. The light blue transparent struts shown in Figure 7 show the unloaded sample, while the third scan in the series is shown here in solid rendering (this is at the first yield dip, the third arrow in Figure 6). The color coding applied to the loaded sample image is a nominal–actual comparison; this quantitatively shows the deformation value (in red was where largest deformation was relative to the unloaded sample). Besides collapse, the largest deformations occurred at the strut junctions. Similar results were reported for Ti6Al4V lattices in Reference [25] where digital image correlation (of external surface of lattice) was used during compression testing of diamond-type lattices. It was found that fracture occurs exclusively at the nodes in this design, and similar layered collapse was observed. The three images are the same region from different viewing angles.

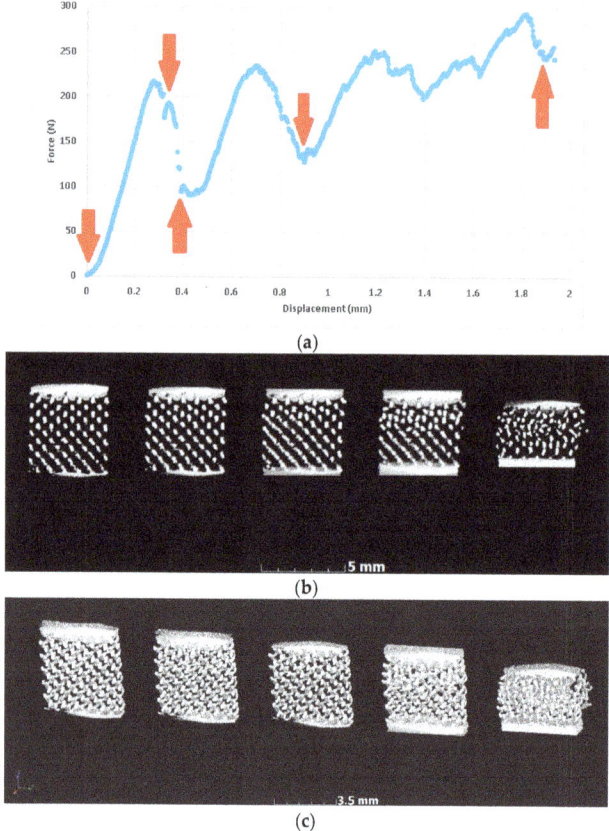

Figure 6. In situ deformation imaging of a lattice (0.8 mm unit cell) at the selected points during yielding; steps are shown as arrows in the force-displacement curve. (**a**) Force-displacement with red arrows indicating steps for microCT scans (stop); (**b**) MicroCT slice images at each step showing collapse; (**c**) The corresponding microCT 3-D images of the lattice at each step (cropped).

Figure 7. Three different angular views of the same internal location showing yielding behavior of individual struts and color coding indicating the largest deformation relative to the unloaded state (deformation upwards in image, where the unloaded state is semi-transparent blue). Loading direction is indicated by arrows.

4. Conclusions

This paper reported the mechanical properties of a series of microlattices with struts near the single-track width of the laser powder bed fusion system used to produce them. The results show that such lattices could be produced successfully but the small strut thickness deviated from the designed value. It was shown that, in the limited range investigated here, the mechanical properties of microlattices produced by LBPF were strongly dependent on actual density and could therefore be predicted with some confidence using this measure alone. Compared to larger lattices, the dependence of the mechanical properties was stronger with density (higher slope in the Ashby–Gibson equation). In situ microCT imaging demonstrated that the largest deformations under compression occurred at the strut junctions. These images represent the first in situ images of a full microlattice structure's yielding behavior.

Author Contributions: Conceptualization, A.D.P.; Methodology, A.D.P., I.Y. (Ina Yadroitsava), and I.Y. (Igor Yadroitsev); Software, A.D.P.; Validation, A.D.P. and D.-P.K.; Formal Analysis, All; Investigation, A.D.P.; Resources, All; Data Curation, All; Writing-Original Draft Preparation, A.D.P.; Writing-Review & Editing, All; Visualization, All.

Funding: The authors thank the South African Research Chairs Initiative of the Department of Science and Technology and National Research Foundation of South Africa (Grant No. 97994), the Collaborative Program in Additive Manufacturing (Contract No. CSIR-NLC-CPAM-15-MOA-CUT-01).

Conflicts of Interest: The authors declare no conflict of interest.

References

1. Schmidt, M.; Merklein, M.; Bourell, D.; Dimitrov, D.; Hausotte, T.; Wegener, K.; Overmeyer, L.; Vollertsen, F.; Levy, G.N. Laser based additive manufacturing in industry and academia. *CIRP Ann.* **2017**, *66*, 561–583. [CrossRef]
2. DebRoy, T.; Wei, H.L.; Zuback, J.S.; Mukherjee, T.; Elmer, J.W.; Milewski, J.O.; Beese, A.M.; Wilson-Heid, A.; De, A.; Zhang, W. Additive manufacturing of metallic components—Process, structure and properties. *Prog. Mater. Sci.* **2017**, *92*, 112–224. [CrossRef]
3. Yadroitsev, I.; Gusarov, A.; Yadroitsava, I.; Smurov, I. Single track formation in selective laser melting of metal powders. *J. Mater. Process. Technol.* **2010**, *210*, 1624–1631. [CrossRef]
4. Du Plessis, A.; Yadroitsev, I.; Yadroitsava, I.; Le Roux, S. X-ray micro computed tomography in additive manufacturing: A review of the current technology and applications. *3D Print. Addit. Manuf* **2018**. [CrossRef]
5. Yadroitsev, I.; Krakhmalev, P.; Yadroitsava, I.; Du Plessis, A. Qualification of Ti6Al4V ELI Alloy Produced by Laser Powder Bed Fusion for Biomedical Applications. *JOM* **2018**, *70*, 372–377. [CrossRef]
6. Tan, X.P.; Tan, Y.J.; Chow, C.S.L.; Tor, S.B.; Yeong, W.Y. Metallic powder-bed based 3D printing of cellular scaffolds for orthopaedic implants: A state-of-the-art review on manufacturing, topological design, mechanical properties and biocompatibility. *Mater. Sci. Eng. C* **2017**, *76*, 1328–1343. [CrossRef] [PubMed]
7. Zhang, X.-Y.; Fang, G.; Zhou, J. Additively Manufactured Scaffolds for Bone Tissue Engineering and the Prediction of their Mechanical Behavior: A Review. *Materials* **2017**, *10*, 50. [CrossRef] [PubMed]
8. Dong, G.; Tang, Y.; Zhao, Y.F. A Survey of Modeling of Lattice Structures Fabricated by Additive Manufacturing. *J. Mech. Des.* **2017**, *139*, 100906. [CrossRef]
9. Gibson, L.; Ashby, M. *Cellular Solids: Structure and Properties*; Cambridge University Press: Cambridge, UK, 1999.
10. Ashby, M.; Evans, T.; Fleck, N.; Hutchinson, J. *Metal Foams: A Design Guide*; Elsevier: New York, NY, USA, 2000.
11. Parthasarathy, J.; Starly, B.; Raman, S.; Christensen, A. Mechanical evaluation of porous titanium (Ti6Al4V) structures with electron beam melting (EBM). *J. Mech. Behav. Biomed. Mater.* **2010**, *3*, 249–259. [CrossRef] [PubMed]
12. Sallica-Leva, E.; Jardini, A.L.; Fogagnolo, J.B. Microstructure and mechanical behavior of porous Ti–6Al–4V parts obtained by selective laser melting. *J. Mech. Behav. Biomed. Mater.* **2013**, *26*, 98–108. [CrossRef] [PubMed]
13. Ahmadi, S.M.; Campoli, G.; Yavari, S.A.; Sajadi, B.; Wauthlé, R.; Schrooten, J.; Weinans, H.; Zadpoor, A.A. Mechanical behavior of regular open-cell porous biomaterials made of diamond lattice unit cells. *J. Mech. Behav. Biomed. Mater.* **2014**, *34*, 106–115. [CrossRef] [PubMed]

14. Ahmadi, S.M.; Yavari, S.A.; Wauthle, R.; Pouran, B.; Schrooten, J.; Weinans, H.; Zadpoor, A.A. Additively Manufactured Open-Cell Porous Biomaterials Made from Six Different Space-Filling Unit Cells: The Mechanical and Morphological Properties. *Materials* **2015**, *8*, 1871–1896. [CrossRef] [PubMed]
15. Bobbert, F.S.; Lietaert, K.; Eftekhari, A.A.; Pouran, B.; Ahmadi, S.M.; Weinans, H.; Zadpoor, A.A. Additively manufactured metallic porous biomaterials based on minimal surfaces: A unique combination of topological, mechanical, and mass transport properties. *Acta Biomater.* **2017**, *53*, 572–584. [CrossRef] [PubMed]
16. Du Plessis, A.; Yadroitsava, I.; Yadroitsev, I.; le Roux, S.; Blaine, D. Numerical comparison of lattice unit cell designs for medical implants by additive manufacturing. *Virtual Phys. Prototyp.* **2018**, 1–16. [CrossRef]
17. Du Plessis, A.; Yadroitsava, I.; Yadroitsev, I. Ti6Al4V lightweight lattice structures manufactured by laser powder bed fusion for load-bearing applications. *Opt. Laser Technol.* **2018**, *108*, 521–528. [CrossRef]
18. Carlton, H.D.; Lind, J.; Messner, M.C.; Volkoff-Shoemaker, N.A.; Barnard, H.S.; Barton, N.R.; Kumar, M. Mapping local deformation behavior in single cell metal lattice structures. *Acta Mater.* **2017**, *129*, 239–250. [CrossRef]
19. Qiu, C.; Yue, S.; Adkins, N.J.; Ward, M.; Hassanin, H.; Lee, P.D.; Withers PJAttallah, M.M. Influence of processing conditions on strut structure and compressive properties of cellular lattice structures fabricated by selective laser melting. *Mater. Sci. Eng. A* **2015**, *628*, 188–197. [CrossRef]
20. Materialise. Available online: https://www.materialise.com/en/software/magics (accessed on 12 August 2018).
21. Du Plessis, A.; le Roux, S.G.; Guelpa, A. The CT Scanner Facility at Stellenbosch University: An open access X-ray computed tomography laboratory. *Nucl. Instrum. Methods Phys. Res. Sect. B Beam Interact. Mater. Atoms.* **2016**, *384*, 42–49. [CrossRef]
22. Volume graphics. Available online: https://www.volumegraphics.com/en/products/vgstudio-max.html (accessed on 12 August 2018).
23. Kouprianoff, D.; du Plessis, A.; Yadroitsava, I.; Yadroitsev, I. Destructive and nondestructive testing on small and intricate SLM components. In Proceedings of the 18th Annual International RAPDASA Conference, Durban, South Africa, 8–10 November 2017.
24. Yadroitsava, I.; Els, J.; Booysen, G.; Yadroitsev, I. Peculiarities of single track formation from TI6AL4V alloy at different laser power densities by SLM. *S. Afr. J. Ind. Eng.* **2015**, *26*, 86–95.
25. Liu, F.; Zhang, D.Z.; Zhang, P.; Zhao, M.; Jafar, S. Mechanical Properties of Optimized Diamond Lattice Structure for Bone Scaffolds Fabricated via Selective Laser Melting. *Materials* **2018**, *11*, 374. [CrossRef] [PubMed]

 © 2018 by the authors. Licensee MDPI, Basel, Switzerland. This article is an open access article distributed under the terms and conditions of the Creative Commons Attribution (CC BY) license (http://creativecommons.org/licenses/by/4.0/).

Article

Two-Scale Tomography Based Finite Element Modeling of Plasticity and Damage in Aluminum Foams

Yasin Amani, Sylvain Dancette *, Eric Maire, Jérôme Adrien and Joël Lachambre

University of Lyon, INSA Lyon, CNRS UMR5510, Laboratoire MATEIS, F-69621 Villeurbanne CEDEX, France; yasin.amani@univ-pau.fr (Y.A.); eric.maire@insa-lyon.fr (E.M.); jerome.adrien@insa-lyon.fr (J.A.); joel.lachambre@insa-lyon.fr (J.L.)
* Correspondence: sylvain.dancette@insa-lyon.fr; Tel.: +33-472-43-85-41

Received: 11 September 2018; Accepted: 6 October 2018; Published: 15 October 2018

Abstract: In this study, finite element (FE) modeling of open-cell aluminum foams in tension was performed based on laboratory X-ray tomography scans of the materials at two different scales. High-resolution stitching tomography of the initial state allowed local intermetallic particles to be distinguished from internal defects in the solid phase of the foam. Lower-resolution scans were used to monitor the deformation and fracture in situ during loading. 3D image-based FE models of the foams were built to simulate the tensile behavior using a new microstructure-informed Gurson–Tvergaard–Needleman model. The new model allows quantitative consideration of the local presence of brittle intermetallic particles in the prediction of damage. It performs well in the discrimination of potential fracture zones in the foam, and can be easily adapted to any type of architectured material where both the global architecture and local microstructural details should be taken into account in the prediction of damage behavior.

Keywords: aluminum foams; intermetallics; X-ray tomography; finite element analysis; damage

1. Introduction

Materials containing gaseous cells are widely found in both nature and engineering applications. Cellular materials can be divided into two different categories based on the continuity of the gaseous phase. The gas phase inside cells can be free (open cell materials) or trapped between cells (closed cell materials). Polymer foams are the most common type of cellular material, but ceramic and metal foams are also produced. Due to their specific structure, cellular materials exhibit a combination of several interesting properties. Mechanically, they hold characteristics like strength, deformability, stiffness, and energy absorption capacity, and are lightweight [1]. Thermally, they are insulators, and some of them are high-temperature-resistant [2]. Acoustically, they are used as effective sound absorbers [3]. In addition, they are frequently applied in other engineering fields, such as packaging, crash-worthiness, and in the production of lightweight sandwich panels [4].

In order to characterize the mechanical properties of cellular materials with complex architecture, an idealized unit cell model assumption was introduced by Ashby [5]. Accordingly, Young's modulus and plastic collapse strength of the foam are related to an exponential power of the foam's relative density [1,2,6]. Andrews et al. [7] noticed that the predicted Young's modulus and strength lie very close to experimental measurements in the case of foams without curvature, corrugation, or internal imperfection.

X-ray micro-computed tomography [8] has been widely used as a non-destructive technique to study yielding mechanism [9] and crack propagation [10]. Finite element (FE) simulations based on tomographic volumes were first used to study the mechanical properties of trabecular bone structures [11]. Maire et al. [12] employed X-ray tomography and morphological granulometry techniques as a generic way to characterize cellular materials to be used for FE calculations. Youssef et al. [13] and Caty et al. [14] developed one of the first methods to build an FE model directly based on the cellular structure obtained by X-ray tomography. Lacroix et al. [15] noted the effect of pore dispersion on the distribution of the FE-computed stress in bone tissue biomaterials. Later, Jeon et al. [16] investigated the deformation and plastic collapse mechanism of closed cell Al foam, and Michailidis et al. [17] determined the stress–strain behavior of open-cell Al and Ni foams. Subsequently, D'Angelo et al. [18] obtained the average Young's modulus and the stress concentrations within the thinnest sections of SiC ceramic foams. Zhang et al. [19] extended the latter method to explain and predict the rupture of the material based on the contour plot of von Mises stress after simulation. Petit et al. [20] developed a method by running FE simulations and qualitatively defining elastoplastic and damage properties of the aluminum and intermetallics phases. However, no study has yet considered the effect of intermetallics on FE simulation results *quantitatively*.

The present paper focuses on the characteristics of open-cell aluminum foams at two different scales: firstly, the macroscopic cellular structure, and secondly, the local microstructure of the 6101 aluminum alloy constituting the cell walls. The uniaxial tensile modulus and strength of several foams produced by ERG Materials and Aerospace Corp. with different cell sizes are discussed based on X-ray tomography analyses combined with the corresponding image-based FE simulations, where the element behavior is enriched by the local image-based fraction of intermetallics. This new procedure allows the influence of intermetallics on the deformation and damage behavior of the foams to be studied.

2. Materials and Experimental Procedures

The studied materials were Duocel® open-cell foams produced and kindly provided by ERG Aerospace Corporation, Oakland, CA, United States. The samples were made of 6101 aluminum alloy, subjected to T6 precipitation-hardening heat treatment. Two foam samples with different cell sizes and testing directions were chosen. Cell sizes of the foam samples were 20 and 30 pores per inch (PPI), corresponding respectively to 0.79 and 1.18 pores per mm. The plasticity and fracture of a 30 PPI foam sample studied in the longitudinal direction were already addressed by Petit [20]. Therefore, the 30 PPI sample was cut in the transverse direction to compare with the study of Petit [20]. The 20 PPI foam sample was cut in the longitudinal direction. The dimensions of the 20 PPI foam sample were 9.4 mm × 6.0 mm × 18.8 mm, and the dimensions of the 30 PPI foam sample were 13.3 mm × 4.8 mm × 10.9 mm.

Table 1 shows the chemical composition of the ERG foam 6101 aluminum alloy characterized by an inductively coupled plasma atomic emission spectrometer by Zhou et al. [21]. Densities of the foam samples were evaluated by weighing on a balance and measuring their dimensions using digital calipers. Afterwards, the relative density of each foam sample was calculated by dividing the global density of the big block of foam by the density of pure Al (2.7 g/cm^{-3}). Relative densities of 0.0733 and 0.0633 were found for the 20 PPI and 30 PPI foam samples, respectively.

Table 1. Chemical composition of the 6101 aluminum foam in weight percent (wt.%). Data from Zhou et al. [21].

Element	Cu	Mg	Mn	Si	Fe	Zn	B
Content	0.03	0.19	0.01	0.27	0.12	0.01	0.03

2.1. Tomography

Dual-scale laboratory X-ray tomography scanning was used is this study to capture both the macroscopic deformation of the foam during loading and the initial local microstructure, as detailed below. The tomograph (phoenix l x-ray v l tome l x s, GE Company, Boston, MA, USA) produces a series of N radiographs corresponding to N angular positions of the sample. Based on the Beer–Lambert law, every line integral of the attenuation coefficient along the beam path corresponds to an element in the recorded projection [22]. The resulting images are superimposed information of a three-dimensional (3D) object in a two-dimensional (2D) plane. The detector is a charge-coupled device (Varian Paxscan, Varian Medical Systems Inc., Palo Alto, CA, USA). It records radiographs passed through the sample, which are imported into a commercial reconstruction software (datos|x, GE Company, Boston, MA, USA). The latter uses a filtered back-projection algorithm [23]. The tomograph was operated at 80 kV acceleration voltage using a tungsten transmission target with a 280 µA current. The spot size was between 2 ∼ 3 µm during all scans, and no filter was used.

First, 3D tomographic images with low resolution (20 µm cubic voxel edge size) were taken to obtain the global structure of the samples. At this resolution, only two phases were imaged: the solid phase and macroscopic void-cells. Each foam sample was then moved toward the X-ray tube to decrease the voxel size so that white intermetallic particles or small cavities could be observed. In the case of the 20 PPI foam sample, the cubic voxel edge size was 4 µm, and it was 6 µm for the 30 PPI foam sample. However, in this high-resolution configuration, the field of view of the detector was not large enough to picture the whole sample. In such cases, the so-called "local tomography" or "region of interest" technique is used. This is a tomography scanning method where portions of the sample are placed in the field of view of the detector during rotation [24].

The 3D images captured by this technique contained high-resolution details of the solid phase, including intermetallic particles, casting defects, and internal cavities of the foam struts, as shown in Figure 1a,b. Note that gray, white, and black colors correspond to aluminum, intermetallic particles, and void-cells, respectively. These intermetallic particles were mainly α-AlFeSi (Al_8Fe_2Si) and β-AlFeSi (Al_5FeSi) precipitates in the grain boundaries [25]. In order to obtain an entire image of the whole foam sample, the local tomography procedure was repeated many times successively by displacing the center of the sample in a plane parallel to the detector plane. The latter was parallel to the (y, z) axes of Figure 2a, where a sketch of the setup is illustrated. Then, these high-resolution 3D images were combined and concatenated to retrieve the whole 3D volume, but this time with a small voxel size. The difference between big and small voxel size tomography is illustrated in Figure 2b,c. It is very clear from this image that the higher resolution allows the local presence of intermetallic particles to be captured. Then, aluminum, intermetallic particles, and void-cells were segmented by standard thresholding based on gray levels distribution and were attributed gray (125), white (255), and black (0) 8-bit values for visualization, respectively. The volume fraction of solid (intermetallic particles and aluminum) and gaseous phases were calculated by counting the number of voxels of each corresponding color using the Fiji software [26]. The results are given in Table 2. The amount of porosity in the solid phase was less than 0.01%, which is negligible. The volume fraction of voids and intermetallic particles in the 20 PPI sample were slightly lower than those in the 30 PPI sample. In addition, the image analysis of the tomographic data revealed that most struts presented a close-to-triangular cross section.

The solid phase was analyzed with the local thickness plugin of the Fiji program. The plugin estimates the local thickness by the largest sphere that fits inside the solid phase and contains its voxels. The result of the analysis is a 3D stack of the foam structure, where the local thicknesses can be represented with a given color map.

The average size and distribution of void-cells in three directions were evaluated by the analysis of the gaseous phase of the 3D binary image using a 3D Watershed plugin implemented in the Fiji program. The plugin splits the continuous gaseous phase into non-overlapping void-cells and assigns them different gray levels. These segmented and labeled void-cells do not contain any strut or node. The so-called "Feret" diameter of each segmented cell could be evaluated with another home-made plugin from the minimum and maximum x, y, and z values of its voxels [27].

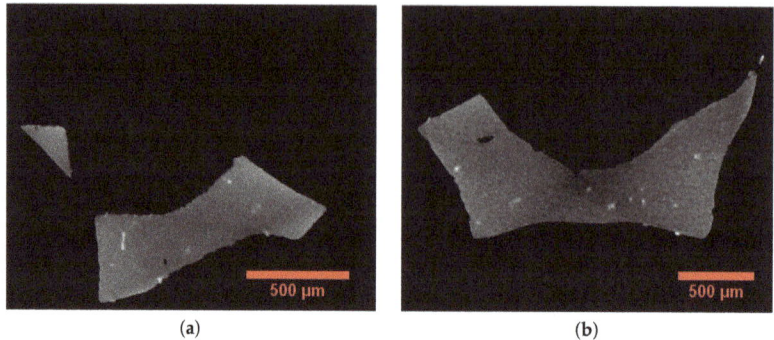

Figure 1. Solid phase defects of foam struts were obtained by local tomography: (**a**) 20 pores per inch (PPI); (**b**) 30 PPI. The light white visible zones are α-AlFeSi (Al_8Fe_2Si)- or β-AlFeSi (Al_5FeSi)-based inclusions.

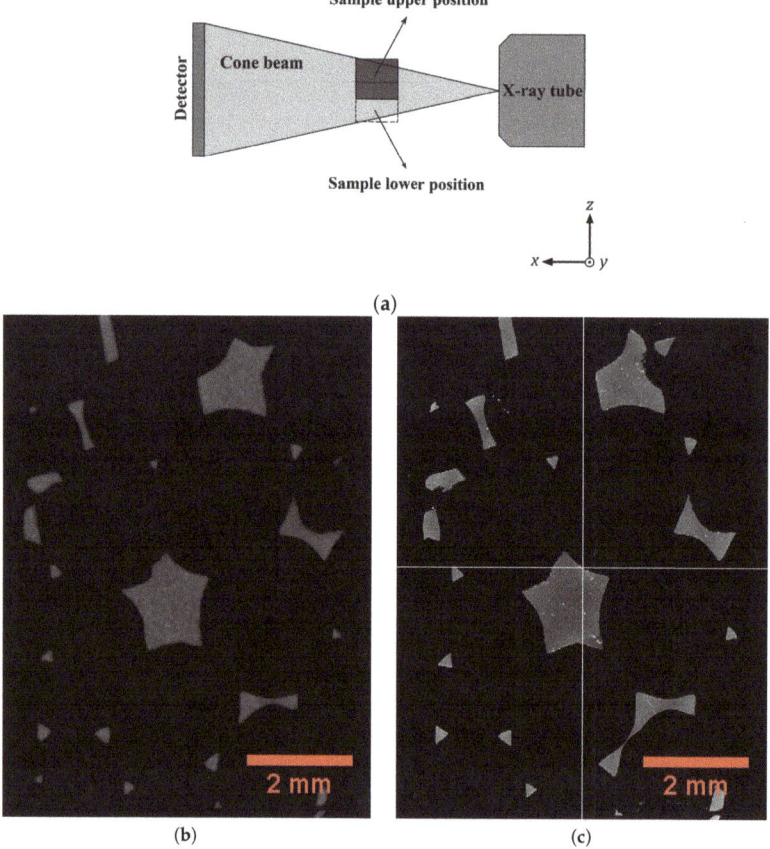

Figure 2. (**a**) Sketch of the stitching tomography setup. Tomography of entire geometry of the 20 PPI sample by: (**b**) global and (**c**) successive local procedure. In (**c**), the intermetallic particles are visible.

Table 2. Volume fractions in percent (%) of the different phases in the foam samples.

Cell Size (PPI)	Void	Aluminium	Intermetallic Particles
20	92.67	7.31	0.02
30	93.07	6.89	0.04

The average thicknesses of the struts, nodes, and average diameters of void-cells in x, y, and z directions are given in Table 3. The direction in which the foam presents the highest average Feret diameter is called longitudinal direction. The two perpendicular directions are called transverse directions. Consequently, the longitudinal direction for the 20 PPI sample was z, and it was x for the 30 PPI sample. It can be noted, however, that the geometrical anisotropy of the foams was quite small, which resulted in an almost isotropic mechanical behavior, as documented for example in Ref. [28] on a similar material.

Table 3. Geometric characteristics of the studied foams.

Cell Size (PPI)	20	30
Strut thickness (mm)	0.16 ± 0.04	0.16 ± 0.04
Node thickness (mm)	0.38 ± 0.10	0.36 ± 0.04
Void-cell dimension in x (mm)	2.60 ± 0.20	2.57 ± 0.49
Void-cell dimension in y (mm)	2.84 ± 0.23	2.36 ± 0.49
Void-cell dimension in z (mm)	3.04 ± 0.55	2.47 ± 0.32

2.2. In Situ Tensile Test

In order to investigate the initiation and subsequent propagation of plastic deformation through the foam structures, the two foam samples were glued to steel M3 screws using an epoxy glue (Araldite 2015) prior to being clamped by the steel machine grips [20]. It is noted that the size of the foam samples should be at least eight times larger than the cell size in order to prevent edge effect on measuring Young's modulus and strength [29]. The samples were then loaded progressively in tension. Tensile tests were performed using an in situ tensile testing machine with a 5 kN load cell. The force transducer (HBM U9B 5kN, Hottinger Baldwin Messtechnik GmbH, Darmstadt, Germany) had a general precision of ±1 N. The loading process was under displacement control at a crosshead displacement speed of 0.001 mm/s in order to ensure quasi-static test conditions and proper control of the in situ loading procedure. In accordance with the sample cutting directions detailed above, the 20 PPI foam sample was stretched in the longitudinal direction, while the 30 PPI sample was stretched in the transverse direction. The loading direction corresponds to z in Table 3 for both foam samples.

The strain was measured based on the displacement of the top and bottom surfaces of the sample in the 3D tomographic images, obtained for the different deformation steps of the tensile test. Low resolution (20 µm cubic voxel edge size) in situ tomographic scans were performed after every 0.2 mm of crosshead displacement. The (nominal) stress was calculated by dividing the measured force on the load cell by the initial rectangular cross-sectional area of the foam sample.

3. Model

3.1. Gurson–Tvergaard–Needleman Damage Model

The plastic behavior of the open-cell alloy foam is directly influenced by the properties of the solid phase. The fracture behavior of the solid aluminum phase is governed by ductile damage, namely the nucleation, growth, and coalescence of cavities. Nucleation is mainly due to microcrack initiation (e.g., from pores or defects in the material [30]), second phase particle decohesion, or fracture in the alloy [31]. The growth of cavities occurs by plastic yielding of the matrix surrounding the cavity. Coalescence occurs when neighboring cavities or cracks merge together [32].

In this study, a standard Gurson–Tvergaard–Needleman (GTN) model [33–37] is used to represent the effect of void nucleation and growth in the simulation. The GTN model renders the effect of porosity on the yield locus and its sensitivity to the hydrostatic component of loading. It reduces to a standard isotropic von Mises yielding criterion in the absence of porosity. The governing equation is defined as:

$$\Phi(\sigma_{eq},\sigma_y,\sigma_H,f) = \left(\frac{\sigma_{eq}}{\sigma_y}\right)^2 + 2fq_1\cosh\left(\frac{3q_2\sigma_H}{2\sigma_y}\right) - (1+q_3f^2) = 0, \quad (1)$$

where Φ is the yield function, q_1, q_2, and q_3 are calibrating parameters, σ_{eq} is the von Mises equivalent stress, σ_y is the yield stress, σ_H is the hydrostatic stress, and f is the void volume fraction (VVF) in the matrix. In addition, starting from an initial void volume fraction f_0, the total change in f, noted $\dot{f}$, is defined as [34]:

$$\dot{f} = \dot{f}_{gr} + \dot{f}_{nucl} = (1-f)\,tr(\dot{\varepsilon}^{pl}) + \frac{f_N}{s_N\sqrt{2\pi}}\exp\left[-\frac{1}{2}\left(\frac{\varepsilon_{eq}^{pl}-\varepsilon_N}{s_N}\right)^2\right]\dot{\varepsilon}_{eq}^{pl}, \quad (2)$$

where $\dot{f}_{nucl}$ is the contribution of nucleating voids; $\dot{f}_{gr}$ is the void growth rate, which is based on mass conservation and directly proportional to the hydrostatic component of plastic strain rate tensor $tr(\dot{\varepsilon}^{pl})$; and ε_{eq}^{pl} is the equivalent plastic strain. ε_N and s_N are the mean value and standard deviation of the normal nucleation distribution. f_N is the volume fraction of the nucleated voids. The power law work hardening used in this study is defined as [38]:

$$\frac{\sigma_y}{\sigma_0} = \left(\frac{\sigma_y}{\sigma_0} + \frac{3G}{\sigma_0}\varepsilon^{pl}\right)^N, \quad (3)$$

where σ_0 is the initial yield stress, N is the hardening exponent, G the elastic shear modulus and ε^{pl} is the plastic strain. In addition, the initiation of necking in tension takes place according to a Considère criterion when the work hardening rate converges to the current yield stress:

$$\frac{\partial\sigma_y}{\partial\varepsilon^{pl}} = \sigma_y. \quad (4)$$

3.2. Mesh Generation

In this study, the initial 3D tomographic volumes of the foam samples were used to generate 3D image-based FE meshes. First, a surface mesh was generated from the solid phase boundaries in the volume. Next, the surface mesh was simplified and remeshed to reduce the number of triangles while preserving a proper description of the surface. Finally, the solid volume was filled by first-order tetrahedra [13] (C3D4 elements in the Abaqus software), allowing for explicit non-linear simulations including damage and fracture. The whole meshing procedure was performed using the commercial Avizo® software [39]. Each foam sample was meshed with four different mesh sizes in order to investigate the mesh size sensitivity of the results, as detailed in Section 3.4. The number of tetrahedra and nodes of the reference volume meshes are detailed in Table 4.

Table 4. Characteristics of the reference volume meshes for the two foam samples.

Cell Size (PPI)	Number of Nodes	Number of Elements	Characteristic Element Size (microns)
20	124,267	419,846	150
30	50,611	146,660	150

Two classes of FE models were considered in this work: (i) *homogeneous* (or microstructure "blind") models, where the local constitutive behavior corresponds to the nominal average behavior of the aluminum alloy everywhere; and (ii) *heterogeneous* models, where the local constitutive behavior depends on the local microstructure (intermetallic fraction in the present case) as informed from the detailed initial 3D tomographic scans of the samples.

The procedure used to generate the *heterogeneous* FE model is described in [40] and briefly recalled here. For each element of the 3D image-based FE mesh, a python script retrieves the voxels of the segmented tomographic volume located in the element interior. The number of white (intermetallic) voxels is then evaluated in order to compute the local intermetallic volume fraction f_{IM} in the element. As a result, the tetrahedra of the volume mesh are tagged depending on the volume fraction of intermetallic particles regrouped into 100 equidistant classes with f_{IM} varying from 0.0 to 1.0. The distributions of the intermetallic volume fraction in the tetrahedra are illustrated in Figure 3 for the 20 PPI and 30 PPI foam samples.

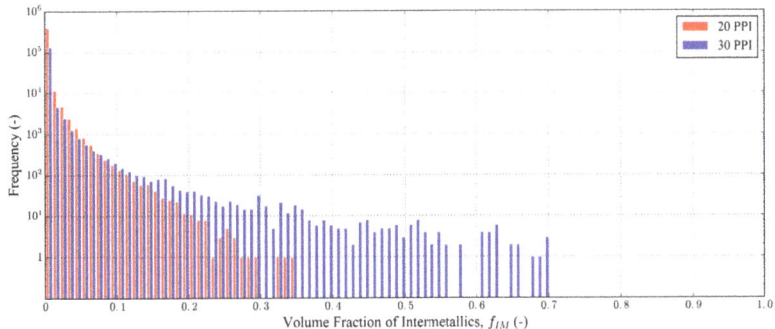

Figure 3. Distributions of intermetallic particles volume fraction in the finite element (FE) tetrahedra for the 20 PPI and 30 PPI foam samples.

3.3. Identification of the Constitutive Model Parameters

The identification of constitutive model parameters is a challenging task, especially when it comes to the local behavior of the solid (metallic) phase in macroporous structures like the present foam samples. Several studies demonstrated that the tensile mechanical behavior of individual struts extracted from the foam might be significantly different from the behavior of the corresponding bulk metal [41–43], even subjected to the same hardening heat treatment. The problem becomes even more complicated if one would like to take the local microstructure (e.g., local intermetallic fraction here) of foam struts into account in the constitutive modeling. The present work proposes to rely on axisymmetric FE unit cell calculation to compute the local constitutive behavior of the different aluminum matrix composites with various intermetallic volume fractions that are available in each tetrahedron of foam sample mesh.

The 2D axisymmetric cell model is illustrated in Figure 4a. Varying the intermetallic (green) radius r allows to cover intermetallic volume fractions f_{IM} from 0.0 to 0.6 in order to compute the resulting composite behavior to be considered for the corresponding elements in the distribution of Figure 3. A linear elastic behavior with a Young's modulus of $E = 160$ GPa and a Poisson's ratio $\nu = 0.33$ was considered for the intermetallic properties [20]. The power-law hardening of Equation (3) was used for the aluminum phase, with $E = 70$ GPa, $\nu = 0.33$, an initial yield stress $\sigma_0 = 97$ MPa, and a hardening exponent $N = 0.052$. This initial yield stress corresponds to the lower range of individual strut yield stresses measured by [41] on a similar foam. The resulting tensile stress–strain curves obtained for increasing intermetallic volume fraction f_{IM} are illustrated in Figure 4b. They were extracted along with the corresponding Young's moduli (ranging from 70 GPa for pure aluminum to 115 GPa for

$f_{IM} = 0.6$) to define the elasto-plastic properties of the foam tetrahedra in the different classes of Figure 3.

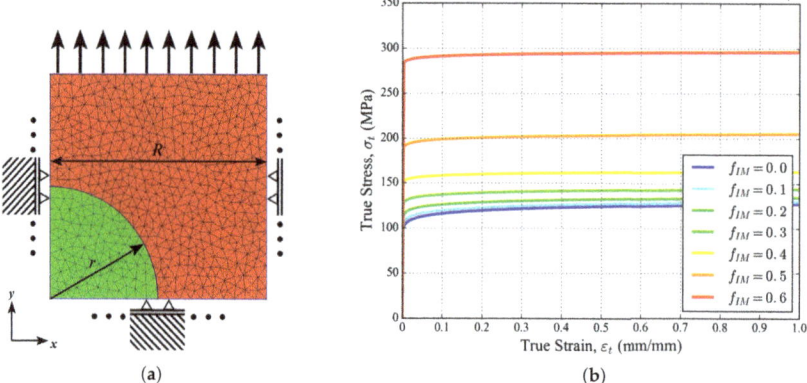

Figure 4. (**a**) 2D axisymmetric model of the aluminum matrix material (red) with a given intermetallic (green) volume fraction f_{IM}. The right exterior line remains straight and vertical during loading, but is free to move in the radial direction. (**b**) Resulting hardening behavior for increasing intermetallic volume fraction.

Damage model parameters were also identified for each element class (each value of f_{IM}). Following [36], the calibrating parameters of the GTN yield function in Equation (1) were chosen as $q_1 = 1.5$, $q_2 = 1$, and $q_3 = 2.25$. In the nucleation part of Equation (2), the volume fraction of nucleating voids was taken as $f_N = 0.04$ [20,44]. The initial void volume fraction f_0 was taken as zero, which is consistent with the negligible initial porosity of the solid phase measured in the tomographic volumes (Section 2.1). ε_N and s_N were gradually decreased with the increasing value of f_{IM} to mimic the transition towards a brittle behavior with the increase in intermetallic fraction in the material (Table 5). Figure 5 illustrates the resulting tensile stress–strain behaviors, including GTN damage of the corresponding aluminum matrix materials. The increase in volume fraction of intermetallic particles clearly results in a stiffer, stronger, but more brittle behavior. This will impact the constitutive response of the tetrahedra with a high intermetallic fraction in the foam FE mesh.

Table 5. Parameters of the Gurson–Tvergaard–Needleman (GTN) nucleation model for increasing volume fraction of intermetallic particles.

Intermetallic Fraction f_{IM}	ε_N	s_N	f_N
0.0	0.18	0.06	0.04
0.1	0.16	0.05	0.04
0.2	0.13	0.04	0.04
0.3	0.10	0.03	0.04
0.4	0.08	0.03	0.04
0.5	0.05	0.02	0.04
0.6	0.03	0.01	0.04

3.4. Simulation Conditions

The FE approach described in the present study consists of a uniaxial monotonous tensile loading applied to the 3D image-based FE meshes generated in the previous section. Two nodes, one on the top surface and another on the bottom surface of the foam samples, were assigned as master nodes. z displacement degree of freedom of each node on the top surface was constrained to be the same as the one of the top master node, which was assigned a positive displacement u_z. Similarly, the z displacement degree of freedom of each node on the bottom surface was constrained to be the same as the one of the bottom master node, which was blocked. Furthermore, nodes located on two central lines on the top and bottom faces were constrained by imposing $u_x = 0$ to prevent the potential rotation of the sample about the tensile axis, which is inhibited experimentally by the machine grips.

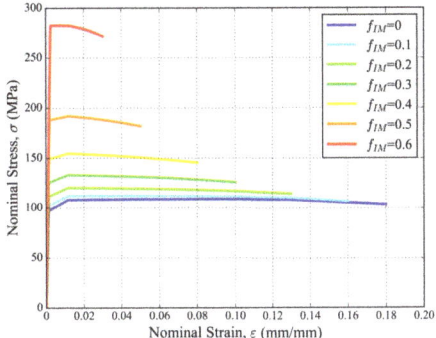

Figure 5. Tensile stress–strain curves of aluminum matrix materials including GTN damage, for increasing fraction of intermetallic particles.

The sensitivity of the simulation results with respect to the element size is illustrated in Figure 6a,b with the predicted apparent Young's modulus obtained with the homogeneous non-porous models. It converged to about 125 MPa with decreasing element size in the case of the 20 PPI foam sample and to about 210 MPa in the case of the 30 PPI foam sample. The characteristic element size of 150 µm was therefore chosen as the reference element size for both samples in Table 4. However, one can note that these predicted values of Young's moduli were overestimated with respect to the experimental measures based on the (3D) image-based macroscopic tensile strain. Two modeling choices in the present study might contribute to this situation: (i) The use of first-order tetrahedral elements imposed by the complexity of the structure and the compatibility with the Explicit solver of the Abaqus FE software. First-order tetrahedra are known to be too stiff in bending-dominated problems [38]. (ii) The uniaxial tension boundary conditions applied in the simulations slightly differ from the effective boundary conditions in the experiments.

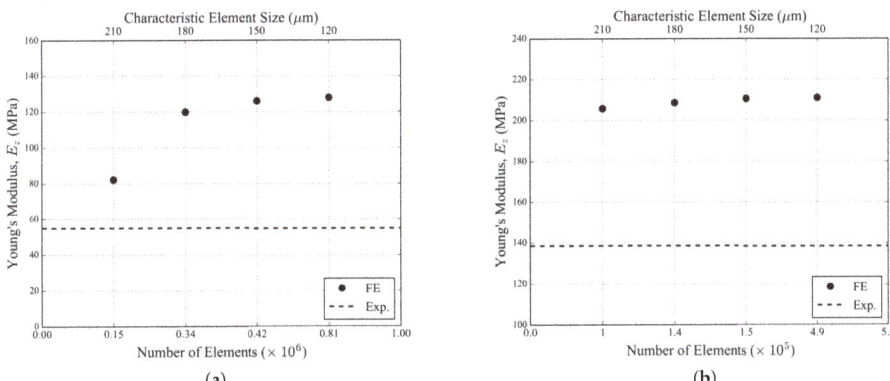

Figure 6. Apparent Young's modulus convergence tests for (**a**) 20 PPI and (**b**) 30 PPI foam samples in the case of the homogeneous non-porous models.

4. Results and Discussion

Figures 7 and 8 respectively illustrate the progressive deformation of the 20 PPI and 30 PPI foam samples during the in situ tensile tests. Spatially uniform straining was observed until the first fracture happened in a strut. Both plastic stretching and plastic bending of the struts might be observed during the tensile deformation of foam blocks. Struts approximately parallel to the loading direction were mostly stretched, while the others were deformed first by bending and then potentially by stretching once they were aligned with the tensile direction. It was observed here that the majority of the struts were not parallel to the loading axis. Therefore, the deformation of the foam blocks was dominated by bending for small levels of elongation. The deformation then shifted to a stretching-dominated mode after significant elongation and re-alignment of the struts. It should be noted that bending-dominated deformation tended to lower the apparent stiffness and strength of the foam samples, while stretching dominated deformation rather tended to increase them.

Fracture in the foam blocks initiated in struts that were parallel to the loading direction and stretched. Subsequent propagation of fracture was achieved in the neighboring struts, mostly reclined with respect to the loading direction and exhibiting first bending and then stretching deformation prior to collapse. Once some struts broke, the fractured area expanded to the adjacent cells.

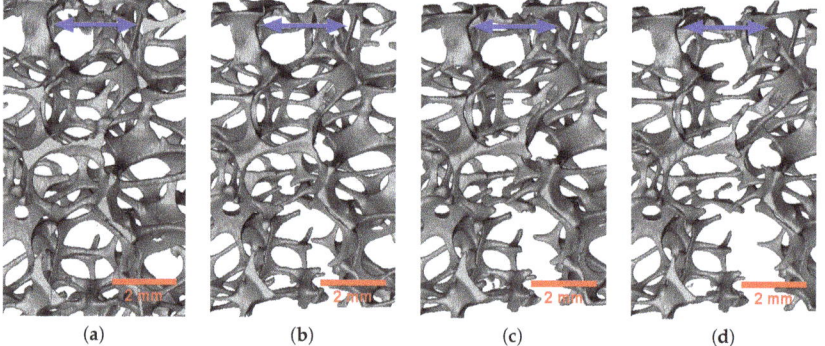

Figure 7. Deformation of the 20 PPI foam sample during the in situ tensile test at nominal strain: (**a**) 0.000; (**b**) 0.019; (**c**) 0.040; (**d**) 0.064. The blue arrow indicates the loading direction.

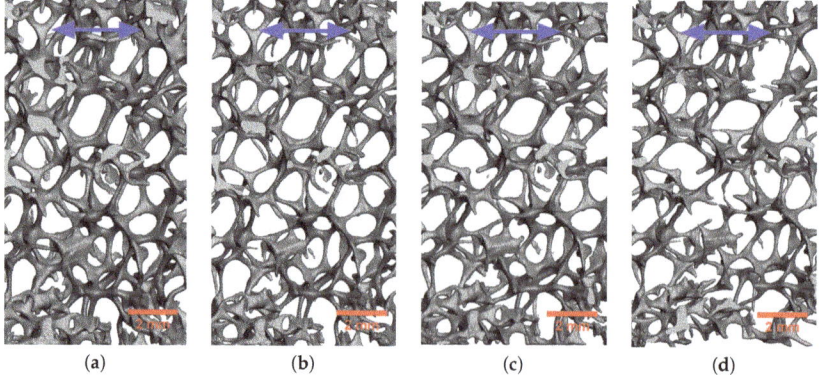

Figure 8. Deformation of the 30 PPI foam sample during the in situ tensile test at nominal strain: (**a**) 0.000; (**b**) 0.031; (**c**) 0.085; (**d**) 0.142. The blue arrow indicates the loading direction.

Figure 9a illustrates the five biggest void-cells of the 20 PPI foam sample, highlighted in red. Figure 9b shows the final step of the fractured foam sample during the in situ tensile test. The fracture area was located in the vicinity of the three big neighboring cells, and passed through two of them (numbers 1 and 3, Figure 9b). In the case of the 30 PPI foam sample, the five biggest cells (Figure 10a) were more scattered than those of the 20 PPI sample. The fracture area was located between the five biggest cells and ended into two of them (numbers 1 and 3, Figure 10b). Furthermore, the 30 PPI foam sample, which was elongated in the transversal direction, was contracted significantly in the x and y directions.

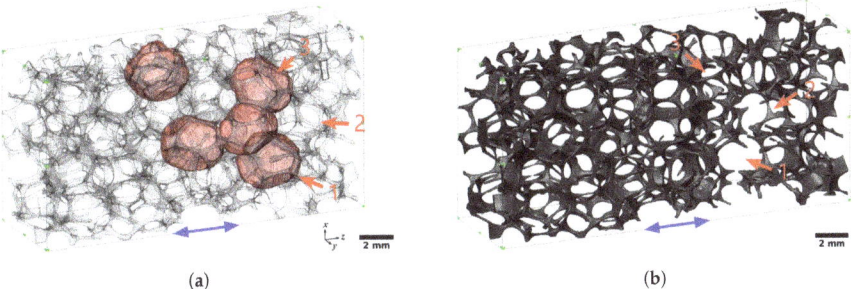

Figure 9. (**a**) The five biggest void-cells and (**b**) the final fractured state of the 20 PPI foam sample. The blue arrow indicates the loading direction. Cells of interest are labeled 1 to 3.

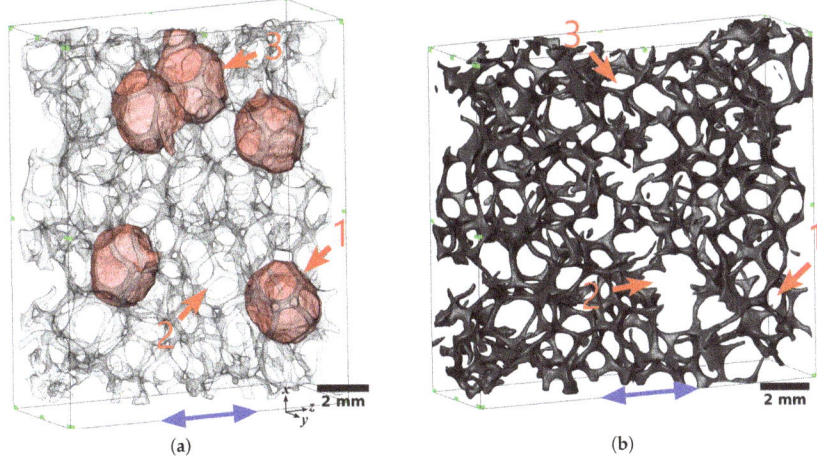

Figure 10. (**a**) The five biggest void-cells and (**b**) the final fractured state of the 30 PPI foam sample. The blue arrow indicates the loading direction. Cells of interest are labeled 1 to 3.

In order to study the effect of intermetallic particles in the solid phase, four FE simulations were performed for each foam sample: (i) *homogeneous* non-porous J2 model, (ii) *heterogeneous* non-porous J2 model, (iii) *homogeneous* porous GTN model, and (iv) *heterogeneous* porous GTN model. J2 models here correspond to standard isotropic (von Mises) plasticity.

Figure 11a,b illustrate the macroscopic tensile curves obtained with the 20 PPI and 30 PPI foam samples in the experiments and simulations. The black points correspond to experimental measurements (note that only the first point laid in the linear elastic regime for each foam sample). The dashed lines correspond to the simulations with isotropic non-porous plasticity (J2 plasticity). The bold lines correspond to the simulations with the GTN damage model. The blue curves correspond to the *homogeneous* simulations with an aluminum matrix without intermetallic particles included. The red curves correspond to the *heterogeneous* simulations where the local presence of intermetallic particles with a given volume fraction f_{IM} is taken into account in the constitutive behavior of the elements. It is clear that taking the presence of intermetallic particles into account in the FE simulations *did not* affect the calculated macroscopic stress–strain curves significantly. However, the non-porous models did not capture the last stages of foam deformation properly, overestimating the reaction stress in the absence of damage.

The tensile curves of Figure 11 exhibited some macroscopic hardening behavior and a peak stress attained after a few percents of macroscopic deformation. The hardening arose from both the geometrical rearrangement of the struts during loading and the constitutive strain hardening of the aluminum solid phase. The 20 PPI foam sample exhibited a 55 MPa Young's modulus and 0.70 MPa tensile strength. The 30 PPI foam sample exhibited a 138 Pa Young's modulus and 0.87 MPa tensile strength, which was both stiffer and stronger than the 20 PPI sample due to the smaller size of the cells. One can note also the higher apparent ductility of the 30 PPI foam sample, accepting a larger level of macroscopic deformation before collapse.

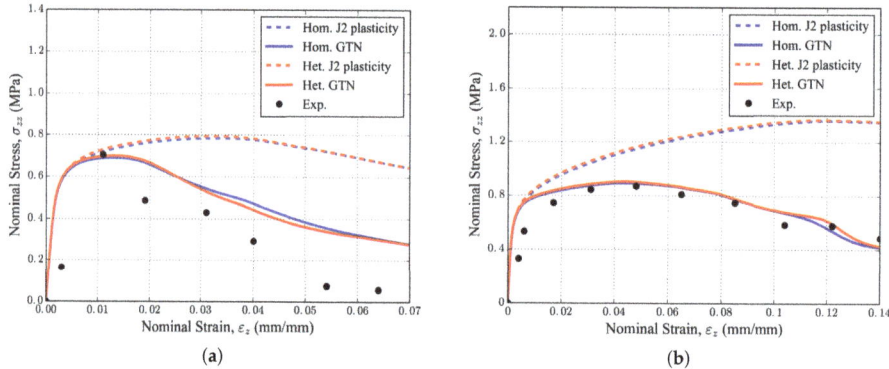

Figure 11. Macroscopic experimental and FE tensile curves of the (**a**) 20 PPI and (**b**) 30 PPI foam samples.

Figure 12a,b illustrate the presence of intermetallic particles in a subregion of the 20 PPI foam sample, before the test and after fracture of the struts. It shows that strut numbers 1 and 5 broke in the vicinity of clusters of intermetallic particles. This was also the case of strut number 4, yet with a smaller amount of particles. On the contrary, other regions with important clusters of intermetallic particles did not fail, illustrating the complexity of the situation leading to fracture. Several factors contributed and competed to trigger fracture, including the loading mode (or misalignment of the struts with the tensile load), local geometrical stress concentration (e.g., due to local reduction in cross section), and the presence of hard and brittle intermetallic particles.

The contour plots of void volume fraction computed in the FE simulations with the homogeneous and the heterogeneous GTN models are illustrated in Figure 12c,d. Figure 12d with the heterogeneous model shows, for example, a similar VVF to the homogeneous model (Figure 12c) in strut number 5 at the place of fracture, despite the presence of a cluster of particles that was not taken into account in the homogeneous model. Strut number 2 illustrates that similar levels of VVF could also be reached in the absence of particles due to geometrical plastic strain localization, which was predicted in both homogeneous and heterogeneous models. Strut number 1 illustrates that such localization of plastic flow due to local cross section reduction can be the critical point even in the presence of particles, in which case both models again provided comparable local VVF.

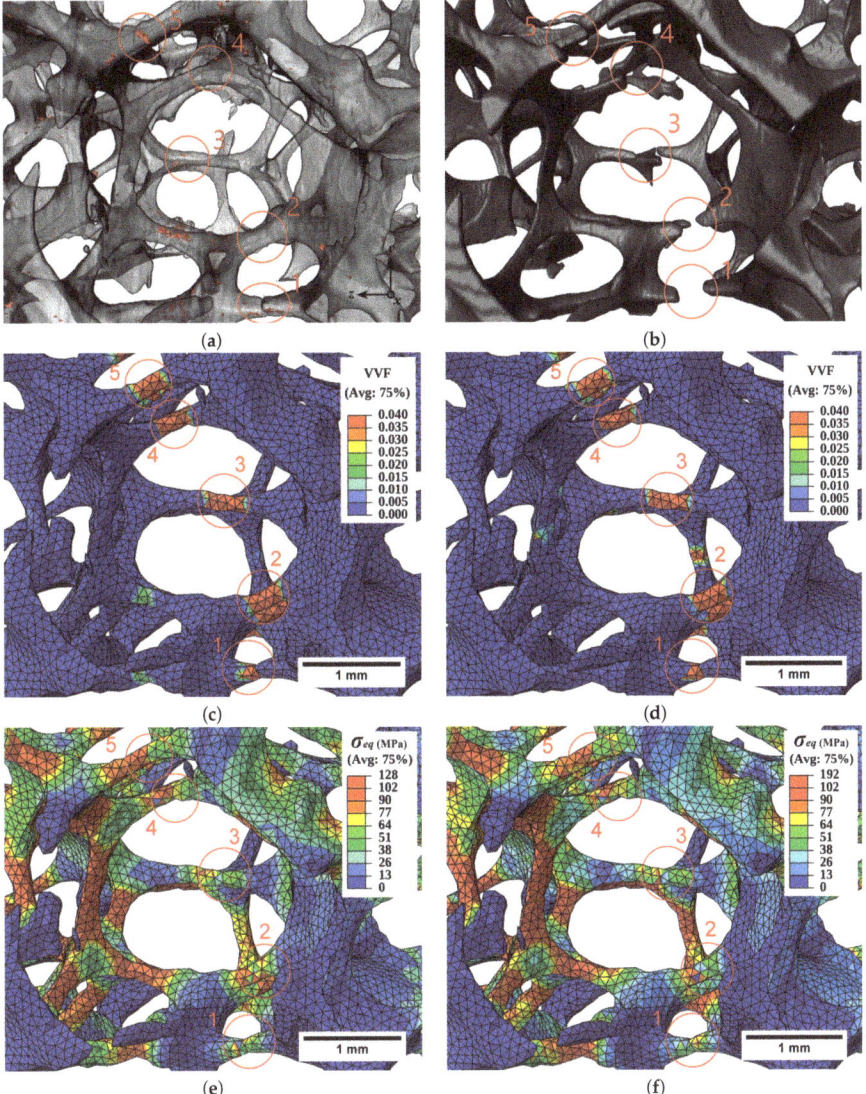

Figure 12. (a) Distribution of intermetallic particles in the solid phase of the 20 PPI sample. (b) Fractured struts of the foam. FE contour plots (at 6.4% strain) of VVF with the (c) homogeneous and (d) heterogeneous GTN models. Corresponding contour plots of equivalent stress with the (e) homogeneous and (f) heterogeneous GTN models. Struts of interest are labeled 1 to 5.

Similar comparisons can be made in terms of equivalent stress distribution in the struts in Figure 12e,f. Stress concentration arose either from local reduction of cross section, in which case both models provided comparable results, or with an additional contribution of the local presence of hard particles. One can note, however, that the sole prediction of stress concentration would fail in systematically discriminating the potential fracture zones at the microscopic level, justifying the use of the GTN porous plasticity model in the present work.

This study illustrates that stress analysis in metal foams is not straightforward due to the non-uniform stress distribution arising from the non-uniform geometry [45] and microstructure.

The distribution of von Mises stress in the foam under loading depends on size, orientation, and spatial arrangement of the cells in a complex manner [17], in addition to the local presence of hard particles. It was however observed here that tensile loading of the foams promoted stress concentration in the struts of the biggest void-cells. The subsequent fracture of these struts formed the fracture plane.

Other studies have also suggested that the fracture mode of ERG foam struts depends on the type of precipitates in the Al matrix. The failure of the struts was observed to begin by realignment and ductile transgranular fracture of struts in the fracture plane [25]. The fracture mode later shifted to major brittle intergranular and minor ductile transgranular failure of the remaining struts due to the presence of α-AlFeSi (Al_8Fe_2Si) and β-AlFeSi (Al_5FeSi) precipitates in the grain boundaries. The experimental protocol of the present study using laboratory tomography does not allow us to easily distinguish these different types of precipitates. The use of in situ synchrotron tomography might be an interesting perspective in this context.

As concerns modeling, current limits of the present approach might be overcome in the future by: (i) using a modified FE simulation environment allowing to alleviate the requirement for first-order tetrahedral elements in highly non-linear simulations with extensive damage development, (ii) using more realistic boundary conditions (e.g., directly mapped from the 3D in situ images), (iii) better identifying the local strut constitutive behavior. Indeed, in this study and in the recent one of Petit et al. [20], direct measures of yield stresses and hardening behavior on single-strut micro-tensile tests [41,43] had to be adjusted to properly reproduce the macroscopic foam response. A perspective might consist of taking the crystallographic orientation of the grains in the struts into account for the calculation of their plastic behavior. This could have significant consequences in this material, since some struts are made of a limited number of grains [21], and could exhibit rather anisotropic plastic behavior. Image-based crystal plasticity FE simulations [46] might be particularly interesting in this context.

As a final comment, the present study shows that *quantitatively* taking the local intermetallic particles into account in the prediction of the foam mechanical behavior does not really affect the macroscopic response, but allows a rather good discrimination of the critical zones for fracture in the structure. This confirms the *qualitative* conclusions and prospects of Petit et al. [20]. Nevertheless, it also tends to indicate in this particular case that accounting for intermetallics in such a homogenized GTN damage simulation framework is only of secondary importance for the prediction of fracture, after the proper accounting for local geometry and loading mode of the nodes/struts. Different conclusions were recently found in the study of Amani et al., [40] where the presence of local heterogeneous microporosity in additively manufactured lattice structures had more significant consequences on the prediction of fracture behavior, yet with a higher global volume fraction and a different type of defect (process-induced cavities) as compared to the present study. Note also that direct full-field modeling of the small intermetallics in the struts would allow a more accurate prediction of local stress heterogeneity and fracture onset. However, such full-field simulations would require a severe increase in computational resources and would hardly be applicable on real-size samples. Moreover, the identification of the local constitutive behavior in the struts remains a challenge. The micromechanical approach developed in this study is based on cell-calculation and the GTN model to derive the constitutive behavior of local particle-rich regions of the foam. It can constitute the basis for a wider range of applications to complement 3D image-based FE studies of macroporous materials, which are now well established. This might include, for example, any multiscale investigation of architectured materials where both (i) a global description of the macroscopic structure, but also (ii) local image-based microstructural details are expected to be required to understand and predict the macroscopic behavior of the material.

5. Conclusions

A method was developed in this study to investigate the tensile behavior of an ERG foam provided in two cell sizes, by taking advantage of in situ tensile tests under microtomography and microstructure-informed FE modeling:

- The internal architecture of the solid phase of the foam was analyzed using high-resolution local tomography, providing elaborate quantitative data of the location of internal defects (e.g., internal microvoids and intermetallic particles inside the sample).
- The deformation and fracture mechanisms of the foam were studied in situ in tension using lower-resolution scans.
- Image-based FE simulation of the tests was performed using a microstructure-informed porous plasticity (Gurson–Tvergaard–Needleman, GTN) model, quantitatively taking the local presence of brittle intermetallic particles into account (the so-called *heterogeneous* GTN model).
- The *heterogeneous* model performed well in the discrimination of potential fracture zones, but did not perform better than the corresponding *homogeneous* (or microstructure "blind") model in the prediction of global stress–strain curves.
- The procedure can be easily utilized for the investigation of other types of architectured materials where both the macroscopic architecture and local microstructural details are expected to be required in order to understand and predict the material behavior.

Author Contributions: Investigation, Y.A., J.A., and J.L.; Methodology, S.D., E.M., and J.A.; Software, S.D. and J.L.; Supervision, S.D. and E.M.; Writing—original draft, Y.A. and S.D.; Writing—review & editing, S.D. and E.M.

Funding: This research received no external funding.

Acknowledgments: The authors would like to thank the ERG Aerospace Corporation for providing the foam samples.

Conflicts of Interest: The authors declare no conflict of interest.

Abbreviations

The following abbreviations are used in this manuscript:

FE	Finite Element
GTN	Gurson–Tvergaard–Needleman
PPI	Pores Per Inch
VVF	Void Volume Fraction

References

1. Gibson, L.J.; Ashby, M.F. *Cellular Solids: Structure and Properties*, 2nd ed., 1st pbk. ed. with corr ed.; Cambridge Solid State Science Series; Cambridge University Press: Cambridge, UK, 1999.
2. Ashby, M.F. *Metal Foams: A Design Guide*; Butterworth-Heinemann: Boston, MA, USA, 2000.
3. Scheffler, M.; Colombo, P. *Cellular Ceramics: Structure, Manufacturing, Properties and Applications*; John Wiley: Hoboken, NJ, USA, 2005.
4. Harte, A. The fatigue strength of sandwich beams with an aluminium alloy foam core. *Int. J. Fatigue* **2001**, *23*, 499–507. [CrossRef]
5. Ashby, M.F. The properties of foams and lattices. *Phil. Trans. R. Soc.* **2006**, *364*, 15–30. [CrossRef] [PubMed]
6. Gibson, L.J.; Ashby, M.F.; Zhang, J.; Triantafillou, T.C. Failure surfaces for cellular materials under multiaxial loads—I. Modelling. *Int. J. Mech. Sci.* **1989**, *31*, 635–663. [CrossRef]
7. Andrews, E.; Sanders, W.; Gibson, L.J. Compressive and tensile behaviour of aluminum foams. *Mater. Sci. Eng. A* **1999**, *270*, 113–124. [CrossRef]
8. Maire, E.; Withers, P.J. Quantitative X-ray tomography. *Int. Mater. Rev.* **2014**, *59*, 1–43. [CrossRef]
9. Bart-Smith, H.; Bastawros, A.F.; Mumm, D.; Evans, A.; Sypeck, D.; Wadley, H. Compressive deformation and yielding mechanisms in cellular Al alloys determined using X-ray tomography and surface strain mapping. *Acta Mater.* **1998**, *46*, 3583–3592. [CrossRef]

10. Caty, O.; Ibarroule, P.; Herbreteau, M.; Rebillat, F.; Maire, E.; Vignoles, G.L. Application of X-ray computed micro-tomography to the study of damage and oxidation kinetics of thermostructural composites. *Nucl. Instrum. Methods Phys. Res. Sect. B* **2014**, *324*, 113–117. [CrossRef]
11. Ulrich, D.; van Rietbergen, B.; Weinans, H.; Rüegsegger, P. Finite element analysis of trabecular bone structure: A comparison of image-based meshing techniques. *J. Biomech.* **1998**, *31*, 1187–1192. [CrossRef]
12. Maire, E. X-ray tomography applied to the characterization of cellular materials. Related finite element modeling problems. *Compos. Sci. Technol.* **2003**, *63*, 2431–2443. [CrossRef]
13. Youssef, S.; Maire, E.; Gaertner, R. Finite element modelling of the actual structure of cellular materials determined by X-ray tomography. *Acta Mater.* **2005**, *53*, 719–730. [CrossRef]
14. Caty, O.; Maire, E.; Youssef, S.; Bouchet, R. Modeling the properties of closed-cell cellular materials from tomography images using finite shell elements. *Acta Mater.* **2008**, *56*, 5524–5534. [CrossRef]
15. Lacroix, D.; Chateau, A.; Ginebra, M.P.; Planell, J.A. Micro-finite element models of bone tissue-engineering scaffolds. *Biomaterials* **2006**, *27*, 5326–5334. [CrossRef] [PubMed]
16. Jeon, I.; Asahina, T.; Kang, K.J.; Im, S.; Lu, T.J. Finite element simulation of the plastic collapse of closed-cell aluminum foams with X-ray computed tomography. *Mech. Mater.* **2010**, *42*, 227–236. [CrossRef]
17. Michailidis, N.; Stergioudi, F.; Omar, H.; Papadopoulos, D.; Tsipas, D.N. Experimental and FEM analysis of the material response of porous metals imposed to mechanical loading. *Colloids Surf. A* **2011**, *382*, 124–131. [CrossRef]
18. D'Angelo, C.; Ortona, A.; Colombo, P. Finite element analysis of reticulated ceramics under compression. *Acta Mater.* **2012**, *60*, 6692–6702. [CrossRef]
19. Zhang, T.; Maire, E.; Adrien, J.; Onck, P.R.; Salvo, L. Local tomography study of the fracture of an ERG metal foam. *Adv. Eng. Mater.* **2013**, *15*, 767–772. [CrossRef]
20. Petit, C.; Maire, E.; Meille, S.; Adrien, J. Two-scale study of the fracture of an aluminum foam by X-ray tomography and finite element modeling. *Mater. Des.* **2017**, *120*, 117–127. [CrossRef]
21. Zhou, J.; Mercer, C.; Soboyejo, W.O. An investigation of the microstructure and strength of open-cell 6101 aluminum foams. *Metall. Mater. Trans. A* **2002**, *33*, 1413–1427. [CrossRef]
22. Herman, G.T. *Fundamentals of Computerized Tomography: Image Reconstruction from Projections*; Springer Science & Business Media: Berlin, Germany, 2009.
23. Buffiere, J.Y.; Maire, E.; Adrien, J.; Masse, J.P.; Boller, E. In situ experiments with X-ray tomography: An attractive tool for experimental mechanics. *Exp. Mech.* **2010**, *50*, 289–305. [CrossRef]
24. Stock, S.R. *MicroComputed Tomography: Methodology and Applications*; CRC Press: Boca Raton, FL, USA, 2009.
25. Amsterdam, E.; Onck, P.R.; Hosson, J.T.M.D. Fracture and microstructure of open cell aluminum foam. *J. Mater. Sci.* **2005**, *40*, 5813–5819. [CrossRef]
26. Schindelin, J.; Arganda-Carreras, I.; Frise, E.; Kaynig, V.; Longair, M.; Pietzsch, T.; Preibisch, S.; Rueden, C.; Saalfeld, S.; Schmid, B.; et al. Fiji: An open-source platform for biological-image analysis. *Nat. Methods* **2012**, *9*, 676–682. [CrossRef] [PubMed]
27. Amani, Y.; Takahashi, A.; Chantrenne, P.; Maruyama, S.; Dancette, S.; Maire, E. Thermal conductivity of highly porous metal foams: Experimental and image based finite element analysis. *Int. J. Heat Mass Transf.* **2018**, *122*, 1–10. [CrossRef]
28. Nieh, T.G.; Higashi, K.; Wadsworth, J. Effect of cell morphology on the compressive properties of open-cell aluminum foams. *Mater. Sci. Eng. A* **2000**, *283*, 105–110. [CrossRef]
29. Andrews, E.W.; Gioux, G.; Onck, P.; Gibson, L.J. Size effects in ductile cellular solids. Part II: Experimental results. *Int. J. Mech. Sci.* **2001**, *43*, 701–713. [CrossRef]
30. Sadowski, T.; Samborski, S. Modeling of Porous Ceramics Response to Compressive Loading. *J. Am. Ceram. Soc.* **2003**, *86*, 2218–2221. [CrossRef]
31. Ferre, A.; Dancette, S.; Maire, E. Damage characterisation in aluminium matrix composites reinforced with amorphous metal inclusions. *Mater. Sci. Technol.* **2015**, *31*, 579–586. [CrossRef]
32. Martin, C.F.; Josserond, C.; Salvo, L.; Blandin, J.J.; Cloetens, P.; Boller, E. Characterisation by X-ray micro-tomography of cavity coalescence during superplastic deformation. *Scr. Mater.* **2000**, *42*, 375–381, doi:10.1016/S1359-6462(99)00355-3. [CrossRef]
33. Gurson, A.L. Continuum theory of ductile rupture by void nucleation and growth: Part I—Yield criteria and flow rules for porous ductile media. *J. Eng. Mater. Technol.* **1977**, *99*, 2, doi:10.1115/1.3443401. [CrossRef]

34. Chu, C.C.; Needleman, A. Void nucleation effects in biaxially stretched sheets. *J. Eng. Mater. Technol.* **1980**, *102*, 249, doi:10.1115/1.3224807. [CrossRef]
35. Tvergaard, V. Influence of voids on shear band instabilities under plane strain conditions. *Int. J. Fract.* **1981**, *17*, 389–407. [CrossRef]
36. Tvergaard, V. On localization in ductile materials containing spherical voids. *Int. J. Fract.* **1982**, *18*, 237–252.
37. Tvergaard, V.; Needleman, A. Analysis of the cup-cone fracture in a round tensile bar. *Acta Metall.* **1984**, *32*, 157–169, doi:10.1016/0001-6160(84)90213-X. [CrossRef]
38. *Abaqus Version 6.13 Documentation Collection*; Dassault Systèmes: Vélizy-Villacoublay, France, 2013.
39. *Avizo® 9 User's Guide*; ThermoFisher Scientific: Waltham, MA, USA, 2016.
40. Amani, Y.; Dancette, S.; Delroisse, P.; Simar, A.; Maire, E. Compression behavior of lattice structures produced by selective laser melting: X-ray tomography based experimental and finite element approaches. *Acta Mater.* **2018**, *159*, 395–407, doi:10.1016/j.actamat.2018.08.030. [CrossRef]
41. Zhou, J.; Allameh, S.; Soboyejo, W.O. Microscale testing of the strut in open cell aluminum foams. *J. Mater. Sci.* **2005**, *40*, 429–439, doi:10.1007/s10853-005-6100-8. [CrossRef]
42. Jung, A.; Wocker, M.; Chen, Z.; Seibert, H. Microtensile testing of open-cell metal foams—Experimental setup, micromechanical properties. *Mater. Des.* **2015**, *88*, 1021–1030, doi:10.1016/j.matdes.2015.09.091. [CrossRef]
43. Amani, Y. Modélisation Basée sur Données de Tomographie aux Rayons x de L'endommagement et de la Conductivité Thermique Dans les Matériaux Cellulaires Métalliques. Ph.D. Thesis, INSA Lyon, Université de Lyon, Lyon, France, 2018.
44. Li, H.; Fu, M.; Lu, J.; Yang, H. Ductile fracture: Experiments and computations. *Int. J. Plast.* **2011**, *27*, 147–180, doi:10.1016/j.ijplas.2010.04.001. [CrossRef]
45. Chan, S.H.; Ngan, A. Statistical distribution of forces in stressed 2-D low-density materials with random microstructures. *Mech. Mater.* **2006**, *38*, 1199–1212, doi:10.1016/j.mechmat.2006.02.007. [CrossRef]
46. Dancette, S.; Browet, A.; Martin, G.; Willemet, M.; Delannay, L. Automatic processing of an orientation map into a finite element mesh that conforms to grain boundaries. *Model. Simul. Mater. Sci. Eng.* **2016**, *24*, 055014. [CrossRef]

© 2018 by the authors. Licensee MDPI, Basel, Switzerland. This article is an open access article distributed under the terms and conditions of the Creative Commons Attribution (CC BY) license (http://creativecommons.org/licenses/by/4.0/).

Article

In-Situ X-ray Tomography Observation of Structure Evolution in 1,3,5-Triamino-2,4,6-Trinitrobenzene Based Polymer Bonded Explosive (TATB-PBX) under Thermo-Mechanical Loading

Zeng-Nian Yuan [1,2], Hua Chen [1,*], Jing-Ming Li [1,*], Bin Dai [1] and Wei-Bin Zhang [1]

[1] Institute of Chemical Materials, China Academy of Engineering Physics, Mianyang 621900, China; znyuan@foxmail.com (Z.-N.Y.); daibin@caep.cn (B.D.); weibinzhang@caep.cn (W.-B.Z.)
[2] Graduate School of China Academy of Engineering Physics, China Academy of Engineering Physics, Mianyang 621900, China
* Correspondence: chenhua9@caep.cn (H.C.); jmli7288@caep.cn (J.-M.L.)

Received: 8 April 2018; Accepted: 3 May 2018; Published: 4 May 2018

Abstract: In order to study the fracture behavior and structure evolution of 1,3,5-Triamino-2,4,6-Trinitrobenzene (TATB)-based polymer bonded explosive in thermal-mechanical loading, in-situ studies were performed on X-ray computed tomography system using quasi-static Brazilian test. The experiment temperature was set from −20 °C to 70 °C. Three-dimensional morphology of cracks at different temperatures was obtained through digital image process. The various fracture modes were compared by scanning electron microscopy. Fracture degree and complexity were defined to quantitatively characterize the different types of fractures. Fractal dimension was used to characterize the roughness of the crack surface. The displacement field of particles in polymer bonded explosive (PBX) was used to analyze the interior structure evolution during the process of thermal-mechanical loading. It was found that the brittleness of PBX reduced, the fracture got more tortuous, and the crack surface got smoother as the temperature rose. At lower temperatures, especially lower than glass transition temperature of binders, there were slipping and shear among particles, and particles tended to displace and disperse; while at higher temperatures, especially above the glass transition temperature of binders, there was reorganization of particles and particles tended to merge, disperse, and reduce sizes, rather than displacing.

Keywords: in-situ X-ray computed tomography; thermal-mechanical loading; polymer bonded explosives; mesoscale characterization; structure evolution

1. Introduction

Polymer bonded explosive (PBX) is a class of heterogeneous composite material that mainly consists of explosive crystal particles and polymer binder matrix. PBX is widely used in weapon systems because of its excellent performance. In the course of the service of the weapon systems, because of various environment temperature and mechanical actions, such as compression and tension, there could be a series of damage occurring, structure evolution, and even fracture behaviors in PBX. These damages, structure evolution, and fractures directly affect the mechanical properties, safety performance, and detonation performance of PBX [1,2]. Therefore, it is important to study the response of PBX under thermo-mechanical loading.

Previous researches mainly studied the single-factor effect on PBX. When considering the mechanical loading, the mechanical properties of PBX were widely studied [3,4]. The deformation [5], creep [6], cohesive [7], and fracture [8] behavior of PBX under mechanical loading were investigated in the digital image correlation (DIC) method. The micro-mechanical evolution was investigated through scanning electron microscopy (SEM). It was found that there was a variety of types of damages in PBX, such as intragranular voids, crystal fractures, interfacial debonding, and deformation twinning [9–11]. The above detection techniques could only obtain the surface morphology, however X-ray micro-computed tomography (μCT) allows for the observation of the internal three-dimensional (3D) structures. The internal deformation of PBX simulant in compression was analyzed in detail by digital volume correlation (DVC) of in-situ μCT [12].

While in the terms of thermal loading effects, state and phase change of both explosives [13] and binders [14] at a wide range of temperatures were studied. The mesoscale structure evolution of PBX during heating was analyzed by ultra-small angle X-ray scattering (USAXS) and μCT [15]. However, mechanical properties and structure evolution of PBX under thermal-mechanical coupling loading still need to be investigated deeply. Because of the technical difficulties of experiments, most of the researches on properties of PBX in thermal-mechanical loading were simulation calculations [16,17]. In respect of experiments, Willamson and others comprehensively studied temperature-time response of an cyclotetramethylenetetranitramine (HMX)-based PBX at a wide temperature range [18]. It was found that the failure strain of PBX is non-sensitive to temperature, so the modules of PBX at different temperatures have a liner relation with failure stress. Other researchers studied the mechanical [19–21] and fracture [22] behavior of PBX under different temperature and loading conditions from different perspectives. The previous researches were mainly focus on the mechanical properties of PBX at high temperatures, while the structure evolution of PBX at both low and high temperatures still needs more investigation. Especially, most of the researches were about HMX-based PBX, while few researches studied 2,4,6-triamino-1,3,5-trinitrobenzene (TATB)-based PBX. Furthermore, quantitative characterization techniques should be built up to describe and analyze the observation result of fracture and structure evolution in mathematical language.

In this paper, in-situ Brazilian test with an improved arc loading head was conducted on a μCT apparatus in order to investigate the interior structure evolution of a TATB-PBX. The test was under quasi-static loading and five different temperatures, ranged from −20 °C to 70 °C. Three-dimensional morphology of cracks was investigated by digital image process. Fracture degree and complexity were defined and used to quantitatively characterize the crack properties. Fractal dimension was used to characterize the roughness of the crack surface. The test samples were also investigated by SEM, and the results of different kinds of detection and analysis were compared. Slice images of μCT were also analyzed by the displacement of particles, and the displacement field of the interior structure of PBX was analyzed.

2. Materials and Methods

The experimental specimen is a kind of TATB-based PBX. The main components are TATB as explosive and F2314 fluororubber as binder. TATB and F2314 are firstly made into molding powders, which are particles with some TATB powder crystals that are wrapped in an F2314 binder. The mass percentage of TATB in this PBX is higher than 90%. The specimen was made into a disk with diameter of 10 mm, and a thickness of 3 mm. The experiment took the way of Brazilian disc quasi-static displacement loading, the loading speed was 0.1 mm/min. As a method of indirect tension, Brazilian disc test is widely used to study the damage behavior of brittle materials. Awaji and Sato [23] improved the loading head in diameter disc compression to reduce the stress concentration at the head and analyzed the stress distribution under this type of loading. Pang [24,25] and others found that when the radius of the arc loading head is 1.35 times of the radius of the specimen, the test result is the closest to that of the direct tensile test. In this experiment, the radius of the arc loading head was designed in this method. Because of the small size of the specimen, and in order to ensure the stability of the specimen, a group of fenders were used in experiment. The test loading diagram was shown in Figure 1.

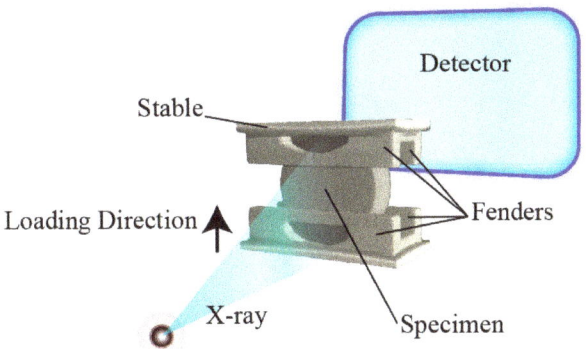

Figure 1. The loading method diagram.

The experiment was carried out with the same scan parameters in every group of tests in order to ensure that the scan results are consistent and comparable. The scan voltage was 60 kV, scan current was 150 mA, exposure time was 0.4 s, and with an image merging number of 5. In this condition, the spatial resolution of this experiment was 21.182 μm/voxel. The loading device was an in-situ CT material testing machine, Deben Microtest CT5000-TEC (Deben UK Ltd, London, UK). In every group of test, the specimen was firstly scanned at room temperature without any loading for structure information of original state. Then, the test temperature, −20 °C, 0 °C, room temperature (22 °C), 55 °C (the glass transition temperature of F2314 is around 50 °C), or 70 °C was set, and we waited for 30 min in order to make the specimen temperature stable. Then, the specimen was loaded to fracture, and was scanned under the fracture state. Replicated tests were taken in order to ensure the repeatability and accuracy of the experiment. The specimen was also loaded to 75% fracture extension at different temperatures, which was obtained in the previous tests, and was scanned to get the structure information of intermediate state of loading at different temperatures. After in-situ CT tests, some of the specimens were also detected by SEM (CamScan Apollo300, CamScan, Cambridgeshire, UK) for more information of more micro-scale structure to compare with the results of CT detection.

3. Results

3.1. Stress-Extension Curves at Different Temperatures

Using Deben Microtest CT5000-TEC device, an extension-force curve in the test can be obtained and recorded automatically, and the stress can be derived by the following expression [10],

$$\sigma_{xx} = \left(1 - \left(\frac{b}{R}\right)^2\right) \frac{2P}{\pi D t} \tag{1}$$

where P is the loading force and D and t are the specimen's diameter and the thickness, respectively, b is the contact half-width of the anvils and R is the radius of the specimen. This b/R ratio should be measured for the specimen in experiments. This ratio is recommended to be greater than 0.27, since this has been found the failure is to be purely tensile [10]. In this paper, all of the b/R ratios in every group of tests are around 0.28 to 0.30.

Through Formula (1), stress-extension curves in every test can be obtained. The curves at different temperature were shown in Figure 2.

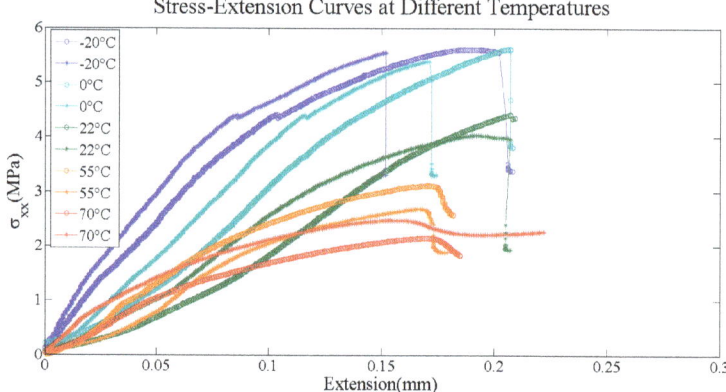

Figure 2. Stress-extension curves at different temperatures, at each temperature there is a repeat group.

As we can see from Figure 2, all of the failure extensions are around 0.15 to 0.20 mm, the difference at the same temperature tests basically equals to the difference among different temperature groups. Therefore, the failure extension is not sensitive to the temperature, which agrees to the previous research [18]: temperature has little effect on tensile failure strain. As the temperature decreases, the binder hardens, the PBX modulus increases, and the failure stress increases accordingly. As the temperature increases, the binder changes to high elastic state, the viscosity increases, the PBX modulus decreases, and the failure stress reduces, accordingly. In low temperature conditions, such as room temperature, 0 °C and −20 °C, the brittleness is significant, the specimen loses its carrying capability at the fracture moment; and, at 55 °C, the load of specimen decreases a little after fracture; when temperature reaches 70 °C, the sample does not lose carrying capability, but continue to load, and an obvious creep occurs.

3.2. CT Images and Digital Image Process

A typical slice image from CT scan and image reconstruction is shown in Figure 3, (a) is the slice of fractured specimen and (b) is the slice of original one. Generally speaking, the grayscale of the image is proportional to the mass density, so different components in the specimen can be identified. However, in this kind of PBX, the mass densities of explosives (TATB molding powder) and binders (F2314) are very close. The density of TATB is 1.93 g/cm^3 and the density of F2314 is around 2.04 g/cm^3. Because of beam hardening and other system errors that are caused by CT, components in the specimen cannot be simply identified and segmented only by grayscale histogram. In this paper, several digital image processing methods, including morphological image processing (such as open and close operations and so on) and various segment techniques (such as Otsu segment and so on) were used to segment explosives (TATB molding powder), binders (F2314), and cracks, and to process the CT slices into binary images. The algorithms and criterion of the grayscale threshold used in segmentation are same in all of the analyses of experiments. The typical cracks at different temperatures are shown in Figure 4a. Three-dimensional crack morphologies can be obtained through reconstruction of binary images, shown in Figure 4b, and more analysis of these binary slices are in the next section.

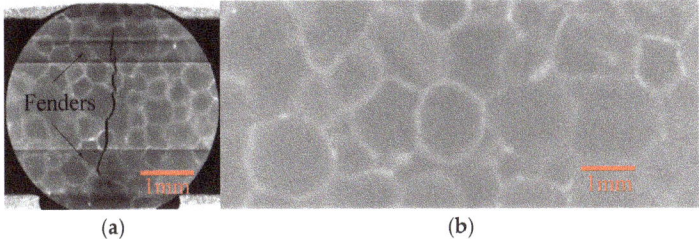

Figure 3. (a) Left one is the computed tomography (CT) slice with cracks. Because the loading head used in Brazilian test in this paper has fenders (as shown in Figure 1) in order to stabilize the specimen, two dark horizontal lines can be seen in the figure; (b) Right one is the CT slice without cracks.

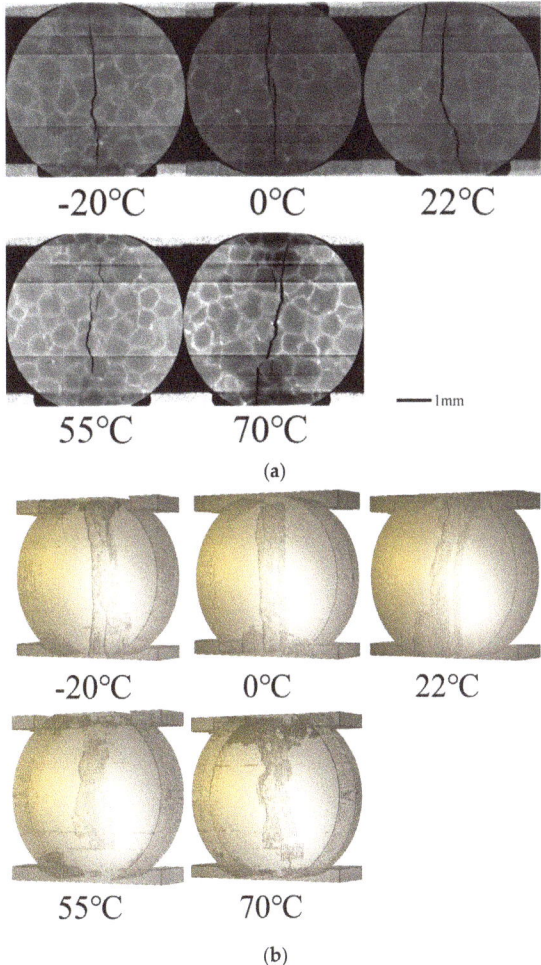

Figure 4. (a) Typical slice images of cracks at different temperatures; (b) Three-dimensional (3D) crack morphology at different temperatures.

As is shown in Figure 4, the cracks are straighter at lower temperatures, while they are more tortuous at higher temperatures. This is because, at lower temperatures, the fracture has significant brittleness; while at higher temperatures, as the binders turn into high elastic state, the viscoelasticity of the specimen enhances, so crack paths are more along the grain boundaries, then a more tortuous crack will occur.

3.3. Fracture Mode Comparison

The fractured specimens at different temperatures were also detected by SEM. The result of SEM detection proves the result of fracture fractal dimension analysis. Figure 5 shows some typical images at different temperatures, −20 °C, 0 °C, 22 °C, 55 °C, and 70 °C, respectively:

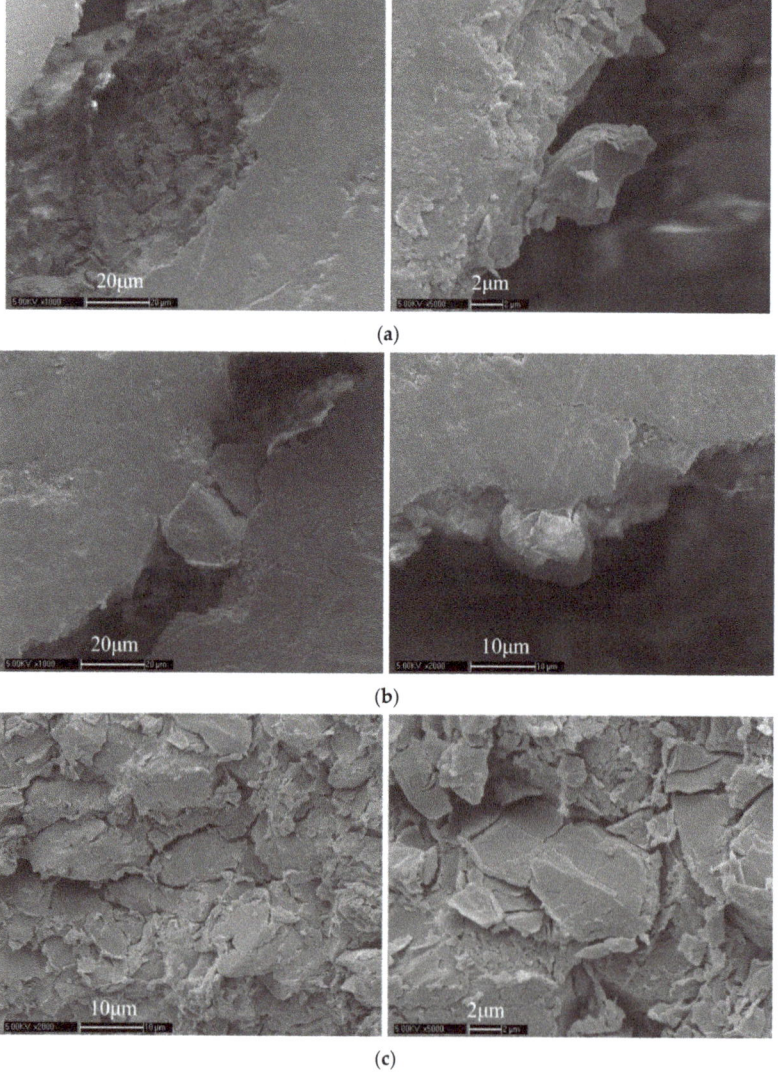

(a)

(b)

(c)

Figure 5. *Cont.*

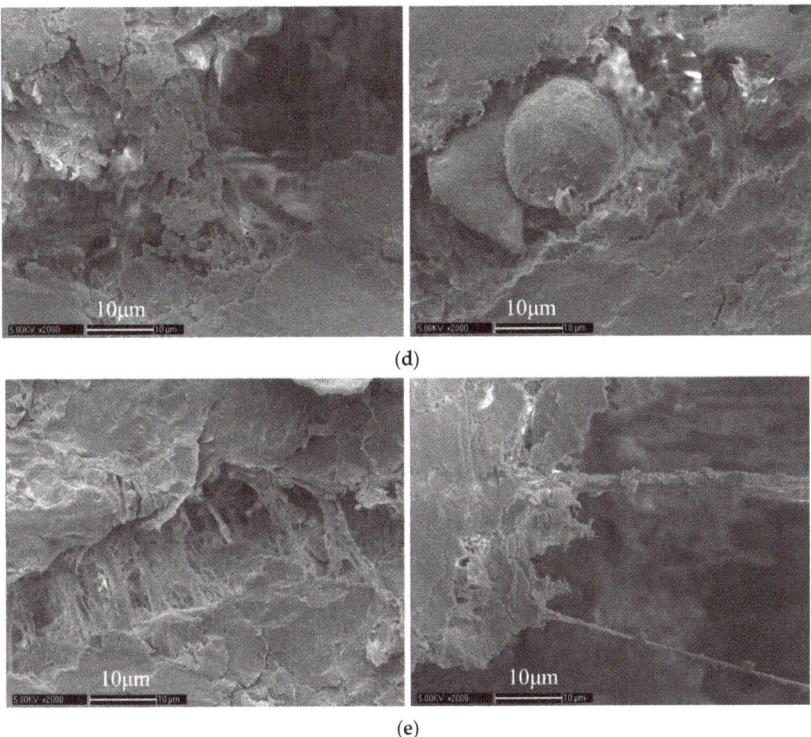

Figure 5. (**a**) SEM images in −20 °C; (**b**) SEM images in 0 °C; (**c**) SEM images in 22 °C; (**d**) SEM images in 55 °C; (**e**) SEM images in 70 °C.

The SEM images show the great difference of the fractures at different temperatures. Figure 5a is SEM images in −20 °C, the rough fracture surface can be clearly seen and the main fracture mode is the break of particles and crystals. Because of the angular broken particles and crystals, the fracture surface is rougher, resulting in a larger fracture fractal dimension. Figure 5b is SEM images in 0 °C, a large amount of fractured particles drop out. Figure 5c is SEM images in 22 °C. These are images of the fracture surface, which can be seen directly that it is rough. Meanwhile, not only broken particles and crystals can be seen, but there are also some broken binders around the fracture surface. Figure 5d is SEM images in 55 °C. The binders bridging the gap of particles in cracks can be seen clearly, as shown in the left image. Some large single particles drop out, a smooth surface that is covered with binders can be seen clearly, which implies a smaller fracture fractal dimension. Because the glass transition temperature of the binder is around 50 °C, in and above this temperature the fracture mode is mainly transgranular and debonding. Figure 5e is SEM images in 70 °C. There are lots of binders bridging the cracks along the crack path. The main fracture mode can be clearly and directly seen through SEM images, which agrees with the previous researches and the result of fracture fractal dimension, which will be discussed in the next section.

4. Discussion

4.1. Degree, Complexity, and Fractal Dimension of Fracture

In order to characterize the fracture morphology characteristics and to link them with mechanical behaviors, quantitative digital characterization techniques are required. In this paper, a characterizing method was defined and used to analyze the experiment results, including fracture degree and complexity, and fractal dimension of the cracks. Through this method, different types of fracture morphology can be distinguished quantitatively, and the results were also proved by SEM detection.

Fracture degree D is defined as the percentage of the volume of the cracks (i.e., the void volume) in the specimen; while the fracture complexity C is defined as the ratio of the fracture surface area to the fracture volume. In order to non-dimensionalize this fracture complexity, the fracture surface area should be multiplied with a voxel width. As shown in Formulas (2) and (3),

$$D = \frac{V_{fracture}}{V_{specimen}} \qquad (2)$$

$$C = \frac{S_{fracture} \cdot W_{voxel}}{V_{fracture}} \qquad (3)$$

where $V_{fracture}$ is the volume of fractures, $V_{specimen}$ is the volume of specimen, $S_{fracture}$ is the area of fracture surface, and W_{voxel} is the voxel width.

Fracture degree is used to characterize the level of the fractures. A larger fracture degree describes a more serious fracture; while, the fracture complexity is used to characterize the tortuosity character of the fractures. A larger fracture complexity describes a more tortuous fracture. In porous media, tortuosity is commonly used to describe diffusion [26], which is defined as the arc-chord ratio: the ratio of the length of the curve to the distance between the ends of it. However, the real fractures are in three-dimensional, the whole volume and surface area of the fractures should be considered. Therefore, the fracture complexity is defined in this way.

The fracture degree and complexity of cracks at different temperatures are shown in Figure 6. The fracture degree shows a slight fluctuation around 1% at different temperatures, while the fracture complexity shows an increasing trend with the rise of temperature. Because, when temperature rises, the brittleness of the specimen will be weakened, while the viscosity will be enhanced. Therefore, the cracks at lower temperature will be more straight, while at higher temperature, it will be more tortuous, which can be also seen from the 3D crack morphology image directly (shown in Figure 4).

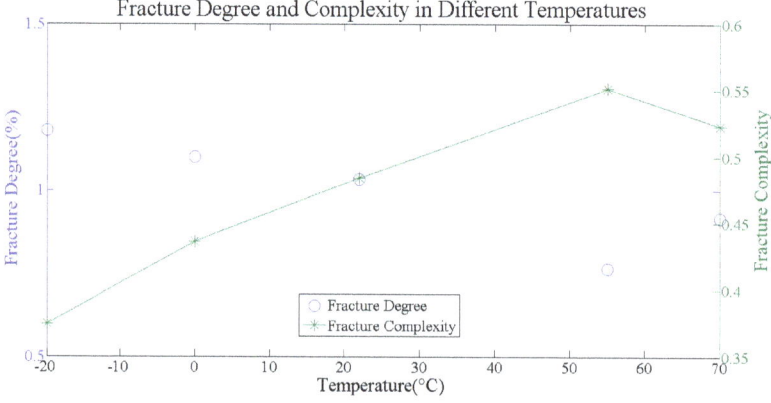

Figure 6. Fracture degree and complexity at different temperatures.

Fractal dimension of fracture is used to describe the roughness of the crack surface. Fractal geometry was firstly formed and defined in order to describe fractal features in 1983 [27]. Fractal feature is a self similarity of geometric characteristics, which means that the object always has a self-similar structure at a smaller scale. Obviously, if the fracture has a fractal feature, then the crack surface is not smooth in a micro-scale. Therefore the fractal dimension of fracture is commonly used to describe the roughness of the crack surface [28]. Previous research in brittle materials found that the fractal dimension of fracture is related to fracture toughness [29–33]. This can be explained as that a larger fracture fractal dimension implies a larger ratio of transgranular or trans-particle fracture, while a smaller fractal dimension implies a larger ratio of intergranular or inter-particle fracture.

In this paper, a box-counting method was used to calculate the fractal dimension of cracks at different temperatures. The box-counting dimension is defined, as following:

$$d = \lim_{\varepsilon \to 0} \left(\frac{\log N(\varepsilon)}{\log(1/\varepsilon)} \right) \quad (4)$$

where d is the fractal dimension, ε is the chosen scale, $N(\varepsilon)$ is the volume of the crack under the scale ε. The algorithm based on box-counting dimension goes as following:

A. choose a series of different sizes of cubes, $\{\varepsilon_i\}$;
B. use these cubes to fill the cracks, count the amount of cubes used, $\{N(\varepsilon_i)\}$; and,
C. fit $\{\log(N(\varepsilon_i))\}$ and $\{\log(1/\varepsilon_i)\}$.

If there is a strong linear correlation, then the fractal characteristic is typical and obvious, and the slope is the fractal dimension of the fracture. In this paper, all of the R^2 in fractal dimension fit is above 0.99, with the p-value being far smaller than 0.01, which means that the cracks have a typical and obvious fractal characteristic. The fracture fractal dimensions at different temperatures are shown in Figure 7.

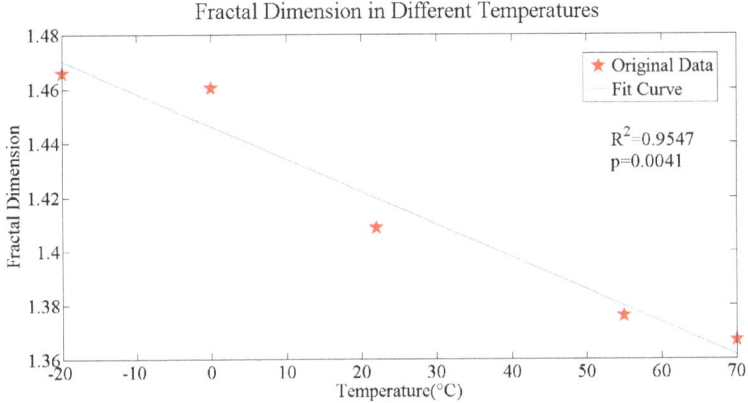

Figure 7. The relation of fracture fractal dimension and temperature.

The fracture fractal dimension showed a decreasing trend with the rise of temperature, which is rational because at a lower temperature, especially lower than the glass transition temperature of the binder, the brittleness of binder is enhanced, so more transgranular and trans-particle fracture will occur; while at a higher temperature, especially higher than the glass transition temperature of the binder, the viscosity of the binder is enhanced so more intergranular and inter-particle fracture will occur. This result agrees with the previous research on TATB-PBX by atomic force microscopy (AFM) [34]. This phenomenon can also be seen directly through SEM images, which was discussed in the previous section.

4.2. Interior Displacement Field of PBX

Through calculating the morphological connectivity of the binary images that were obtained from CT slices, the "center of pixel mass" of each particle can be obtained by using the coordinate distribution of the particles' pixels, which is the position coordinate of the center of the particle, as shown below.

$$(x,y)_{particle} = \frac{particle(x_i, y_i)_{particle} dS_i}{S_{particle}} \tag{5}$$

In the Formula (5), $(x,y)_{particle}$ is the position coordinate of the particle; $(x_i, y_i)_{particle}$ is the ith pixel's coordinate in the particle; dS_i is the area of the ith pixel, which equals 1 here; and, $S_{particle}$ is the area of the whole particle, which equals the amount of the pixels that are included in the particle.

Because of the little loading displacement and the advantage of in-situ experiment, the thickness and the position of the specimen barely changed after loading. As a result, the particular middle slice can be accurately identified and located, and the particles barely moved in the thickness position of the specimen. When comparing the particle distribution morphology before and after loading, the interior displacement field of PBX at different temperatures can be obtained, as shown in Figure 8. The red ones and red circle markers show the original positions of particles, while the blue ones and blue star markers show the positions of particles after loading.

Through the above analysis, it can be concluded that: 1. at $-20\,^\circ$C to room temperature, which are all under the glass transition temperature of the binders, there are slipping and shear among the particles. Dispersion of particles can also be seen, which is proved as particle break by SEM detection; 2. at $0\,^\circ$C, it can be observed that left half of particles move top-left, while right half of particles move top-right, which is exactly same as the test loading method, the top loading head is stable while the bottom loading head move upwards. This implies that the micro-structure evolution is the same as the macro-mechanical behavior; 3. at room temperature, shear along the loading direction is observed, which is because the limitation and the slight unbalance of the loading heads; 4. particles in the middle part of the specimen tend to disperse, which agrees with the fracture behavior under Brazilian test loading method; 5. when the temperature is up to $55\,^\circ$C, which is above the glass transition temperature of binders, the binders begin to transform from glass state to high elastic state, so the binders soften. Therefore, the binder's module reduces while the viscosity enhances, and it has better liquidity. So, the displacement of most particles is smaller, while the particles tend to disperse rather than displacing. The volume of binders tends to expending, accordingly, it can be seen that the particle size tends to reduce; 6. when temperature is higher and up to $70\,^\circ$C, the binders are totally on high elastic state, the viscosity and the liquidity are more significant, resulting in a greater tendency of binder expending and particle reorganization. Particles tend to reshape rather than displacing, and there is complex structure evolution inside the specimen; and, 7. when comparing with the SEM images, it can be concluded that at lower temperatures, the dispersion of particles is mainly the break of particles, while at higher temperatures, is mainly debondings and break of binders.

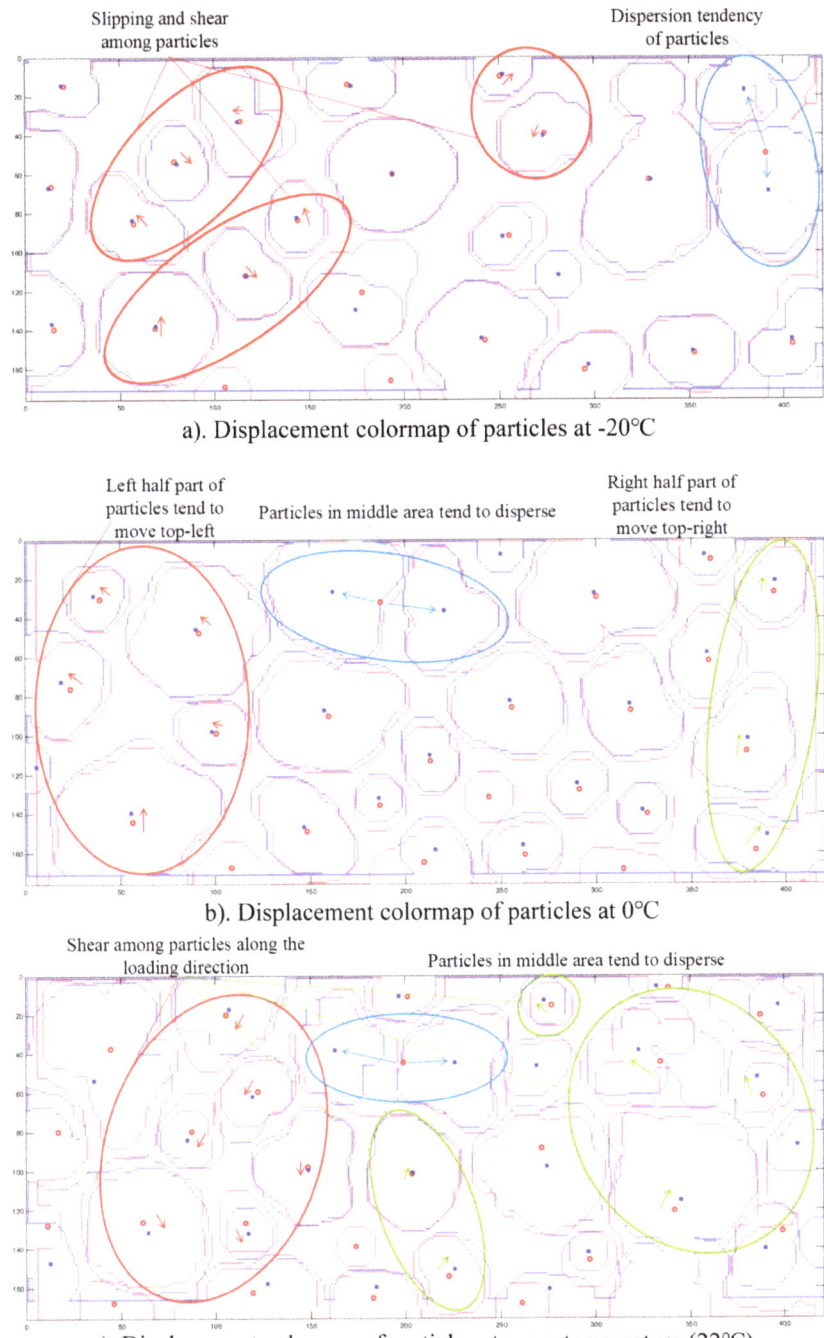

a). Displacement colormap of particles at -20°C

b). Displacement colormap of particles at 0°C

c). Displacement colormap of particles at room temperature (22°C)

Figure 8. *Cont.*

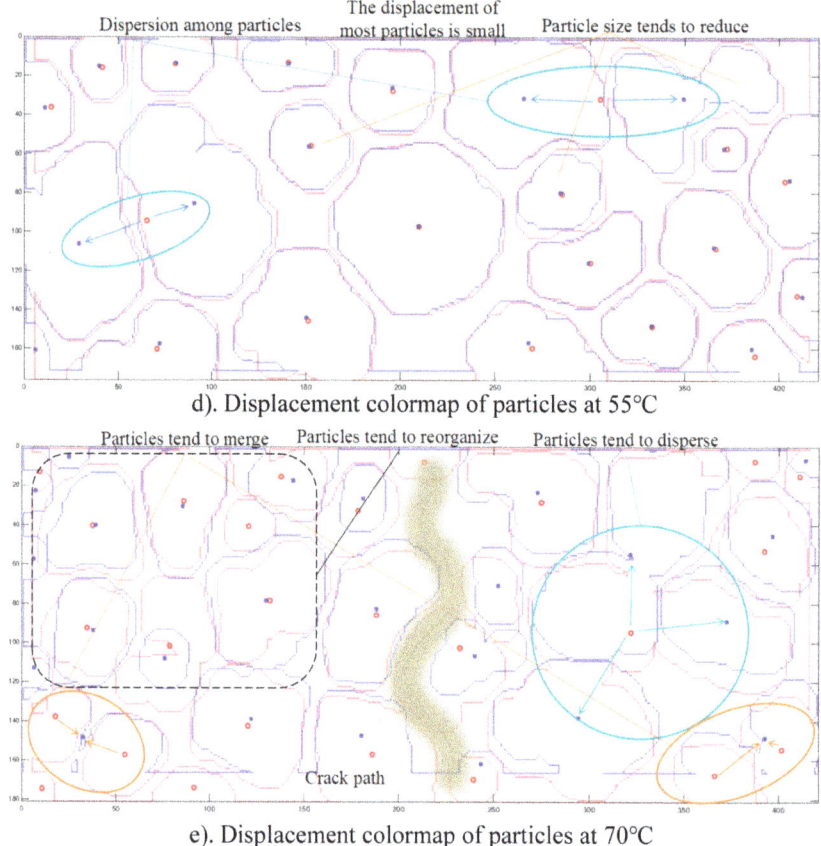

Figure 8. Displacement field of particles at (**a**) −20 °C; (**b**) 0 °C; (**c**) 22 °C; (**d**) 55 °C; and (**e**) 70 °C, respectively. The red ones and red circle markers show the original positions of particles, while the blue ones and blue star markers show the positions of particles after loading.

5. Conclusions

A series of Brazilian tests of TATB-PBX at different temperatures were conducted, while the evolution of micro-structure inside the specimens was obtained through in-situ µCT observation. Through the digital imaging process, CT slice images were analyzed. Fracture degree and complexity were defined and were used to quantitatively describe the fracture characteristics at different temperatures. The fractal characteristic of cracks was also analyzed by using box-counting method. The interior displacement fields of particles were also studied and all of the above results were also proved by SEM detection.

The 3D crack morphology was obtained, and it could be seen clearly and directly at lower temperatures that the cracks were straighter while at higher temperatures, especially above the glass transition temperature of binders, the cracks were more tortuous. The fracture degree was used to characterize the level of the fractures; while the fracture complexity was used to characterize the tortuosity character of the fractures. It could be seen that the fracture complexity tended to increase as the temperature rose, which implied a more tortuous crack. The fractal dimension of cracks showed a great minus linear relation with the temperature, which meant that as the temperature rose, the fracture fractal dimension decreased and the fracture surface was smoother. The interior displacement fields of particles showed the micro-structure evolution inside the specimens, which agreed with the macro-mechanical behavior at different temperatures. In SEM images, the typical morphology of fractures at different temperatures was obtained and proved the above conclusions. It was found that at lower temperatures, especially lower than the glass transition temperature of binders, there was slipping and shear among particles, and particles tended to displace and disperse; while at higher temperatures, especially above the glass transition temperature of binders, there was reorganization of particles and particles tended to merge, disperse, and reduce sizes, rather than displacing.

Author Contributions: J.-M.L. conceived the whole research; H.C. and Z.-N.Y. designed the experiments in detail; Z.-N.Y. and H.C. performed the experiments and analyzed the data; B.D. gave some advice in experiments and W.-B.Z. gave some advice in analysis. Z.-N.Y. and H.C. wrote the paper.

Acknowledgments: This work was founded by NSAF (U1330202), National Natural Science Foundation of China (11572294), and National Key Scientific Instrument and Equipment Development Project (2013YQ03062907).

Conflicts of Interest: The authors declare no conflict of interest. The founding sponsors had no role in the design of the study; in the collection, analyses, or interpretation of data; in the writing of the manuscript, and in the decision to publish the results.

References

1. Balzer, J.E.; Siviour, C.R.; Walley, S.M.; Proud, W.G.; Field, J.E. Behaviour of ammonium perchlorate-based propellants and a polymer-bonded explosive under impact loading. *Proc. R. Soc. Lond. Ser. A* **2004**, *460*, 781–806. [CrossRef]
2. Barua, A.; Horie, Y.; Zhou, M. Energy localization in HMX-Estane polymer-bonded explosives during impact loading. *J. Appl. Phys.* **2012**, *111*, 399–586. [CrossRef]
3. Tang, M.-F.; Li, M.; Lan, L.-G. Review on the mechanical properties of cast PBXs. *Chin. J. Energ. Mater.* **2013**, *21*, 812–817.
4. Ellis, K.; Leppard, C.; Radesk, H. Mechanical properties and damage evaluation of a UK PBX. *J. Mater. Sci.* **2005**, *40*, 6241–6248. [CrossRef]
5. Grantham, S.; Siviour, C.; Proud, W.; Field, J. High-strain rate Brazilian testing of an explosive simulant using speckle metrology. *Meas. Sci. Technol.* **2004**, *15*, 1867–1870. [CrossRef]
6. Guo, B.-Q.; Xie, H.-M.; Chen, P.-W.; Zhang, Q.-M. Creep properties identification of PBX using digital image correlation. In Proceedings of the Fourth International Conference on Experimental Mechanics, Singapore, 18–20 November 2009.
7. Tan, H.; Liu, C.; Huang, Y.; Geubelle, P. The cohesive law for the particle/matrix interfaces in high explosives. *J. Mech. Phys. Solids* **2005**, *53*, 1892–1917. [CrossRef]
8. Zhou, Z.; Chen, P.-W.; Duan, Z.-P.; Huang, F.-L. Study on fracture behaviour of a polymer-bonded explosive simulant subjected to uniaxial compression using digital image correlation method. *Strain* **2012**, *48*, 326–332. [CrossRef]
9. Chen, P.-W.; Huang, F.-L.; Ding, Y.-S. Microstructure, deformation and failure of polymer bonded explosives. *J. Mater. Sci.* **2007**, *42*, 5257–5280. [CrossRef]
10. Rae, P.J.; Goldrein, H.T.; Palmer, S.J.P.; Field, J.E.; Lewis, A.L. Quasi-static studies of the deformation and failure of β-HMX based polymer bonded explosives. *Proc. R. Soc. A* **2002**, *458*, 743–762. [CrossRef]
11. Palmer, S.J.P.; Field, J.E.; Huntley, J.M. Deformation, strengths and strains to failure of polymer bonded explosives. *Proc. R. Soc. Lond. A* **1993**, *440*, 399–419. [CrossRef]
12. Hu, Z.; Luo, H.; Bardenhagen, S.G. Internal deformation measurement of polymer bonded sugar in compression by digital volume correlation of in-situ tomography. *Exp. Mech.* **2015**, *55*, 289–300. [CrossRef]
13. Xue, C.; Sun, J.; Kang, B.; Liu, Y.; Liu, X.-F.; Song, G.-B.; Xue, Q.-B. The β−δ-phase transition and thermal expansion of octahydro-1,3,5,7-tetranitro-1,3,5,7-tetrazocine. *Propellants Explos. Pyrotech.* **2010**, *35*, 333–338. [CrossRef]
14. Zhou, H.P.; Pang, H.Y.; Wen, M.P.; Li, J.-M.; Li, M. Comparative studies on the mechanical properties of three kinds of binders. *Mater. Rev.* **2009**, *23*, 34–36.

15. Willey, T.M.; Lauderbach, L.; Gagliardi, F.; van Buuren, T.; Glascoe, E.A.; Tringe, J.W.; Lee, J.R.I.; Springer, H.K.; Ilavsky, J. Mesoscale evolution of voids and microstructural changes in HMX-based explosives during heating through the β-δ phase transition. *J. Appl. Phys.* **2015**, *118*, 055901. [CrossRef]
16. Wang, X.-J.; Wu, Y.-Q.; Huang, F.-L.; Jiao, T.; Clifton, R.J. Mesoscale thermal-mechanical analysis of impacted granular and polymer-bonded explosives. *Mech. Mater.* **2016**, *99*, 68–78. [CrossRef]
17. Wang, X.-J.; Wu, Y.-Q.; Huang, F.-L. Thermal–mechanical–chemical responses of polymer-bonded explosives using a mesoscopic reactive model under impact loading. *J. Hazard. Mater.* **2017**, *321*, 256–267. [CrossRef] [PubMed]
18. Williamson, D.M.; Siviour, C.R.; Proud, W.G.; Palmer, S.J.P.; Govier, R.; Ellis, K.; Blackwell, P.; Leppard, C. Temperature–time response of a polymer bonded explosive in compression (EDC37). *J. Phys. D Appl. Phys* **2008**, *41*, 085404. [CrossRef]
19. Çolak, Ö.Ü. Mechanical behavior of PBXW-128 and PBXN-110 under uniaxial and multiaxial compression at different strain rates and temperatures. *J. Test. Eval.* **2004**, *32*, 390–395. [CrossRef]
20. Thompson, D.G.; Deluca, R.; Brown, G.W. Time–temperature analysis, tension and compression in PBXs. *J. Energ. Mater.* **2012**, *30*, 299–323. [CrossRef]
21. Tang, M.-F.; Wen, M.-P.; Tu, X.-Z.; Lan, L.-G.; Dai, X.-G. Influence and mechanism of high temperature and mechanical stress on the mechanical behaviors of PBXs. *Chin. J. Energ. Mater.* **2018**, *26*, 150–155.
22. Liu, Z.-W.; Xie, H.-M.; Li, K.-X.; Chen, P.-W.; Huang, F.L. Fracture behavior of PBX simulation subject to combined thermal and mechanical loads. *Polym. Test.* **2009**, *28*, 627–635. [CrossRef]
23. Awaji, H.; Sato, S. Diametral compressive testing method. *J. Eng. Mater. Technol.* **1979**, *101*, 139–147. [CrossRef]
24. Pang, H.-Y.; Li, M.; Wen, M.-P.; Lan, L.-G.; Jing, S.-M. Comparison on the Brazilian test and tension test of the PBX. *Chin. J. Energ. Mater.* **2011**, *34*, 42–45.
25. Pang, H.-Y.; Li, M.; Wen, M.-P.; Lan, L.-G.; Jing, S.-M. Different loading methods in Brazilian test for PBX. *Chin. J. Energ. Mater.* **2012**, *20*, 205–209.
26. Epstein, N. On tortuosity and the tortuosity factor in flow and diffusion through porous media. *Chem. Eng. Sci.* **1989**, *44*, 777–779. [CrossRef]
27. Mandelbrot, B.B.; Wheeler, J.A. The fractal geometry of nature. *Am. J. Phys.* **1983**, *51*, 468. [CrossRef]
28. Mecholsky, J.J.; Passoja, D.E.; Feinberg-Ringel, K.S. Quantitative analysis of brittle fracture surfaces using fractal geometry. *J. Am. Ceram. Soc.* **1989**, *72*, 60–65. [CrossRef]
29. Horovistiz, A.L.; de Campos, K.A.; Shibata, S.; Prado, C.C.S.; de Oliveira Hein, L.R. Fractal characterization of brittle fracture in ceramics under mode I stress loading. *Mater. Sci. Eng A* **2010**, *527*, 4847–4850. [CrossRef]
30. Khanbareh, H.; Wu, X.; Zwaag, S. Analysis of the fractal dimension of grain boundaries of AA7050 aluminum alloys and its relationship to fracture toughness. *J. Mater. Sci.* **2012**, *47*, 6246–6253. [CrossRef]
31. Hilders, O.; Zambrano, N. The effect of aging on impact toughness and fracture surface fractal dimension in SAF 2507 super duplex stainless steel. *J. Microsc. Ultrastruct.* **2014**, *2*, 236–244. [CrossRef]
32. Xie, H.-P.; Chen, Z.-D. Fractal geometry and fracture of rock. *Acta Mech. Sinica* **1988**, *4*, 255–264.
33. Zheng, G.-M.; Zhao, J.; Li, L.; Cheng, X.; Wang, M. A fractal analysis of the crack extension paths in a Si_3N_4 ceramic tool composite. *Int. J. Refract. Met. Hard Mater.* **2015**, *51*, 160–168. [CrossRef]
34. Cheng, K.-M.; Liu, X.-Y.; Guan, D.-B.; Xu, T.; Wei, Z. Fractal analysis of TATB-based explosive AFM morphology at different conditions. *Propellants Explos. Pyrotech.* **2007**, *32*, 301–306. [CrossRef]

© 2018 by the authors. Licensee MDPI, Basel, Switzerland. This article is an open access article distributed under the terms and conditions of the Creative Commons Attribution (CC BY) license (http://creativecommons.org/licenses/by/4.0/).

Article
3D Imaging of Indentation Damage in Bone

Tristan Lowe [1,*,†], Egemen Avcu [1,2,†], Etienne Bousser [1,3], William Sellers [4] and Philip J. Withers [1]

1. Henry Moseley X-ray Imaging Facility, Henry Royce Institute, School of Materials, The University of Manchester, Manchester M13 9PL, UK; egemen.avcu@manchester.ac.uk (E.A.); etienne.bousser@manchester.ac.uk or etienne.bousser@polymtl.ca (E.B.); p.j.withers@manchester.ac.uk (P.J.W.)
2. Ford Otosan Ihsaniye Automotive Vocational School, Machine and Metal Technologies, Kocaeli University, 41680 Kocaeli, Turkey
3. Engineering Physics Department, Polytechnique Montréal, Montreal H3T1J4, QC, Canada
4. School of Earth and Environmental Sciences, The University of Manchester, Manchester M13 9PL, UK; William.Sellers@manchester.ac.uk
* Correspondence: Tristan.Lowe@manchester.ac.uk; Tel.: +44-161-306-2250
† These authors contributed equally to this work.

Received: 12 November 2018; Accepted: 10 December 2018; Published: 13 December 2018

Abstract: Bone is a complex material comprising high stiffness, but brittle, crystalline bio-apatite combined with compliant, but tough, collagen fibres. It can accommodate significant deformation, and the bone microstructure inhibits crack propagation such that micro-cracks can be quickly repaired. Catastrophic failure (bone fracture) is a major cause of morbidity, particularly in aging populations, either through a succession of small fractures or because a traumatic event is sufficiently large to overcome the individual crack blunting/shielding mechanisms. Indentation methods provide a convenient way of characterising the mechanical properties of bone. It is important to be able to visualise the interactions between the bone microstructure and the damage events in three dimensions (3D) to better understand the nature of the damage processes that occur in bone and the relevance of indentation tests in evaluating bone resilience and strength. For the first time, time-lapse laboratory X-ray computed tomography (CT) has been used to establish a time-evolving picture of bone deformation/plasticity and cracking. The sites of both crack initiation and termination as well as the interconnectivity of cracks and pores have been visualised and identified in 2D and 3D.

Keywords: aging; in situ; crack initiation and propagation; damage modes; osteoporosis; osteogenesis imperfecta; porosity; bone matrix quality

---

1. Introduction

Bone is one of the major innovations of vertebrates, allowing a rigid endoskeleton that can both grow to a genetically controlled shape and size and also remodel dynamically in response to load. Structurally it is a composite material consisting of high stiffness, but brittle, crystalline bio-apatite combined with compliant, but tough, collagen fibres [1]. Physiologically, it accommodates growth, remodelling, repair, metabolism and sensation. The material components are arranged in a highly organised fashion with repeated and nested functional units. The structure of bone has evolved to provide the necessary rigid framework for effective locomotion as well as the protection of vital organs [2]. Typical limb bones have an overall shape reflecting the joints and muscle attachment points with an internal structure consisting of spongy trabecular bone at the epiphyses and metaphysis and forming a rigid, hollow tube of cortical bone in the shaft. Within the cortical bone lies an organised microstructure of laminated bone forming multi-layer, multi-scale tubes which allow the physiological functions of bone to occur alongside its biomechanical role. However, this microstructure varies between individual bones, evolves with age, and differs between different vertebrate species [3].

Bone is highly mass optimised, with peak strains due to normal activity that are approximately half the yield strain, providing a safety factor of about two according to strain gauge studies [4]. Unfortunately, bone fractures in humans are all too common, particularly in aging populations where falls are more frequent and bone quality is lower. Conditions that reduce bone loading, such as bed rest, spaceflight, and a reduction in physical activity associated with growing old, lead to a weakening of the bone due to remodelling that results in an increased risk of traumatic injury [5].

Indentation techniques have become some of the most widely used methods to characterise the fracture behaviour of bone tissue over the last decade [6–8]. These techniques allow for the determination of mechanical properties such as hardness and elastic modulus via the acquired load-displacement curves [9]. Since indentation testing can be applied at different length scales to evaluate the mechanical resistance of bone to plastic deformation [8], it has a great potential for investigating and assessing bone quality, plasticity and fracture properties [10,11]. Bone fails due to cracks that propagate through the crystalline matrix. However, the microstructure of bone is such that crack propagation is often stopped at an early stage, and these micro-cracks can be quickly repaired by the body's internal repair processes [12]. It is only in situations where cracks can propagate through most of the shaft, either because of a succession of small injuries without sufficient time for repair (so-called fatigue fractures [13]), or because the traumatic event is sufficiently large to overcome the individual crack blunting mechanisms, that these cracks lead to catastrophic fracture [14,15]. Thus, a detailed understanding of crack propagation within bone is essential to better understand the process of bone fracture, and the risk factors associated with developmental processes, disease and activity. As such this area of biology has been intensely studied over the last few decades. The advent of high-resolution CT has allowed cracks and microcracks to be accurately visualised and quantified (e.g., [16,17]), however previous work in this area has concentrated on visualising cracks after they have formed. There has been very limited work on time-lapse imaging of loaded bone [18] to measure and track crack growth directly. As a result, considerably more work is merited, particularly with regard to indentation methods that could potentially allow in vivo assessment of bone quality. Dynamic visualisation of a crack whilst it is forming requires both high resolution to allow the crack tip to be adequately imaged, and also fast acquisition rates to allow multiple scans to be taken over many loading increments, whilst minimising time-dependent creep. It also requires specialised loading equipment to apply the load or strain without unduly obstructing the X-rays within the scanner. On the one hand, recent advances in laboratory-based micro CT imaging mean that, from a spatial resolution viewpoint, a synchrotron is no longer necessary for much of the imaging. On the other hand, for time-lapse experiments, synchrotrons significantly outperform laboratory-based systems in terms of acquisition rate and signal to noise [19–21]. Nevertheless, limited access to synchrotron beamtime means that if X-ray imaging is to become a useful research tool in the study of bone mechanics as a function of aging, diet, osteoporosis or repair, then it is important to consider laboratory-based X-ray CT methods.

The purpose of this study is to examine what can be achieved using state-of-the-art X-ray CT systems and customised indentation rigs in terms of visualising and tracking bone plastic deformation and crack propagation under indentation loading. This study questions whether laboratory-based systems are adequate for applications such as non-linear bone fracture modelling, bone property measurement, and the experimental identification of bone plasticity and toughening mechanisms.

2. Materials and Methods

A dried, de-fleshed mouse femur was chosen for this study (supplementary document S1): it is a suitable size to fit within the available chambers; the current rig is not humidity controlled so a wet specimen would not be suitable, and the mouse is a commonly used species for studying bone physiology and pathology [22]. The specimen was mounted on a magnetic plate using wax to ensure the head of the femur was positioned vertically for indentation of the fine trabecular region (Figure 1).

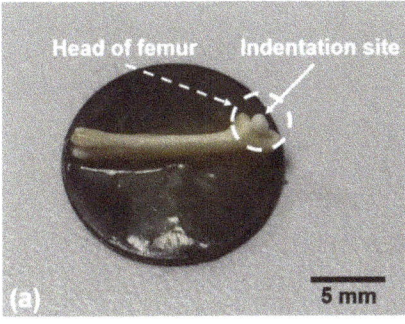 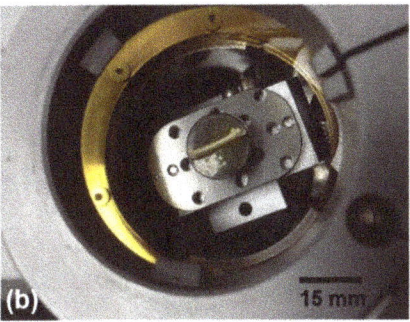

Figure 1. (a) Bone mounted on a steel plate with the femoral head pointing vertically for indentation, (b) Specimen mounted within the Hysitron in-situ nanomechanical testing rig.

Indentation was performed using a specially designed Hysitron in-situ nanomechanical testing rig (Bruker Hysitron IntraSpect 360 indentation rig) developed in association with the Henry Moseley X-ray Imaging Facility. It was mounted in a Zeiss Xradia 520 Versa X-ray microscope system (Figure 2) to allow imaging while the indentation was taking place. The Hysitron rig uses a piezoelectric load cell design with capacitive depth sensing. It can provide a maximum force of 10 N and maximum displacement of 80 μm and will run in load- or displacement-controlled testing modes using a Performech digital controller. This allows it to capture transient events, such as fracture initiation, whilst using a wide variety of probe materials and geometries. The indentation was performed using a three-sided pyramidal (Berkovich) indenter under load-controlled indentation mode. The Zeiss Xradia 520 Versa was operated at 110 kV and 91 mA with a specimen-to-source distance of 94 mm and a specimen-to-detector distance of 43.4 mm for optimum imaging geometry to optimise the resolution of the images obtained by minimising both pixel size and focal spot blurring [23]. The whole indentation area could be viewed at 9.77× magnification corresponding to a voxel size of 0.95 μm under these settings. Each projection was collected over 120 s to minimise image noise in the projections and maximise spatial resolution, resulting in a 2D spatial resolution in the 2–3 μm range which was validated using a JIMA spatial resolution chart. The specimen was rotated over a 360° rotation range, collecting 1601 projections and giving an overall scan time of 52 h.

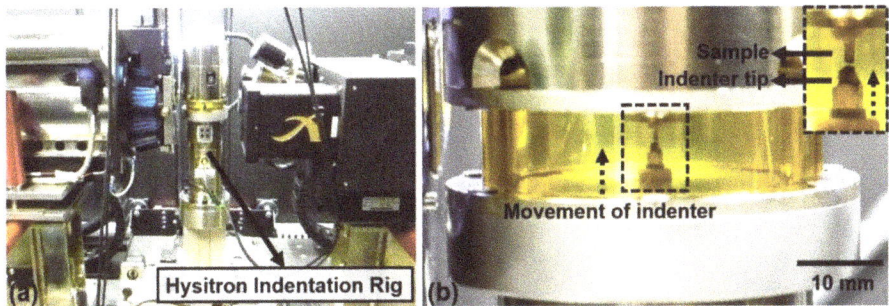

Figure 2. (a) The Hysitron indentation rig mounted within the Zeiss-Xradia VersaXRM-520 system during initial specimen alignment, (b) Magnified image showing the indentation tip position.

After image acquisition, the data sets were uploaded into the Zeiss Xradia XMReconstructor software for reconstruction of the 3D virtual slices using a filtered back projection algorithm. The reconstructed data was then analysed using the Avizo 9.2 (Thermo Fisher Scientific, Waltham, MA, USA) visualisation software to segment and image the virtual slices and 3D volume renderings.

After the indentation experiment, the residual imprint location on the femoral head was investigated using an FEI Magellan scanning electron microscope (SEM), and secondary electron (SE) images of the indentation site were collected at 1 kV with a tilt angle of 45°.

3. Results and Discussions

The indentation sequence is illustrated in Figure 3a with the load and displacement curves recorded shown in Figure 4. The load was linearly incremented from 0 to a maximum of 2.5 N, since at this point the displacement sensor limit was reached and the sensor saturated, leading to a significant drop in load and a slight increase in displacement during the hold segment (Figure 4).

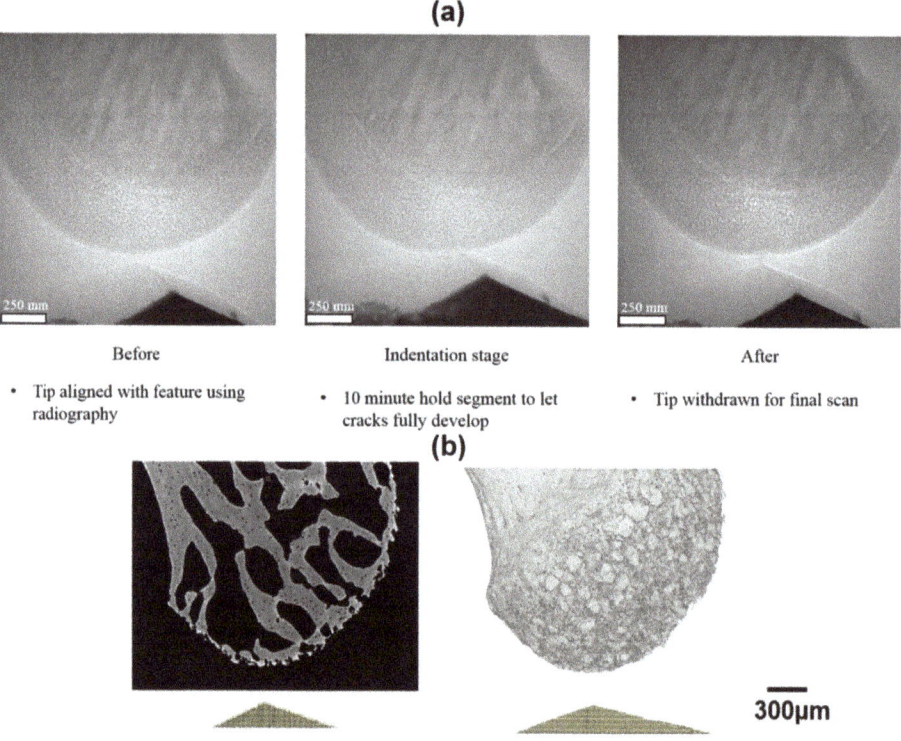

Figure 3. (a) Radiographs showing the indentation sequence using a Berkovich indenter geometry, (b) Virtual slice and 3D reconstruction of the head of the mouse femur.

The load-displacement curve of the full indentation and the corresponding contact stiffness (dP/dh) during loading are illustrated in Figure 5. The fracture of the bone can be identified in the load-displacement curve through pop-in events in the loading segment, which translate to significant drops in contact stiffness accompanying local fractures (Figure 5). The residual imprint does not have the usual shape for a Berkovich indentation and seems to be slightly elongated (Figure 6). This could indicate the movement of the tip during the indentation. In addition to the sensor saturation effects, this could be one of the reasons for the drop in load and displacement drift during the test holding segment (Figure 4).

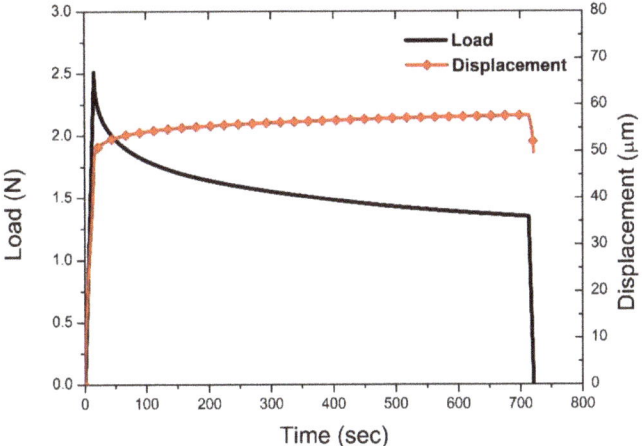

Figure 4. The load and displacement curves recorded as a function of time.

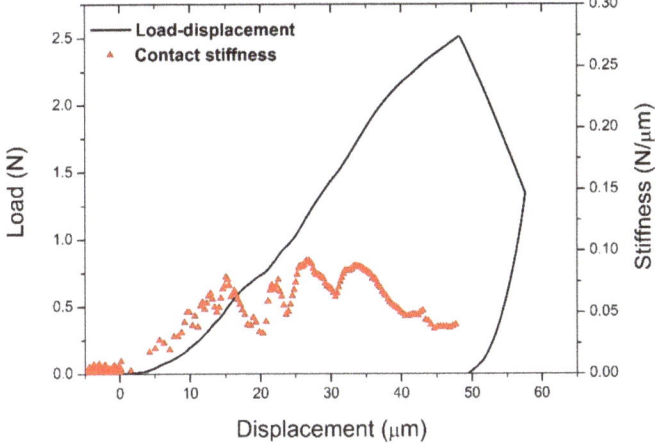

Figure 5. The load-displacement curve of the full indentation and variation of the contact stiffness during the loading segment. The load drop during the hold stage is evident.

Due to the heterogeneous microstructure of bone [24], as illustrated in Figure 3b, the selection of the indentation site is essential [25] in determining the mechanical response and the inferred properties such as hardness and elastic modulus [26,27]. One other factor that needs to be considered in terms of indentation testing is the size of the indenter [28]. The inferred mechanical properties of the bone are likely to depend on the indent size due to the scale of the microstructural features such as pores and cracks within the bone structure sampled during the indentation process [29]. In the present study, the indent size is not a critical concern since the main focus of the study is to generate and visualise sub-surface crack networks to understand the crack propagation and fracturing behaviour of bone rather than to examine the mechanical properties (e.g., hardness and modulus) by implementing indentation under X-ray CT.

Figure 7 shows a 3D image of the indentation site and also the geometry of the contact area formed during the indentation. This indicates a material pile-up behaviour [30]. It has been reported that pile-up of material on the indentation site may occur when using a sharp Berkovich indenter [9,31,32].

However, the amount of pile-up is limited, as seen in Figure 7, since bone has a relatively low effective modulus-to-yield stress ratio in compression [32]. Although the plastically deformed indentation site and limited pile-up can be seen through the volumetric rendering of the indentation site given in Figure 7, it is not possible to identify the sub-surface damage mechanisms such as splitting, cracking and plastic deformations induced by the indentation. Therefore, individual virtual cross-sectional slices (orthoslices) at different distances have been used to identify these damage modes and to understand the fracture mechanics of bone. This is challenging due to the complex and hierarchical structure of bone [33]. Figure 8 shows the individually rendered cross sections at large distances away from the indentation site and near the indentation site. Small dimensional changes through the movement of small pores in the bone microstructure are visible at a distance of 214 µm from the indentation site (slice 60) while no visible changes can be observed at a distance of 250 µm (slice 40).

Figure 6. Secondary electron SEM image of the indentation residual imprint location on the femoral head and higher-magnification of the indentation (inset) at 1 kV with a tilt angle of 45°.

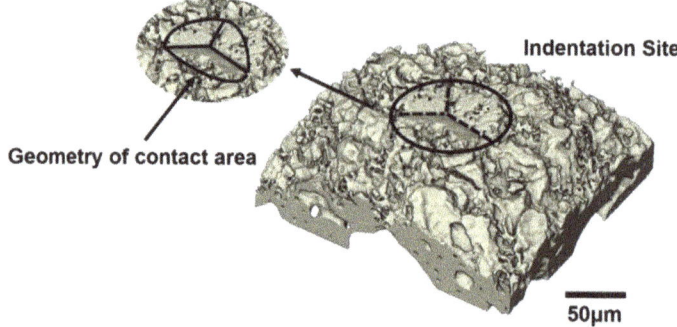

Figure 7. Volumetric rendering of the indentation site and geometry of the contact area after removal of the indenter.

Figure 8 shows various damage mechanisms including visible plastic strain and crack initiation and propagation through the virtual cross-sectional slices near the indentation site (slices 117 and 125). Fine cracks and the displacement of pores can be identified at a distance of 113 µm (slice 117) while crack formation and growth at an angle of approximately 45° to the indentation direction can be observed nearer to the indentation site (slice 125). Crack bridging is one of the major mechanisms that inhibits crack extension in bone tissue [34] and can occur when a crack grows from the indent [27]. Here the cracks along the direction 45° to the indentation direction show some slight evidence of crack bridging (Figure 8 slice 125) which may lead to a decrease in the driving force at the crack tip through the interaction between the cracks and the microstructural features [35]. Some interactions between the propagating cracks and the pre-existing pores and micro-cracks are also visible in slices 117 and 125. Although it is clear that the pore networks in the bone structure affect the mechanical behaviour of bone [36], it is currently unknown how important these pores are in preventing crack propagation and increasing the fracture resistance by contributing to toughening mechanisms such as microcracking, crack bridging, and crack deflection [10,15,16,35]. Haversian canals have an important role in the crack propagation behaviour of human cortical bone (specifically crack deflection) [15], however, they have been reported as 'not found' in mouse bone, and in any case not in trabecular bone [7]. Thus, the interconnectivity of pores and cracks may have a relatively strong effect on the fracturing of mouse bone compared to human bone. In order to better understand the effects of the pre-existing pore networks on the crack propagation, the interaction of cracks with pores and fine trabecular structure is visualised and discussed in detail through 2D and 3D visualisations in the following sections.

The propagation of micro-cracks local to the indentation site and the interaction between the growing cracks and the pores found in the trabecular bone can be seen in Figures 8 and 9. It is evident that there is no significant orientation of the pore structure, and the sizes of pores range between 5 and 30 µm. The micro-crack formation might be dependent on these pre-existing pore networks illustrated in both Figures 8 and 9. On the one hand, pores may detrimentally affect the mechanical properties of bone due to the decrease in the total load bearing area [37] as well as acting as stress concentration sites. On the other hand, the internal pore structure may have a positive effect on suppressing crack propagation during indentation. Thus, it is important to understand the effects of crack-pore interactions on the toughening mechanisms of bone structure. Unfortunately, the existing literature is limited regarding the analysis of the interaction of cracks with the existing pore networks in the bone microstructure, and no work has been conducted on how cracks develop and interact with the microstructural features of bone over time during indentation loading.

In Figure 9, we can see how a crack interacts with an existing pore within the bone structure, and the growth of the crack appears to be terminated at this point. However, the 2D view of cracks and their interactions with the pores illustrated in these figures can be highly misleading. In fact, the cracks themselves have complex 3D shapes that propagate through the bone, and the level of detail is therefore clearly necessary if this approach is to be used to validate non-linear crack growth models and their complex interactions with the pore networks in the bone structure. Figure 10 illustrates the individually rendered slice at a distance of 135 µm from the opposite side of the indenter tip and the 3D rendering of the crack structure illustrating the interaction between the growing crack and the pores found within the trabecular bone. The 2D visualisation suggests that the pore structure has completely stopped the crack from propagating, but the 3D view showing a significant flexural crack and its interaction with an existing pore implies that the real story is much more complex (Figure 10). To the best of our knowledge, this is the first time that indentation of bone has been visualised in high resolution to analyse the initiation and propagation of cracks and their interactions with the microstructural features of bone, specifically with pre-existing pore networks.

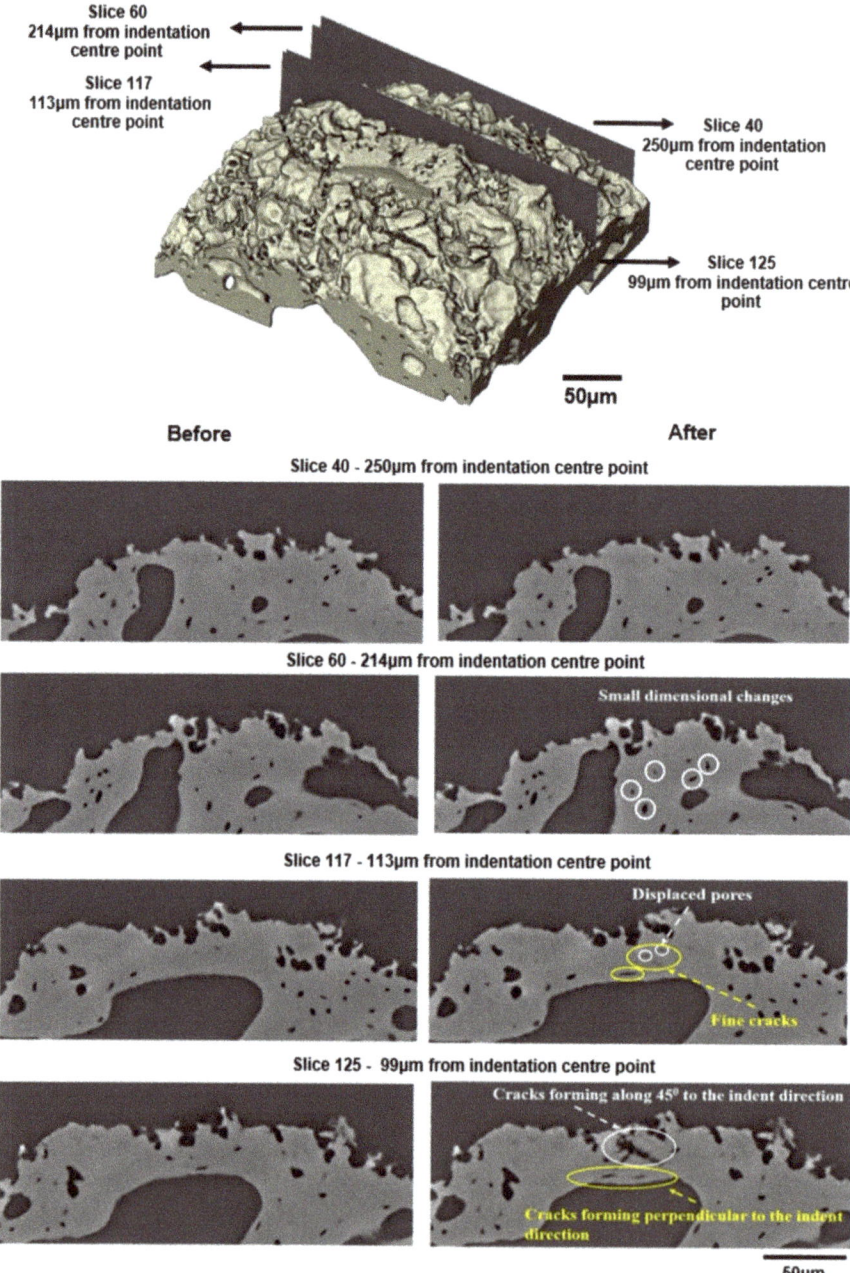

Figure 8. Individual virtual slices before (**left**) and after (**right**) indentation at distances of 250 µm, 214 µm, 113 µm and 99 µm from the indenter tip (see also animated sequences in supplementary video online).

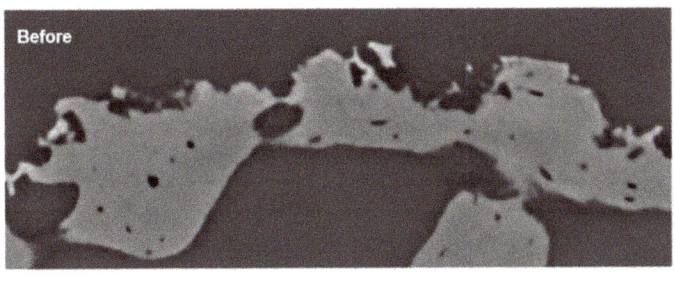

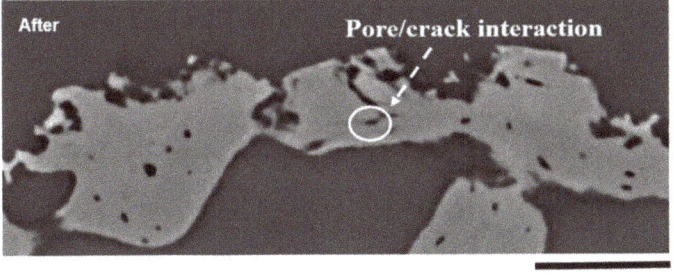

Figure 9. Individual slice images before and after indentation showing plastic deformation, collapse and cracking as a result of the indenter at a distance of 66 µm from the indenter tip (see also animated sequences in supplementary video online).

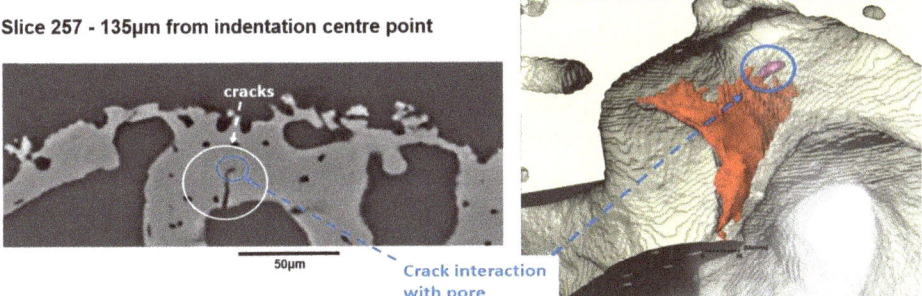

Figure 10. 2D and 3D visualisations of a larger flexural crack from the opposite side of the indenter tip showing how the 2D image can be misleading in terms of crack complexity and propagation.

Figure 11 shows the damage at the centre of the indentation site. Some evidence of plastic deformation beneath the surface is visible which could provide resistance to crack propagation by blunting the crack tip through the formation of plastically deformed zones [7]. However, this intrinsic toughening mechanism seems to be very limited, preventing the formation of large shear cracks just beneath the vicinity of the indent corner (Figure 11c). The rotation of the indentation site around the pivot point (rotation point) is identified through the angular movement of the indented microstructure as highlighted in Figure 11b,c. It can be inferred that the bending moment induced by this identified rotation may be the underlying reason for the severe plastic deformation and shear crack propagation at the centre of the indentation site. A limited pile-up of material can be seen at the right edges of the impression, proving that plastic deformation occurred during indentation while displacement of materials is visible at the left edge due to the generated high strain at the surface of the indentation zone (Figure 11).

Previously indentation studies have yielded only information on the surface crack geometries, while sub-surface crack behaviour has only been predicted, or investigated through cross-sectional scanning electron microscope (SEM) images [27]. Here we have successfully visualised and linked different surface and sub-surface damage mechanisms in the bone structure, such as localized plastic deformation, pile-up of materials and propagation of different crack types by fracturing the head of a mouse femur using load-controlled indentation testing within the X-ray CT scanner. Since mouse bone has been widely used as a model for human bone diseases [7,24] this approach provides a novel way of exploring the bone quality and the fracture mechanics of bone as a function of aging and/or bone disease. While we are currently limited to using dried bone, future work will include the incorporation of a humidity controlled chamber so that freshly prepared bone can be used, allowing its well known differences in material properties to be assessed [38].

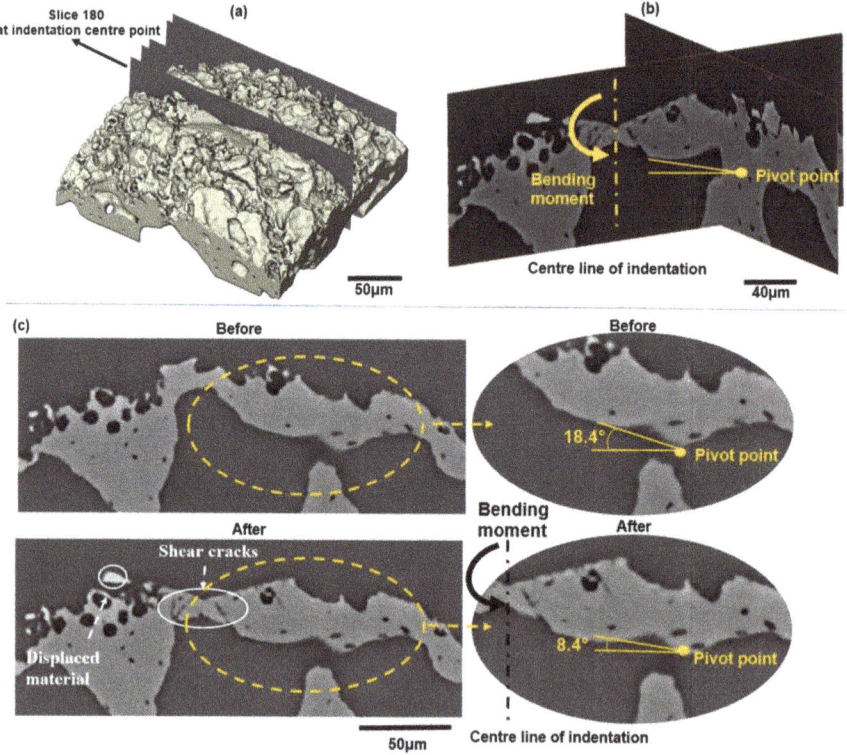

Figure 11. (a) Volumetric rendering of the indentation site, (b) Orthoslice from the other orientation showing the pivot point, (c) Individual slice images showing the damage immediately below the indentation tip (see also animated sequences in supplementary video online).

4. Conclusions

The present paper demonstrates the use of an indentation rig within an X-ray CT scanner to visualise and characterise the cracking of bone material at an unprecedented level of detail for the first time. This has revealed a range of surface and sub-surface damage mechanisms in the bone structure, such as localised plastic deformation and the propagation of different types of cracking. The results show that indentation coupled with X-ray CT has the potential to quantify the bone fracture mechanics associated with aging and disease process, such as osteoporosis and osteogenesis imperfecta.

The interactions of cracks with the existing pore networks and micro-cracks have been visualised in high resolution in 2D and 3D to understand the fracturing process of the bone structure. Further research is recommended on the 3D visualisation of these interactions to understand the fracture mechanics of bone. Time-lapse 3D imaging is a promising tool to better understand in-vivo indentation of bone to evaluate bone quality, the effects of normal development, and the impacts of disease processes and potential treatments.

The results illustrate that the 3D propagation of cracks and other deformation effects within the bone under the indentation load can be visualised with a resolution of ~2 µm by laboratory-based X-ray CT. This is sufficient to observe early crack formation while enabling us to see the microstructural features that might help prevent crack propagation. However, the current acquisition rate means that each scan took 52 h to acquire. This limited the number of frames that could be acquired in a time-lapse sequence. Furthermore, it introduces the risk of creep and relaxation affecting the interpretation of the results. Therefore, in-situ indentation tests under synchrotron X-ray CT imaging would be desirable to observe in-situ crack development as the indenter is progressively loaded. This would also allow the use of CT data to better understand the strain-strain features that occur during indentation in terms of the plastic hinging and crack propagation sub-surface. Such an approach will enable better interpretation and of indentation curves in terms of the effects of aging or disease. Nevertheless lab. X-ray CT is demonstrated to be a useful tool for characterising and quantifying sub-surface indentation damage for bone indentation testing.

Supplementary Materials: The following are available online at http://www.mdpi.com/1996-1944/11/12/2533/s1, Video S1: Animated sequences showing indentation damage in bone, Document S1: Materials and methods.

Author Contributions: Conceptualization, T.L., E.A., P.J.W. and W.S.; Funding Acquisition, P.J.W.; Investigation, T.L. and E.B.; Methodology, T.L. and E.B.; Project Administration, P.J.W.; Recourses, W.S.; Supervision, P.J.W.; Visualisation, T.L., E.B., and E.A.; Writing—Original Draft; E.A., W.S. and T.L.; Writing—Review & Editing; E.A., E.B., W.S., T.L., and P.J.W.

Funding: The X-ray kit was funded by the Higher education Funding Council for England UK (HEFCE). We would like to acknowledge the UK Research Partnership Investment Fund (UKRPIF) and the Engineering and Physical Sciences Research Council (ESPRC) for funding the Henry Moseley X-ray Imaging Facility through grants (EP/F007906/1, EP/F001452/1 and EP/M010619/1). P.J.W. also acknowledges support from the European Research Council (ERC) grant No. 695638 CORREL-CT. E.B. would like to acknowledge the Canadian Natural Sciences and Engineering Research Council's support through the Postdoctoral Fellowship program.

Acknowledgments: The authors would like to thank Xun Zhang and Douglas Stauffer for advising on the use of the Hysitron indentation rig, and Andrew Chamberlain for providing the bone samples.

Conflicts of Interest: The authors declare no conflict of interest. The funders had no role in the design of the study; in the collection, analyses, or interpretation of data; in the writing of the manuscript, or in the decision to publish the results.

References

1. Currey, J.D. *Bones: Structure and Mechanics*; Princeton University Press: Princeton, NJ, USA, 2002; ISBN 9780691128047.
2. Alexander, R.M. *Principles of Animal Locomotion*; Princetown University Press: Princeton, NJ, USA, 2006; ISBN 9781400849512.
3. Quemeneur, S.; De Buffrenil, V.; Laurin, M. Microanatomy of the amniote femur and inference of lifestyle in limbed vertebrates. *Biol. J. Linn. Soc.* **2013**, *109*, 644–655. [CrossRef]
4. Biewener, A.A. Biomechanics of Mammalian Terrestrial Locomotion. *Science* **1990**, *250*, 1097–1103. [CrossRef] [PubMed]
5. LeBlanc, A.D.; Spector, E.R.; Evans, H.J.; Sibonga, J.D. Skeletal responses to space flight and the bed rest analog: A review. *J. Musculoskelet. Neuronal Interact.* **2007**, *7*, 33–47. [PubMed]
6. Dall'Ara, E.; Schmidt, R.; Zysset, P. Microindentation can discriminate between damaged and intact human bone tissue. *Bone* **2012**, *50*, 925–929. [CrossRef] [PubMed]

7. Carriero, A.; Zimmermann, E.A.; Shefelbine, S.J.; Ritchie, R.O. A methodology for the investigation of toughness and crack propagation in mouse bone. *J. Mech. Behav. Biomed. Mats* **2014**, *39*, 38–47. [CrossRef] [PubMed]
8. Hunt, H.B.; Donnelly, E. Bone quality assessment techniques: Geometric, compositional, and mechanical characterization from macroscale to nanoscale. *Clin. Rev. Bone Miner. Metab.* **2016**, *14*, 133–149. [CrossRef]
9. Oliver, W.C.; Pharr, G.M. Measurement of hardness and elastic modulus by instrumented indentation: Advances in understanding and refinements to methodology. *J. Mater. Res.* **2011**, *19*, 3–20. [CrossRef]
10. Nyman, J.S.; Granke, M.; Singleton, R.C.; Pharr, G.M. Tissue-Level Mechanical Properties of Bone Contributing to Fracture Risk. *Curr. Osteoporos. Rep.* **2016**, *14*, 138–150. [CrossRef]
11. Chang, A.; Easson, G.W.; Tang, S.Y. Clinical Measurements of Bone Tissue Mechanical Behavior Using Reference Point Indentation. *Clin. Revn Bone Miner. Metab.* **2018**, *16*, 87–94. [CrossRef]
12. Bruce Martin, R. Fatigue Damage, Remodeling, and the Minimization of Skeletal Weight. *J. Theor. Biol.* **2003**, *220*, 271–276. [CrossRef]
13. Lee, T.C.; O'Brien, F.L.; Parkesh, R.; Gunnlaugsson, T. Visualisation and Quantification of Fatigue Induced Microdamage in Bone: Histology and Radiology. In Proceedings of the 11th International Conference on Fracture 2005 (ICF 11), Turin, Italy, 20–25 March 2005.
14. Alexander, R.M. *Animal Mechanics*; Sidgwick & Jackson: London, UK, 1968.
15. Mohsin, S.; O'Brien, F.J.; Lee, T.C. Osteonal crack barriers in ovine compact bone. *J. Anat.* **2006**, *208*, 81–89. [CrossRef] [PubMed]
16. Larrue, A.; Rattner, A.; Peter, Z.A.; Olivier, C.; Laroche, N.; Vico, L.; Peyrin, F. Synchrotron radiation micro-CT at the micrometer scale for the analysis of the three-dimensional morphology of microcracks in human trabecular bone. *PLoS ONE* **2011**, *6*, e21297. [CrossRef] [PubMed]
17. O'Brien, F.J.; Taylor, D.; Dickson, G.R.; Lee, T.C. Visualisation of three-dimensional microcracks in compact bone. *J. Anat.* **2000**, *197*, 413–420. [CrossRef] [PubMed]
18. Thurner, P.J.; Wyss, P.; Voide, R.; Stauber, M.; Stampanoni, M.; Sennhauser, U.; Muller, R. Time-lapsed investigation of three-dimensional failure and damage accumulation in trabecular bone using synchrotron light. *Bone* **2006**, *39*, 289–299. [CrossRef] [PubMed]
19. Lee, T.C.; Mohsin, S.; Taylor, D.; Parkesh, R.; Gunnlaugsson, T.; O'Brien, F.J.; Giehl, M.; Gowin, W. Detecting microdamage in bone. *J. Anat.* **2003**, *203*, 161–172. [CrossRef] [PubMed]
20. Metscher, B.D. MicroCT for comparative morphology: Simple staining methods allow high-contrast 3D imaging of diverse non-mineralized animal tissues. *BMC Physiol.* **2009**, *9*, 11. [CrossRef] [PubMed]
21. Maire, E.; Withers, P.J. Quantitative X-ray tomography. *Int. Mater. Rev.* **2013**, *59*, 1–43. [CrossRef]
22. Pogoda, P.; Priemel, M.; Schilling, A.F.; Gebauer, M.; Catala-Lehnen, P.; Barvencik, F.; Beil, F.T.; Munch, C.; Rupprecht, M.; Muldner, C.; et al. Mouse models in skeletal physiology and osteoporosis: Experiences and data on 14,839 cases from the Hamburg Mouse Archives. *J. Bone Miner. Metab.* **2005**, *23* (Suppl. 1), 97–102. [CrossRef]
23. Feser, M.; Gelb, J.; Chang, H.; Cui, H.; Duewer, F.; Lau, S.H.; Tkachuk, A.; Yun, W. Sub-micron resolution CT for failure analysis and process development. *Meas. Sci. Technol.* **2008**, *19*, 094001. [CrossRef]
24. Carriero, A.; Bruse, J.L.; Oldknow, K.J.; Millan, J.L.; Farquharson, C.; Shefelbine, S.J. Reference point indentation is not indicative of whole mouse bone measures of stress intensity fracture toughness. *Bone* **2014**, *69*, 174–179. [CrossRef]
25. Zwierzak, I.; Baleani, M.; Viceconti, M. Microindentation on cortical human bone: Effects of tissue condition and indentation location on hardness values. *Proc. Inst. Mech. Eng. Part H J. Eng. Med.* **2009**, *223*, 913–918. [CrossRef] [PubMed]
26. Johnson, W.M.; Rapoff, A.J. Microindentation in bone: Hardness variation with five independent variables. *J. Mater. Sci. Mater. Med* **2007**, *18*, 591–597. [CrossRef] [PubMed]
27. Kruzic, J.J.; Kim, D.K.; Koester, K.J.; Ritchie, R.O. Indentation techniques for evaluating the fracture toughness of biomaterials and hard tissues. *J. Mech. Behav. Biomed. Mater.* **2009**, *2*, 384–395. [CrossRef] [PubMed]
28. Oyen, M.L.; Shean, T.A.V.; Strange, D.G.T.; Galli, M. Size effects in indentation of hydrated biological tissues. *J. Mater. Res.* **2011**, *27*, 245–255. [CrossRef]
29. Boughton, O.R.; Ma, S.; Zhao, S.; Arnold, M.; Lewis, A.; Hansen, U.; Cobb, J.P.; Giuliani, F.; Abel, R.L. Measuring bone stiffness using spherical indentation. *PLoS ONE* **2018**, *13*, e0200475. [CrossRef] [PubMed]

30. Hardiman, M.; Vaughan, T.J.; McCarthy, C.T. The effects of pile-up, viscoelasticity and hydrostatic stress on polymer matrix nanoindentation. *Polym. Test.* **2016**, *52*, 157–166. [CrossRef]
31. Tang, Z.; Guo, Y.; Jia, Z.; Li, Y.; Wei, Q. Examining the Effect of Pileup on the Accuracy of Sharp Indentation Testing. *Adv. Mater. Sci. Eng.* **2015**, *2015*, 1–10. [CrossRef]
32. Zysset, P.K. Indentation of bone tissue: A short review. *Osteoporosis* **2009**, *20*, 1049–1055. [CrossRef]
33. Ritchie, R.O.; Nalla, R.K.; Kruzic, J.J.; Ager, J.W.; Balooch, G.; Kinney, J.H. Fracture and Ageing in Bone: Toughness and Structural Characterization. *Strain* **2006**, *42*, 225–232. [CrossRef]
34. Diez-Perez, A.; Guerri, R.; Nogues, X.; Caceres, E.; Pena, M.J.; Mellibovsky, L.; Randall, C.; Bridges, D.; Weaver, J.C.; Proctor, A.; et al. Microindentation for in vivo measurement of bone tissue mechanical properties in humans. *J. Bone Miner. Res.* **2010**, *25*, 1877–1885. [CrossRef]
35. Ager, J.W.; Balooch, G.; Ritchie, R.O. Fracture, aging, and disease in bone. *J. Mater. Res.* **2011**, *21*, 1878–1892. [CrossRef]
36. Hauthier, R. Crack Propagation Mechanisms in Human Cortical Bone on Different Paired Anatomical locations: Biomechanical, Tomographic and Biochemical Approaches. Ph.D. Thesis, Université de Lyon, Lyon, France, 2017.
37. Zimmermann, E.A.; Busse, B.; Ritchie, R.O. The fracture mechanics of human bone: Influence of disease and treatment. *BoneKEy Rep.* **2015**, *4*, 743. [CrossRef] [PubMed]
38. Currey, J. Measurement of the mechanical properties of bone: A recent history. *Clin. Orthop. Relat. Res.* **2009**, *467*, 1948–1954. [CrossRef] [PubMed]

© 2018 by the authors. Licensee MDPI, Basel, Switzerland. This article is an open access article distributed under the terms and conditions of the Creative Commons Attribution (CC BY) license (http://creativecommons.org/licenses/by/4.0/).

Article

3D Analysis of Deformation and Porosity of Dry Natural Snow during Compaction

Lavan Kumar Eppanapelli *, Fredrik Forsberg, Johan Casselgren and Henrik Lycksam

Division of Fluid and Experimental Mechanics, Luleå University of Technology, 971 87 Luleå, Sweden; fredrik.forsberg@ltu.se (F.F.); johan.casselgren@ltu.se (J.C.); henrik.lycksam@ltu.se (H.L.)
* Correspondence: lavan.eppanapelli@ltu.se; Tel.: +46-920-49-3098

Received: 28 September 2018; Accepted: 8 March 2019; Published: 13 March 2018

Abstract: The present study focuses on three-dimensional (3D) microstructure analysis of dry natural snow during compaction. An X-ray computed microtomography (micro-CT) system was used to record a total of 1601 projections of a snow volume. Experiments were performed in-situ at four load states as 0 MPa, 0.3 MPa, 0.6 MPa and 0.8 MPa, to investigate the effect of compaction on structural features of snow grains. The micro-CT system produces high resolution images (4.3 µm voxel) in 6 h of scanning time. The micro-CT images of the investigated snow volume illustrate that grain shapes are mostly dominated by needles, capped columns and dendrites. It was found that a significant number of grains appeared to have a deep hollow core irrespective of the grain shape. Digital volume correlation (DVC) was applied to investigate displacement and strain fields in the snow volume due to the compaction. Results from the DVC analysis show that grains close to the moving punch experience most of the displacement. The reconstructed snow volume is segmented into several cylinders via horizontal cross-sectioning, to evaluate the vertical heterogeneity of porosity distribution of the snow volume. It was observed that the porosity (for the whole volume) in principle decreases as the level of compaction increases. A distinct vertical heterogeneity is observed in porosity distribution in response to compaction. The observations from this initial study may be useful to understand the snow microstructure under applied stress.

Keywords: tomography; micro-CT; snow grains; digital volume correlation; snow microstructure; snow properties

---

1. Introduction

The three-dimensional (3D) arrangement of ice crystals and pores, i.e., the microstructure of snow, changes with time due to exchanges of matter between the ice crystals. Although the link between the snow microstructure and its physical properties has been addressed for a long time [1,2], it is still difficult to characterize the snow microstructure and its evolution over time. Two common approaches used to characterize the 3D microstructure of snow are serial sectioning [3,4] and X-ray computed microtomography (micro-CT) imaging [5,6]. The micro-CT is a technique for non-destructive 3D imaging of internal microstructures [7]. Based on the acquired 3D data it is possible to make a quantitative analysis of internal features such as porosity, cracks, grains, fibres etc., as well as material deformation and strain [8]. In addition, the micro-CT enables the evolution of a material microstructure to be studied in both temporal and spatial domain.

The micro-CT method has been used by many researchers for more than ten years to visualize the snow microstructure. However, the 3D quantitative analysis of displacement of snow grains during compaction is limited. Schleef and Löwe [9] addressed the influence of external mechanical stress on isothermal densification and specific surface area (SSA) of snow, using the micro-CT measurements. They reported that evolution of the snow SSA is independent of the snow density, while snow densification increases with increasing external stress. Kaempfer and Schneebeli [10] investigated the isothermal metamorphism of fresh snow at different temperatures for nearly one year. They deduced snow microstructural parameters from the tomographic images, which describe the structural information related to grain boundaries. Pinzer et al. [11] performed time-lapse micro-CT experiments on snow metamorphism under a static temperature gradient. They observed the structural evolution and mass transfer within snow through ice grains. Ebner et al. [12] further observed the snow metamorphism exposed to an advective airflow and reported that saturated airflow has no influence over grains coarsening rate. Wang and Baker [13] investigated the snow microstructure evolution under compression tests, based on the X-ray micro-CT imaging. They also performed analysis of SSA, structure model index and structure thickness. One of their findings from interrupted compression tests was that the SSA of snow decreased more rapidly than the determined values of SSA. Kerbrat et al. [14] and Hagenmuller et al. [15] proposed image processing techniques to determine snow properties such as density and SSA based on the micro-CT measurements. They emphasized that the retrieval method of these snow properties is sensitive to the voxel size (10 µm for their experiments), especially for fresh snow. Wiese and Schneebeli [8] performed the micro-CT measurements of snow microstructure under the influence of settlement at constant temperature gradient. Their observations show an increase in density, strain and viscosity over time due to settlement induced via external loading. All these investigations have larger focus on microstructural parameters, and smaller focus on structural changes in a snowpack at granular level due to compaction. Further, Schleef et al. [16] observed the impact of various levels of compression on the microstructure of snow, which is similar to the presented study. The major focus of their study was on the dependence of density and SSA to the applied stress using a microcompression device. However the presented study focuses on the characteristics of individual ice crystals under external loading. A high quality tomographic data with spatial resolution of 4.3 µm was carried out in this study to observe the microstructural features of snow grains with respect to applied stress.

The purpose of this initial study is to analyze the 3D images of snow grains during compression tests induced via in-situ uniaxial load. The presented analysis of the micro-CT measurements focuses on displacement and strain fields based on digital volume correlation (DVC) [17] and porosity distribution of the investigated snow volume. Microstructural parameters such as density and SSA have been calculated for various load stages from the tomography data. The investigated snow volume is prepared from a freshly fallen low density snow that was collected right after precipitation. Section 2 focuses on experimental arrangement and measurement procedure. Section 3 describes the applied techniques, which are DVC and porosity analysis. Section 4 details the observations of the study and discussions of the observed results are given in Section 5. A summary including conclusions are presented in Section 6.

2. Experimental Procedure

2.1. Micro-CT System with In-Situ Load Module

Snow sampling and scanning were performed at the micro-CT lab, Luleå University of Technology, Sweden. The 3D images of snow microstructure were obtained using a ZEISS Xradia 510 Versa (Carl Zeiss X-ray Microscopy, Pleasanton, CA, USA). The Xradia 510 Versa can achieve 0.7 µm of true spatial resolution with minimum achievable voxel size of 0.07 µm. The components of the micro-CT system are a sealed microfocus X-ray tube, 4-axis sample stage, a photo detector and a load stage, see Figure 1a. The system is equipped with a Deben CT5000TEC temperature (Deben UK Limited, Bury Saint Edmunds, UK) controlled load stage with a 500 N load cell. The Xradia 510 Versa can be operated at tube voltage range of 30–160 kV with maximum output of 10 W.

A sample holder (Figure 1b) was specifically designed to visualize material properties for the micro-CT experiments. This holder has a fixed punch made of brass, a Polymethyl methacrylate (PMMA) tube with inner and outer diameter of 6 mm and 10 mm, respectively. Further, a moving punch with diameter of 5.95 mm was made of aluminium. The diameter of the moving punch is slightly smaller than the inner diameter of the PMMA tube, in order to ensure a smooth and frictionless compression. The in-situ load stage was used to apply uniaxial stress to compact a snow volume via the moving punch, which means the compression is applied from the bottom, and both the punch plates have smooth and flat surface. The investigated snow sample was small enough such that it was not required to move the sample vertically during data acquisition.

Scout-and-Scan[TM] Control Software Control Software (Carl Zeiss X-ray Microscopy, Pleasanton, CA, USA) was used for reconstructing the scanned images. Quantitative analysis of microstructure (shape of crystals, porosity etc.) was obtained from 3D image analysis, using the software Dragonfly Pro (Carl Zeiss X-ray Microscopy, Pleasanton, CA, USA) (ORS).

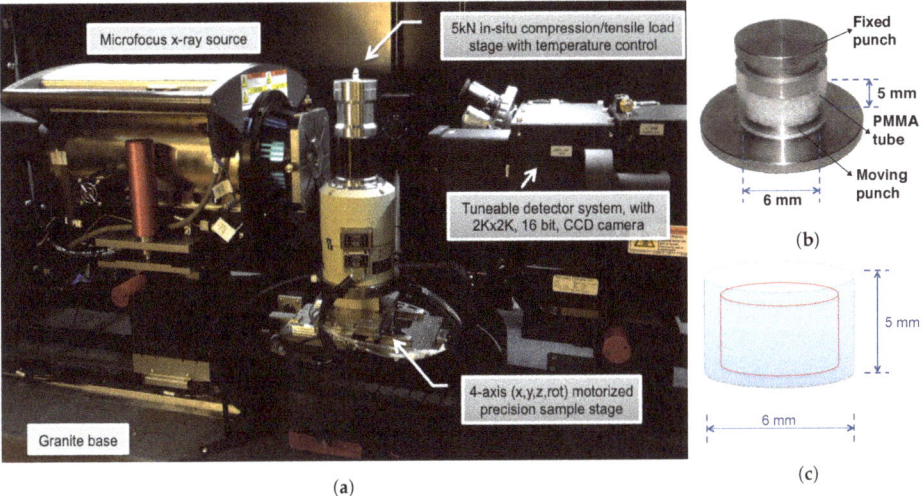

Figure 1. Experimental arrangement of the micro-CT system: (**a**) the micro-CT consists of a sealed microfocus X-ray tube, 4-axis sample stage, a photo detector and a temperature controlled in-situ load stage; (**b**) the sample holder was 6 mm in diameter and 5 mm in height, note that the material visible in the sample holder is sugar; (**c**) selection of volume of interest (VOI) in the snow volume.

2.2. Snow Sampling

An undisturbed natural dry snow block was collected right after latest precipitation outside of Luleå University of Technology (LTU), Sweden and ambient temperature was about −4 °C at the time of collection. The acquired snow block was immediately placed in a freezer held at −18 °C. Tools such as sample holder, spatula etc., that were required to prepare a final snow sample were also kept in the freezer. After the sample holder and tools were cooled down, a cylindrical snow sample was prepared using thermal insulated gloves, still being inside the freezer. Thereafter, the sample holder with snow sample was transferred quickly from the freezer to the in-situ load stage that was held at −15 °C. Surface of the snow sample was flattened using the moving punch to minimize the skewness of the applied load.

2.3. Data Acquisition

The X-ray source voltage and current were set to 50 kV and 80 µA, respectively. Temperature of the in-situ load stage was maintained at −15 °C for all the measurements. To avoid the microstructural damage close to the edges, an interior VOI was defined for the DVC analysis, which corresponds to 4.4 mm in diameter and 3.9 mm in height, see Figure 1c. A total of 1601 projections of the investigated snow volume were recorded as the sample rotated over 360° in high resolution setting for a period of 6 h.

Prior to the scan, the snow sample was placed in the micro-CT chamber for 30 min to ensure thermal equilibrium for the sample and mount. Initially, a low resolution scan was carried out for 30 min to improve stability of the source and to check if the reconstruction is as expected. Thereafter, higher resolution scans of the snow volume were performed using the 4× objective and 4.4 mm field of view (FOV). In this case, the micro-CT reconstructs the spatial distribution of ice crystals and pore space with a resolution in terms of voxel size 4.3 µm at an exposure time of 12 s per projection.

Figure 2a shows a schematic representation of the loading profile, where micro-mechanical uniaxial compression tests on the snow sample were performed at four load states such as 0 MPa, 0.3 MPa, 0.6 MPa and 0.8 MPa. The load stage was used in a continuous mode where the compression was applied until the user-defined load was reached. The snow volume was scanned first at 0 MPa (unloaded state) and then subsequent scans were conducted at three load states. After each loading state, the snow sample was allowed to rest for 30 min prior to the scan.

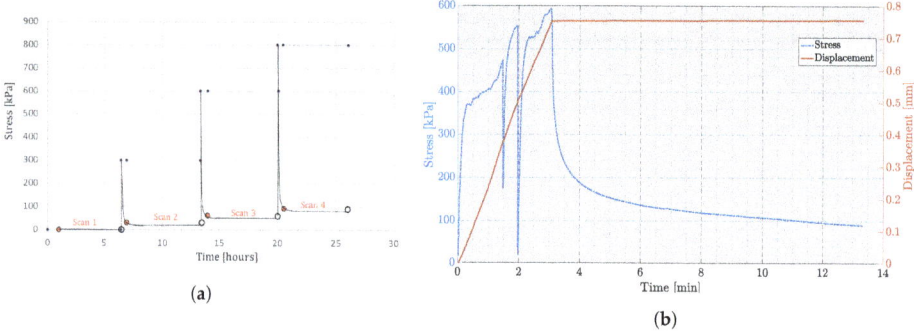

Figure 2. Graphical representation of compression tests: (**a**) schematic representation of the applied loading profile; (**b**) real-time force and displacement data for 0.6 MPa loading profile. In Figure 2a, red circles represent the beginning of scanning, white circles represent the end of scanning and blue circles represent the relaxation time during respective load cycles.

Figure 2b shows the loading and displacement data for the load state of 0.6 MPa. Figure 2b shows rapid and large deformation of snow during the loading period due to grain rearrangements, however the load profile curve was steeper up to 0.35 MPa as the snow sample was already compacted up to 0.3 MPa during the previous load state. After 0.35 MPa, snow tends to resist the force up to 0.45 MPa where the snow structure breaks due to failure of cohesive bonds and recrystallization. The similar behavior can be observed when the force starts to increase up to 0.55 MPa. After reaching the desired load, the snow sample is set to relax for 30 min prior to the scan. During the load relaxation, the stress relaxed to a residual value of about 64 kPa after 30 min. Apparently the described global structural failures (see Figure 2b) during the load profile allow the ice crystals to move into more compact arrangements, which tend to strengthen the bonds between grains [18,19]. Furthermore, there was no scan performed during the relaxation phase, as this would create motion artefacts due to high resolution.

3. Methods

3.1. DVC

The DVC technique is used to quantify the internal displacements and strain field throughout a sample volume. This technique focuses on movement and re-distribution of microstructural features of the sample volume in response to compaction via e.g., uniaxial stress. A complete description of the DVC technique can be found in [20–22].

The DVC technique requires volume images of a sample in reference (unloaded) and deformed (loaded) states. These volume images are then divided into sub-volumes that are independently correlated, and mapped into global deformation and strain fields. The DVC analysis was carried out using the software LaVision Davis (LaVision Inc., Ypsilanti, MI, USA) 8.4, where a multi-grid differential correlation approach was used, with final sub-volume size of 32 × 32 × 32 voxels and 75% overlap.

3.2. Porosity, Density and SSA

The reconstructed 3D images were first filtered by a median filter to remove noise in the images. After filtering, the volume data was segmented using a threshold obtained with the Otsu method [23]. The filtering and thresholding algorithms segment the grayscale images to binary images containing the voxels composed of either ice or air. Furthermore, these binary images were then used to determine density, SSA and porosity distribution of the snow volume. The Otsu segmentation method was previously shown to be subject to systematic bias in the estimation of density and SSA [15].

Porosity can be defined as the ratio between the volume of the pore space (V_e) to the total volume of the sample (V_t). In order to determine the volume of pore space, first volume of ice crystals (V_i) is determined based on the segmented grayscale images [24,25]. Subtraction of V_i dataset from V_t dataset then gives the required V_e dataset and calculation of porosity is in the form

$$\varphi = \frac{V_e}{V_t} * 100 = \frac{V_t - V_i}{V_t} * 100 \text{ [in \%]}. \qquad (1)$$

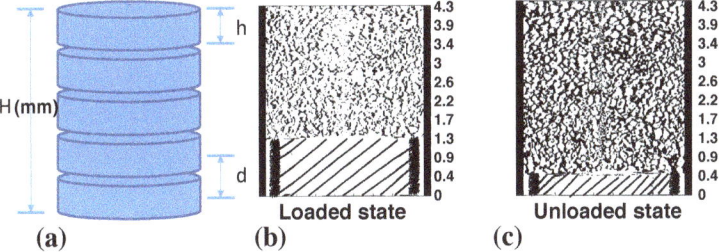

Figure 3. Concepts of porosity calculation: (a) in this study, H (height of bed) is 4.23 mm at an unloaded state while h (section height) and d (section distance) are 0.47 mm; (b) an example of loaded state; (c) an example of unloaded state. Note that area with lines in Figure 3b,c represents the moving punch and the gap between the moving punch and sample holder is 0.05 mm.

Porosity distribution of the investigated volume was calculated for the whole volume and for discretized sections of the volume. In this study, the number of discretized sections varies with the applied load, see Figure 3a. For example, snow volume at unloaded state was divided into 9 sections. In addition, section height and distance were maintained constant during the discretization at all load states. Equation (2) was then used to calculate porosity within each of these sections and for the whole volume.

The density of the snow sample is calculated from the volume of the ice crystals (V_i) and the total volume of the snow sample (V_t) [13,26]. It is given in the form:

$$\rho_{snow} = \frac{V_i}{V_t} * \rho_{ice} \text{ [in kg m}^{-3}\text{]}, \qquad (2)$$

where ρ_{ice} = 917 kg m^{-3}.

Specific surface area (SSA) is calculated from the volume fraction of ice crystals (V_i) and area of ice crystals (A_i). It is given in the form:

$$SSA = \frac{A_i}{V_i} \text{ [in mm}^{-1}\text{]}. \tag{3}$$

The calculated SSA, density and porosity values from the micro-CT data at four loading states are presented in Section 4.3.

4. Results

The experimental observations from the micro-CT analysis are presented in this section. Distribution of snow grains is given in Section 4.1. Section 4.2 presents the changes in snow microstructure with respect to the applied load. Microstructural parameters for the discretized sections of the snow volume are presented in Section 4.3.

4.1. Distribution of Snow Grains

Some examples of the complex 3D morphology of the investigated natural dry snow are presented in Figure 4. The snow sample was composed of ice crystals with significant variations in shape and size. However, moving through the cross sections, the most dominated shapes are needles, capped columns and dendrites, see Figure 4a–c.

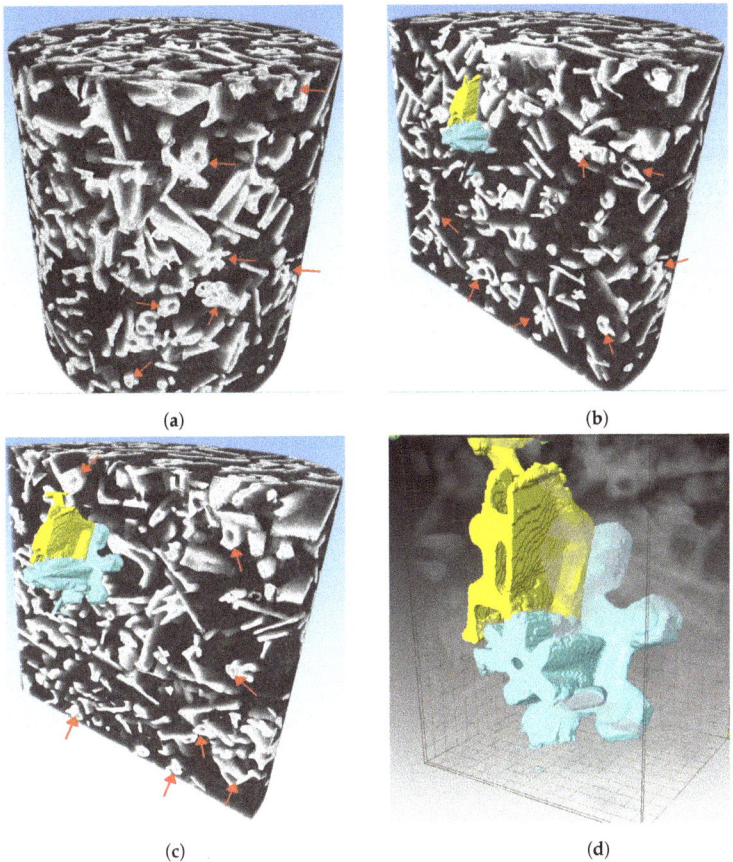

Figure 4. 3D images of snow grains distribution: (**a**) scan of the whole investigated snow volume; (**b**) scan of the snow volume across a cross-section and selection of two individual grains; (**c**) slightly deeper cross-section than the one shown in Figure 4b, to focus on the individual grains; (**d**) structure of two individual snow grains. Red arrows in Figure 4 represent an example of the described hollow core structure.

There was no possibility for snow grains to grow into different shapes during the scanning time as the snow temperature was kept constant at −15 °C. In addition, there was no presence of water in the pore space between ice crystals. These natural ice crystals observed to be completely non-isotropic in shape and size. A number of the ice crystals observed to have a deep hollow core (see-through tunnel) with variable dimensions (Figure 4d). To study the nature of this hollow core further, two individual ice crystals of different shapes are selected. The structure of these two crystals can be seen at two successive cross-sections in Figures 4b,c. The presented crystals (Figure 4) appeared to have hollow core with closed tip at one end (Figure 4d, the tip was cut). The width of the presented ice crystals was approximately 0.85 mm, and the diameter of the tunnel ranges from 0.06 mm to 0.22 mm Moreover, pore volume of blue and yellow colored crystals (from Figure 4d) was approximately 0.092 mm^3 and 0.051 mm^3, respectively. One can observe in Figure 4 that a significant number of snow grains have the described hollow core structure, and some examples are shown by red arrows in Figures 4a–c.

4.2. Displacement of Snow Grains

Figures 5 and 6 show the displacement and strain fields between unloaded state (0 MPa) and the loaded states at 0.3 MPa (Figures 5b and 6b), 0.6 MPa (Figures 5c and 6c) and 0.8 MPa (Figures 5d and 6d). The spatial resolution of the DVC analysis is limited to the size of the correlation window (sub volume), which is 32 × 32 × 32 voxels. Hence, the DVC analysis is not carried out at granular level, for individual grain tracking.

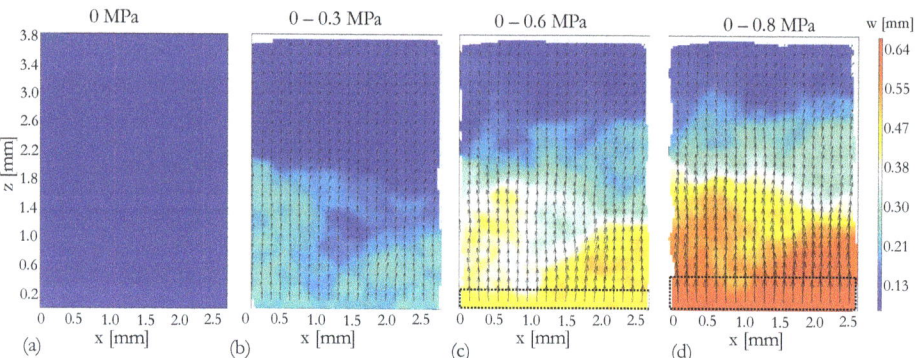

Figure 5. The DVC results for the snow sample at three loaded states, 0.3 MPa, 0.6 MPa and 0.8 MPa. w refers to the displacement in z-direction within the investigated volume. Figure 5a represents the displacement field at unloaded state. Figure 5b,c,d represent the displacement fields between unloaded state and loaded states 0.3 Mpa, 0.6 Mpa and 0.8 Mpa, respectively. Arrows indicate the direction of the displacement field and dashed black lines in Figure 5c,d represent the moving punch.

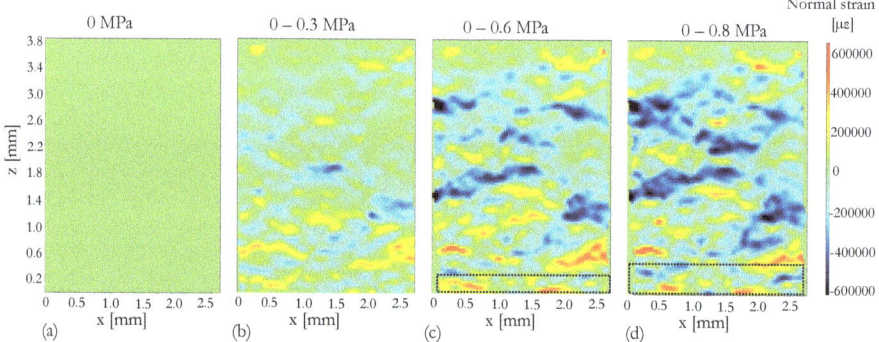

Figure 6. Normal strain fields with respect to the applied stress and dashed black lines in Figure 6c,d represent the moving punch. Figure 6a represents the strain field at unloaded state. Figure 6b,c,d represents the strain fields between unloaded state and loaded states 0.3 Mpa, 0.6 Mpa and 0.8 Mpa, respectively.

One can observe in Figure 5 that the measured displacement field at all the loaded states exhibit similar feature of upward compression (in z-direction). This is expected due to the position of the moving punch. In addition, overall displacement of snow volume appeared to be increasing as the applied stress increases, especially close to the moving punch. However, larger amount of deformation can be observed at the center of the sample (in radial direction) compared to the boundaries. This phenomenon may be due to the smaller punch diameter than the sample holder.

Snow grains close to the fixed punch appeared to be insensitive to the applied stress at least up to 0.8 MPa, as no significant displacement is observed, see Figure 5. This may be due to the boundary condition imposed by the fixed punch. At 0.8 MPa of applied stress, significant deformations may be due to breakage of bonds between ice crystals and re-distribution of ice crystals. However, strain fields in Figure 6 shows that the deformations were fairly distributed through out the whole volume where parts of the snow volume experience positive strain field (tensile) while other parts experience negative stain field (compression). One can observe that there is also a part of the snow volume which experiences almost zero deformation. However, tracking of individual grains due to compression needs to be studied further.

Figure 7 shows the reconstructed grayscale images of the investigated snow volume at four loading states. Four cross-cuts per loading state are presented so that, behavior of ice crystals can be observed close to the fixed punch (top layer, XY-plane, Figure 7a), in the center of the snow volume (XY-plane, Figure 7b), close to the moving punch (bottom layer, XY-plane, Figure 7c), and center cross-section in z-direction (XZ-plane, Figure 7d). The internal features of the snow volume in Figure 7 are very extensive, therefore four crystals from each image are selected to describe general observations. Note that, the selected cross-cuts for a respective loading were correspond to the same cross-cut (slice), to avoid the rotational effects and to ensure that the observed solid movements of grains are solely due to compression tests.

Figure 7e,i,m,q represent a cross-cut close to the fixed punch (top layers, XY-plane) at load states 0 MPa, 0.3 MPa, 0.6 MPa and 0.8 MPa, respectively. The selected crystals in these four images are named as A, B, C and D. Crystals A and B show that not all the crystals in this cross-cut experience displacement. The dendrite crystals in this case were almost unchanged in terms of shape and size. Crystals C and D experience grains breakage. Analysis of displacement field in Figure 5 show that this part of the snow volume experienced very small displacement. However, observation of individual ice crystals shows a small degree of grain displacement close to the fixed punch, which can be also seen from the strain fields in Figure 6.

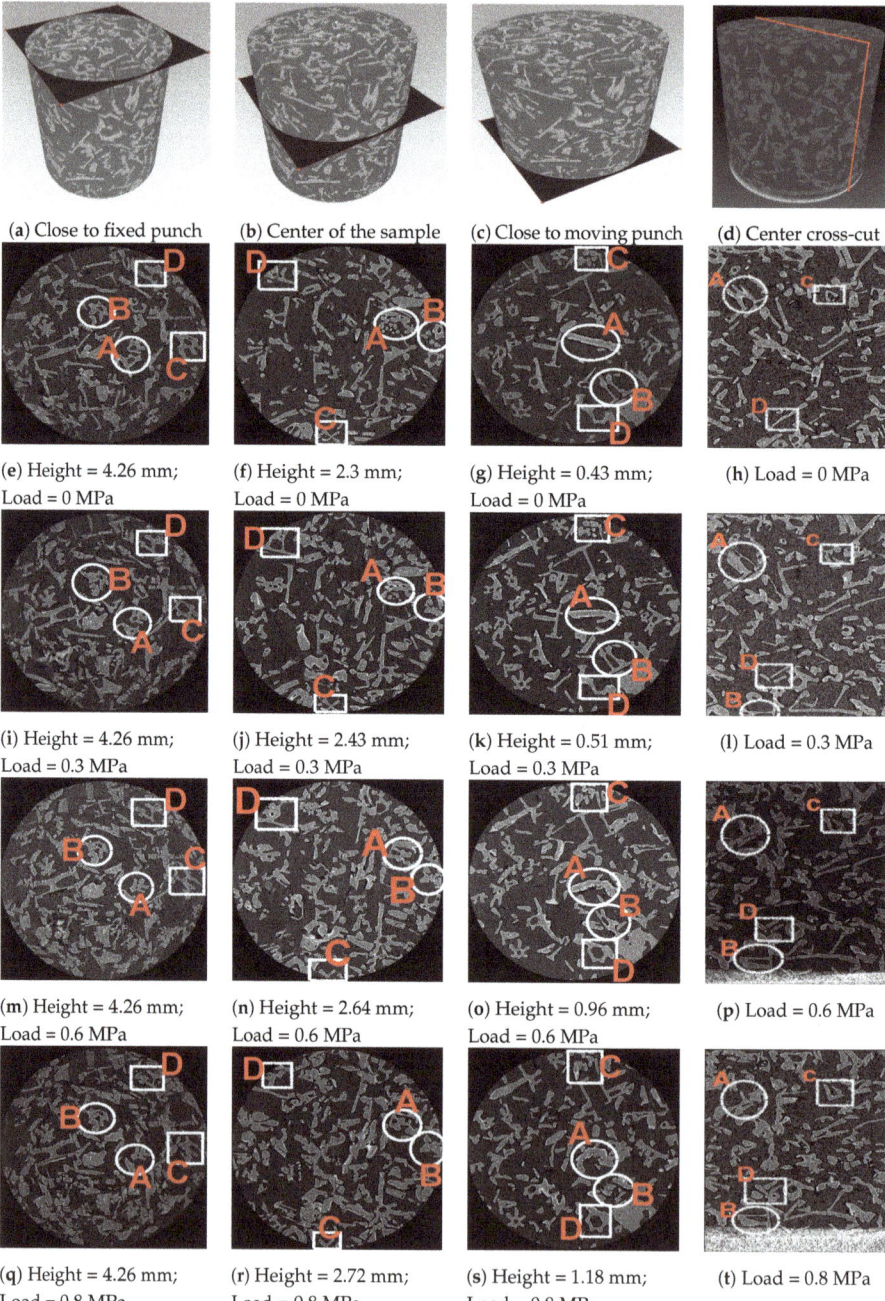

Figure 7. Distribution of ice crystals at four slices of the investigated snow volume at a given load. Note that the compaction is applied from the bottom of the snow volume via moving punch and it is visible in Figure 7p,t.

Figure 7f,j,n,r represent a cross-cut (XY-plane) at the center of the snow volume at all load states. The selected crystals in these four images are named as A, B, C and D. Crystals A and B tend to move closer to each other and form well connected grains. Crystals C and D, experience re-distribution due to the formation of new crystals, and one can observe that crystals C and D moves out of the scan volume as the compaction increases.

Figure 7g,k,o,s represent a cross-cut (bottom layers, XY-plane) close to the moving punch at all load states. The selected crystals in these four images are named as A, B, C and D. Crystals A and B tend to experience breakage of grains bond due to compaction, while crystals C and D tend to form into new crystals. As this part of the snow volume experienced the compaction directly from the punch, crystals observed to be deformed significantly via bond breakage. This was already observed from the displacement field in Figure 5.

Figure 7h,l,p,t represent a center cross-cut (XZ-plane) in z-direction at all load states. The selected crystals in these four images are named as A, B, C and D. Crystals A, C and D tend to experience recrystalization as the applied stress increases. Moreover, one can observe the moving punch in Figure 7p,t. As the punch compact the snow surface, crystals tend to appear within the VOI (for example, crystal B).

4.3. Porosity Measurements

Figure 8 shows the calculated porosity distribution for the discretized sections of the snow volume and the whole volume. The straight lines in Figure 8 represent the average porosity values for the whole volume at a given load state. Note that the moving punch applied compaction from the bottom, as shown in Figure 3b,c. The calculated density and SSA are given in Table 1.

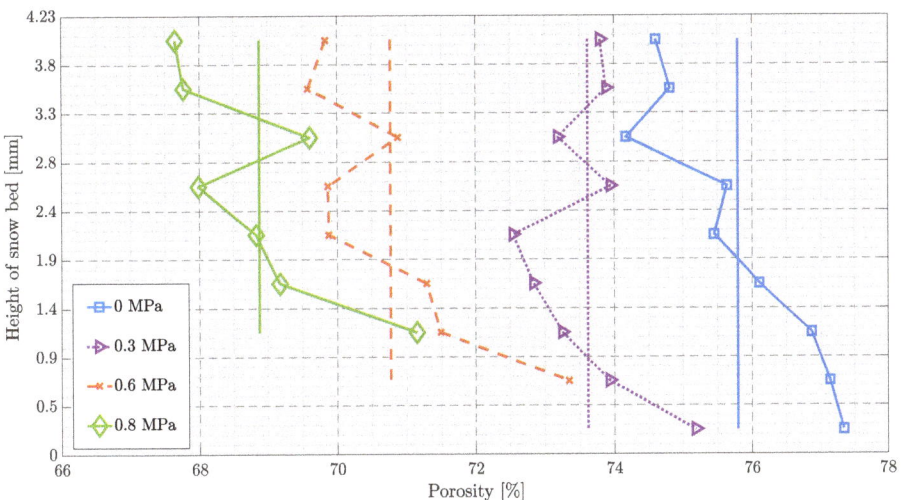

Figure 8. Porosity distributions for the discretized sections of the investigated snow volume. The calculated density values of the snow volume at each load state are given (in units kg m^{-3}).

Table 1. Microstructural parameters under loading conditions.

Stress (MPa)	Density (kg m^{-3})	SSA (mm^{-1})
0	224	66
0.3	243	64
0.6	265	62
0.8	284	60

One can observe that the porosity of the whole snow volume linearly decreases as the applied compression increases. Under compaction, the ice crystals tend to move closer resulting in densification where the ice crystals re-distribute into the pore space. Note that the moving punch is excluded from the data shown in Figure 8, which can be observed from the reduction of data points as the investigated snow volume experienced compaction.

There is a distinct vertical heterogeneity can be observed for the porosity distribution in the axial direction, see Figure 8. At an unloaded state, the initial snow sample exhibits a higher porosity (77.5%) at the bottom compared to the top (74.5%) of the sample. This may be due to the preparation of the snow sample. The sample holder was pushed straight into the dry natural snow and shaved off the excess snow from the top with a sharp spatula, producing a flat and smooth surface. The described sample preparation appeared to force the grains to move closer at the top of the sample holder, which resulted in around 3% of porosity variation between top and bottom layers.

At 0.3 MPa loading stage, this vertical heterogeneity in porosity distribution between the top (74%) and bottom (75%) layers was reduced to around 1%. This may be due to the initial compression, which forces the grains to re-arrange. Furthermore, there was more than 6 hours of settling process during the scan at unloaded state. It may also be possible that the bottom layers experience the weight of top layers due to settlement thus forcing the distribution of ice crystals into the pore space.

At 0.6 MPa and 0.8 MPa load stages, the vertical heterogeneity in porosity distribution between the top (73.5% & 71%, respectively) and bottom (70% & 67.5%, respectively) layers was then re-increased to around 3%. As described in Figure 7, snow grains at higher loads experience bond breakage and grain re-arrangement. Table 1 further shows that the overall density increases with compression while the SSA decreases.

5. Discussions

General observations can be deduced from the results of this initial study. The present study shows that displacement of snow's internal structures during compaction can be investigated from the micro-CT data coupled with the DVC and porosity analysis. Embodied in this study are the vital observations that (1) the majority of snow grains have deep hollow core, (2) displacement of snow grains due to grains breakage and recrystallization, and (3) vertical heterogeneity in porosity distribution of the snow sample in response to compaction.

The growth of ice crystals depends on the temperature and humidity [27]. Shimada and Ohtake [28], Wergin et al. [29] and Riche et al. [30] reported that ice crystals grow as columns and needles around −5 °C and as plates and dendrites around −15 °C. This may explain the complex structure of the investigated dry natural snow as colder temperatures higher up in the atmosphere produces dendritic snow and needles are formed during the precipitation close to ground.

The complex microstructure of the investigated snow volume shows the inter-granular structural changes due to compaction. Compaction of snow by applying micro-mechanical uniaxial load forces the individual ice crystals to move closer via grain sintering and bond breakage. This process is known as pressure sintering [31], which results in rounded and less complex microstructure of snow. Wilkinson [32] detailed the influence of pressure sintering on snow microstructure and its mechanical properties. The sintering process in a snow pack was observed previously [33,34], where snow coarsening was forced due to loading and liquid water content. A common observation of the snow compaction between the tomography data and our previous works, is that the majority of displacement of snow grains occur at the near-surface layers. Gubler [35], Szabo and Schneebeli [36] reported that they observed similar behavior of snow microstructure during compaction with respect to temperature and time. Moreover, Schleef et al. [16] observed the sintering effect during the relaxation phase of a snow sample, which has similar microstructural characteristics as the one presented in this study. Major observations from their work are based on variations in strain rates and microstructural properties during the relaxation stage. They also discussed about a critical point during higher compression rates where a snow structure can be collapsed and rearranges into more compact structure.

The tomography data presented in this initial study was useful in terms of higher resolution measurements, observation of hollow core structure and analysis of porosity distribution. However the study is also limited by the long measurement time, lack of stress relaxation measurements, correlation window size for DVC analysis and the sample size. In principle, this initial study can be further extended in future to analyze different snow types under the influence of loading and isothermal metamorphism conditions. In addition, the selected VOI (refer to Section 2.2) must consider the whole volume to observe the edge effects. Figure 7 shows that some of ice crystals move out of the scan volume during compaction due to the finite size of the VOI, where the relevant information about the re-distribution of these ice crystals is missing.

Furthermore, the sample preparation appears to play a major role especially when it comes to tomography measurements. Ebner et al. [12], Zermatten et al. [37] performed structural analysis of snow samples by exposing to an advective airflow, and further observed porosity distribution, coarsening rate and settling of snow. In addition, Ebner et al. [12] reported that there is no correlation between the porosity distribution and settling of snow. However, a clear correlation is observed in this study where, porosity distribution at 0.3 MPa load stage appeared to be due to the snow settlement, see Figure 8.

The elastic characteristics of the described hollow core feature can be further investigated in future experiments. The observations from this study to understand the microstructural changes in snow grains and properties in response to compression tests, may be helpful to investigate the snow mechanics. Further experiments considering the limitations, can be advantageous in various fields of snow research. In specific, the observations in this study can be useful to understand snow mechanics for winter tire testing, ski track preparation and avalanche prediction. When snow is compressed, friction between the snow and tire decreases, which can be investigated further based on the presented study. For the ski tracks and avalanche, the investigation in this study helps to understand the snow quality and snow pack stability.

6. Conclusions

An X-ray micro-tomographic (micro-CT) measurements coupled with digital volume correlation and microstructural study are performed to investigate the microstructure of dry natural snow. The three-dimensional (3D) reconstruction of snow microstructure is essential to understand its metamorphism, and physical and mechanical properties of a snowpack. This initial study allows for an observation of the changes in internal microstructure during compression tests. Moreover, displacement and strain fields of the snow volume and microstructural properties such as porosity, density and specific surface area (SSA) are determined. The porosity distribution of the whole snow volume decreases with an increase in level of compaction. In addition, the analysis of vertical heterogeneity in porosity distribution shows the characteristics of snow sample preparation and settling of snow. The observations in this study further showed that the majority of ice crystals in the investigated snow volume has a deep hollow core, irrespective of the shape and size of the crystal. The DVC analysis showed the displacement field in response to compaction, while the two-dimensional (2D) cross-cuts of the snow volume showed the grain re-arrangement and bond breakage due to compaction. The presented techniques may be useful to determine stress-strain response of snow for better understanding of the various snow layer transitions.

Author Contributions: Conceptualization, F.F. and J.C.; methodology, F.F.; software, F.F. and H.L.; validation, F.F., H.L. and L.E.; formal analysis, F.F. and L.E.; investigation, F.F. and L.E.; resources, F.F., H.L. and L.E.; data curation, F.F.; writing–original draft preparation, L.E. and F.F.; writing–review and editing, L.E. F.F., J.C., and H.L.; visualization, F.F. and L.E.; supervision, F.F. and J.C.; project administration, J.C.; funding acquisition, J.C.

Funding: This research was founded by European Regional Development Fund, Region Norrbotten and others, grant number 20201424.

Conflicts of Interest: The authors declare no conflict of interest.

References

1. Dozier, J.; Davis, R.E.; Perla, R. On the objective analysis of snow microstructure. *Avalanche Form. Mov. Eff.* **1987**, *162*, 49–59.
2. du Roscoat, S.R.; King, A.; Philip, A.; Reischig, P.; Ludwig, W.; Flin, F.; Meyssonnier, J. Analysis of Snow Microstructure by Means of X-ray Diffraction Contrast Tomography. *Adv. Eng. Mater.* **2011**, *13*, 128–135. [CrossRef]
3. Perla, R.; Dozier, J.; Davis, R. Preparation of serial sections in dry snow specimens. *J. Microsc.* **1986**, *142*, 111–114. [CrossRef]
4. Good, W. Thin sections, serial cuts and 3-D analysis of snow. *IAHS Publ.* **1987**, *162*, 35–48.
5. Flin, F.; Brzoska, J.B.; Lesaffre, B.; Coléou, C.; Pieritz, R.A. Full three-dimensional modelling of curvature-dependent snow metamorphism: First results and comparison with experimental tomographic data. *J. Phys. D Appl. Phys.* **2003**, *36*, A49. [CrossRef]
6. Schneebeli, M.; Sokratov, S.A. Tomography of temperature gradient metamorphism of snow and associated changes in heat conductivity. *Hydrol. Process.* **2004**, *18*, 3655–3665. [CrossRef]
7. Maire, E.; Withers, P.J. Quantitative X-ray tomography. *Int. Mater. Rev.* **2014**, *59*, 1–43. [CrossRef]
8. Wiese, M.; Schneebeli, M. Snowbreeder 5: A Micro-CT device for measuring the snow-microstructure evolution under the simultaneous influence of a temperature gradient and compaction. *J. Glaciol.* **2017**, *63*, 355–360. [CrossRef]
9. Schleef, S.; Löwe, H. X-ray microtomography analysis of isothermal densification of new snow under external mechanical stress. *J. Glaciol.* **2013**, *59*, 233–243. [CrossRef]
10. Kaempfer, T.U.; Schneebeli, M. Observation of isothermal metamorphism of new snow and interpretation as a sintering process. *J. Geophys. Res. Atmos.* **2007**, *112*, 2156–2202. [CrossRef]
11. Pinzer, B.; Schneebeli, M.; Kaempfer, T. Vapor flux and recrystallization during dry snow metamorphism under a steady temperature gradient as observed by time-lapse micro-tomography. *Cryosphere* **2012**, *6*, 1141–1155. [CrossRef]
12. Ebner, P.P.; Schneebeli, M.; Steinfeld, A. Tomography-based monitoring of isothermal snow metamorphism under advective conditions. *Cryosphere* **2015**, *9*, 1363–1371. [CrossRef]
13. Wang, X.; Baker, I. Observation of the microstructural evolution of snow under uniaxial compression using X-ray computed microtomography. *J. Geophys. Res. Atmos.* **2013**, *118*, 12371–12382. [CrossRef]
14. Kerbrat, M.; Pinzer, B.; Huthwelker, T.; Gäggeler, H.; Ammann, M.; Schneebeli, M. Measuring the specific surface area of snow with X-ray tomography and gas adsorption: Comparison and implications for surface smoothness. *Atmos. Chem. Phys.* **2008**, *8*, 1261–1275. [CrossRef]
15. Hagenmuller, P.; Matzl, M.; Chambon, G.; Schneebeli, M. Sensitivity of snow density and specific surface area measured by microtomography to different image processing algorithms. *Cryosphere* **2016**, *10*, 1039–1054. [CrossRef]
16. Schleef, S.; Löwe, H.; Schneebeli, M. Hot-pressure sintering of low-density snow analyzed by X-ray microtomography and in situ microcompression. *Acta Mater.* **2014**, *71*, 185–194. [CrossRef]
17. Bay, B.K.; Smith, T.S.; Fyhrie, D.P.; Saad, M. Digital volume correlation: Three-dimensional strain mapping using X-ray tomography. *Exp. Mech.* **1999**, *39*, 217–226. [CrossRef]
18. St Lawrence, W.; Bradley, C.C. The deformation of snow in terms of structural mechanism. In *Snow Mechanics Symposium*; International Association of Hydrological Sciences: Washington, DC, USA, 1975; pp. 155–170.
19. Theile, T.; Szabo, D.; Luthi, A.; Rhyner, H.; Schneebeli, M. Mechanics of the ski–snow contact. *Tribol. Lett.* **2009**, *36*, 223–231. [CrossRef]
20. Forsberg, F.; Sjödahl, M.; Mooser, R.; Hack, E.; Wyss, P. Full Three-dimensional strain measurements on wood exposed to three-point bending: Analysis by use of digital volume correlation applied to synchrotron radiation micro-computed tomography image data. *Strain* **2010**, *46*, 47–60. [CrossRef]
21. Buljac, A.; Jailin, C.; Mendoza, A.; Neggers, J.; Taillandier-Thomas, T.; Bouterf, A.; Smaniotto, B.; Hild, F.; Roux, S. Digital Volume Correlation: Review of Progress and Challenges. *Exp. Mech.* **2018**, *58*, 661–708. [CrossRef]
22. Bay, B. Methods and applications of digital volume correlation. *J. Strain Anal. Eng. Des.* **2008**, *43*, 745–760. [CrossRef]

23. Otsu, N. A threshold selection method from gray-level histograms. *IEEE Trans. Syst. Man Cybern.* **1979**, *9*, 62–66. [CrossRef]
24. Taud, H.; Martinez-Angeles, R.; Parrot, J.; Hernandez-Escobedo, L. Porosity estimation method by X-ray computed tomography. *J. Pet. Sci. Eng.* **2005**, *47*, 209–217. [CrossRef]
25. Nikishkov, Y.; Airoldi, L.; Makeev, A. Measurement of voids in composites by X-ray Computed Tomography. *Compos. Sci. Technol.* **2013**, *89*, 89–97. [CrossRef]
26. Lundy, C.C.; Edens, M.Q.; Brown, R.L. Measurement of snow density and microstructure using computed tomography. *J. Glaciol.* **2002**, *48*, 312–316. [CrossRef]
27. Libbrecht, K.G. The physics of snow crystals. *Rep. Prog. Phys.* **2005**, *68*, 855–895. [CrossRef]
28. Shimada, W.; Ohtake, K. Three-Dimensional Morphology of Natural Snow Crystals. *Cryst. Growth Des.* **2016**, *16*, 5603–5605. [CrossRef]
29. Wergin, W.P.; Rango, A.; Erbe, E.F. Observations of snow crystals using low-temperature scanning electron microscopy. *Scanning* **1995**, *17*, 41–50. [CrossRef]
30. Riche, F.; Montagnat, M.; Schneebeli, M. Evolution of crystal orientation in snow during temperature gradient metamorphism. *J. Glaciol.* **2013**, *59*, 47–55. [CrossRef]
31. Lehning, M.; Bartelt, P.; Brown, B.; Fierz, C.; Satyawali, P. A physical SNOWPACK model for the Swiss avalanche warning: Part II. Snow microstructure. *Cold Reg. Sci. Technol.* **2002**, *35*, 147–167. [CrossRef]
32. Wilkinson, D. A pressure-sintering model for the densification of polar firn and glacier ice. *J. Glaciol.* **1988**, *34*, 40–45. [CrossRef]
33. Eppanapelli, L.K.; Casselgren, J.; Wåhlin, J.; Sjödahl, M. Investigation of snow single scattering properties based on first order Legendre phase function. *Opt. Lasers Eng.* **2017**, *91*, 151–159, doi:10.1016/j.optlaseng.2016.11.013. [CrossRef]
34. Eppanapelli, L.K.; Lintzén, N.; Casselgren, J.; Wåhlin, J. Estimation of Liquid Water Content of Snow Surface by Spectral Reflectance. *J. Cold Reg. Eng.* **2018**, *32*, 05018001. [CrossRef]
35. Gubler, H. Strength of bonds between ice grains after short contact times. *J. Glaciol.* **1982**, *28*, 457–473. [CrossRef]
36. Szabo, D.; Schneebeli, M. Subsecond sintering of ice. *Appl. Phys. Lett.* **2007**, *90*, 151916. [CrossRef]
37. Zermatten, E.; Schneebeli, M.; Arakawa, H.; Steinfeld, A. Tomography-based determination of porosity, specific area and permeability of snow and comparison with measurements. *Cold Reg. Sci. Technol.* **2014**, *97*, 33–40. [CrossRef]

© 2019 by the authors. Licensee MDPI, Basel, Switzerland. This article is an open access article distributed under the terms and conditions of the Creative Commons Attribution (CC BY) license (http://creativecommons.org/licenses/by/4.0/).

Article

Time-Resolved Tomographic Quantification of the Microstructural Evolution of Ice Cream

Jingyi Mo [1,2,*], Enyu Guo [3], D. Graham McCartney [1,2], David S. Eastwood [2,4], Julian Bent [5], Gerard Van Dalen [5], Peter Schuetz [5], Peter Rockett [1] and Peter D. Lee [1,2,*]

1. Department of Mechanical Engineering, University College London, London WC1E 7JE, UK; graham.mccartney@nottingham.ac.uk (D.G.M.); peterrockett33@gmail.com (P.R.)
2. Research Complex at Harwell, RAL, Didcot OX11 0FA, UK; david.eastwood@manchester.ac.uk
3. School of Materials Science and Engineering, Dalian University of Technology, Dalian 116024, China; eyguo@dlut.edu.cn
4. The Manchester X-ray Imaging Facility, School of Materials, The University of Manchester, Manchester M13 9PL, UK
5. Unilever R&D, Colworth MK44 1LQ, UK; Julian.Bent@unilever.com (J.B.); Gerard-van.Dalen@unilever.com (G.V.D.); Peter.Schuetz@unilever.com (P.S.)
* Correspondence: j.mo@ucl.ac.uk (J.M.); peter.lee@ucl.ac.uk (P.D.L.)

Received: 10 September 2018; Accepted: 15 October 2018; Published: 19 October 2018

Abstract: Ice cream is a complex multi-phase colloidal soft-solid and its three-dimensional microstructure plays a critical role in determining the oral sensory experience or mouthfeel. Using in-line phase contrast synchrotron X-ray tomography, we capture the rapid evolution of the ice cream microstructure during heat shock conditions in situ and *operando*, on a time scale of minutes. The further evolution of the ice cream microstructure during storage and abuse was captured using ex situ tomography on a time scale of days. The morphology of the ice crystals and unfrozen matrix during these thermal cycles was quantified as an indicator for the texture and oral sensory perception. Our results reveal that the coarsening is due to both Ostwald ripening and physical agglomeration, enhancing our understanding of the microstructural evolution of ice cream during both manufacturing and storage. The microstructural evolution of this complex material was quantified, providing new insights into the behavior of soft-solids and semi-solids, including many foodstuffs, and invaluable data to both inform and validate models of their processing.

Keywords: ice cream; microstructure; tomography; ice crystals; coarsening; soft solids

1. Introduction

Ice cream is a widely consumed dairy product whose complex microstructure determines its texture and oral sensory perception. The main constituents of ice cream are air cells, ice crystals, finely dispersed (emulsified) fat globules, and a continuous unfrozen aqueous solution phase (containing sugars and other additives) and, as shown in Figure 1a, the ice crystal fraction varies markedly with temperature. Ice cream is known to have a complex microstructure which is a function of its formulation but also its thermal history during manufacture, storage, and handling. An important aim of any manufacturer is to ensure their process produces ice cream with a smooth texture which does not degrade to a grainy and coarse texture before the product is consumed and enjoyed. The size of the ice crystals is widely recognized to be a key factor in the perception of texture by consumers; larger crystals being, in general, undesirable [1].

It is thus crucial to be able to observe how the complex three-dimensional structure of ice cream evolves over time with changes in temperature. The traditional method of examination using cryo-SEM of extracted samples (Figure 1b) is capable of providing excellent spatial resolution. However, the contrast between the different phases can be limited and only a two-dimension image is captured of the complex three dimensional (3D) interconnected network of air cells, ice crystals, and unfrozen matrix. During the past decade, there has been a significant increase in the use of synchrotron computed tomography (sCT) to examine the evolution of structures over time in three dimensions as the temperature is changed (termed 4D, 3D plus time, tomography) and so sCT is, in principle, well suited to the study of ice cream microstructures. Indeed, very recent studies have demonstrated the power of 3D sCT for studying the complex microstructure of ice cream; Figure 1c,d. sCT permits quantitative measurements of different phases and allows the visualization of 3D connectivity.

In the present paper, we will describe the application of 4D sCT to reveal the microstructural evolution in ice cream over a range of different timescales, namely that which takes place during the manufacturing process (short timescales, minutes) and storage over long timescales (days/weeks). We apply in situ, time-resolved sCT to reveal the microstructure evolution mechanisms during manufacturing; whereas we use an ex situ sCT methodology to provide time-resolved data during long-term ice cream storage.

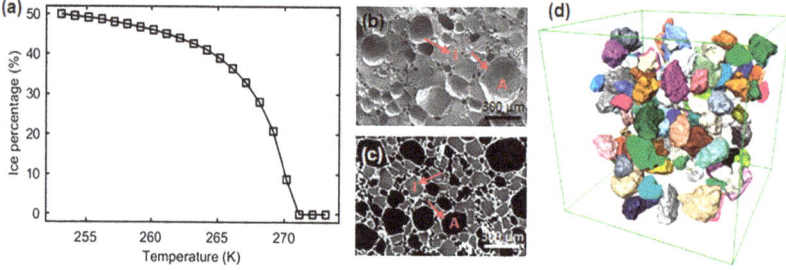

Figure 1. The characteristics of ice crystals in ice cream samples: (**a**) ice crystal volume fraction as a function of temperature; (**b**) a typical cryo-SEM image, showing air cells, A, and ice crystals, I; (**c**) 2D X-ray tomographic image of an ice cream sample. Note, the contrast was enhanced via edge-constrained diffusion filtering (300 iterations). Dark features are air cells, grey features are ice crystals and white regions are the unfrozen matrix; (**d**) 3D renderings of the ice crystals from X-ray tomographic images, providing a 3D view of the features. Here, the ice crystals are individually color rendered (after Reference [2]).

The manufacturing process for ice cream introduces around 50 vol% of air cells so the typical make up of frozen ice cream, by volume, is 50% air cells, 30% ice crystals, 10–15% sugar solution matrix and the balance dispersed fat droplets and non-fat solids [1,3]. Ice cream is commonly manufactured in a scraped-surface freezer (SSF) in either a batch or continuous mode as described, for example, in Cook and Hartel and others [4–8]. The initial processes occurring in the SSF are likely to be the most important in determining ice crystal size and size distribution. In a typical SSF, the mix is rapidly cooled at the wall of a highly chilled cylindrical barrel and ice crystals nucleate and grow at, or close to, the wall and potentially into an undercooled liquid [9]. Rotating scraper blades break up the initially formed crystals and sweep them into the bulk of the mix (where the mean temperature is closer to 269 K to 267 K) causing the fragmentation and partial melting of the initial ice crystals. This process has clear similarities to the metallurgical process known as rheocasting or thixocasting [10]. At the exit of the SSF, about 50% of the water is frozen and the mix is cooled to about 267 K. The semi-solid product is typically pumped into a container for supply to the consumer and hardened in an air blast freezer. It is cooled to around 255 K over a period of around 120 min (0.1 K/min) [9] and then stored at or below this temperature. Ideally, the ice crystals will be kept as small as possible at the exit from the SSF but as the temperature drops in the freezer the original ice crystals grow and their volume fraction increases (Figure 1a).

The texture and sensorial quality of ice cream are related to the size and connectivity of the air cells and ice crystals [1,11,12]. It is reported that the microstructural features vary in size considerably with typical values being air cells 20–150 µm, ice crystals 10–75 µm, fat globules ~0.5 µm with the unfrozen matrix phase being a continuous matrix [9]. Once ice cream is distributed to consumers it can undergo significant temperature variations typically in the range 258 K to 268 K and these will affect the ice crystal size and hence perceived texture [7]. Empirically, it is found that long-term storage above 258 K and/or thermal cycling degrades the ice cream's quality due to coarsening of ice crystals and air cells. Below about 243 K, the matrix phase is transformed to a glass which stabilizes the microstructure but above these temperature, ice crystals are thermodynamically unstable due to the curvature effects and the kinetics of ice crystal coarsening can be quite rapid. The well-known Ostwald ripening phenomenon is a major factor for ice crystal coarsening although other processes such as physical agglomeration also affect the size distribution. Measurements of crystal size distribution have shown that crystal coarsening is very significant in the temperature range of 258 K to 268 K and also that temperature oscillations of only around ±1 K can significantly enhance the coarsening rate [4].

Whilst significant progress has been made in elucidating the process-structure-quality relationships of ice cream, a major limitation is the complex procedures needed to perform microstructural observations by cryo-SEM or transmission electron microscopy (TEM). These are post-mortem techniques [13], and provide only two-dimensional (2D) information on the surface that sliced through the sample [14–17]. Consequently, interpretation of microstructure evolution is subject to assumptions and opens to ambiguity. Recently, X-ray microtomography techniques have been employed to study 3D structures in food products and frozen food, including ice cream, but with somewhat limited success [7,18,19]. Pinzer et al. used a laboratory X-ray microCT in a cold room to investigate the long-term microstructural evolution of ice cream and quantified changes in air cell and ice crystal size during thermal cycles between −5 to −16 °C over a period of 24 h [18]. However, the resolution of the structure was a limitation with this instrument.

Ice cream falls into the category of soft solids (of which there are a number found in nature, including food products) and has similar behavior and properties to many man-made and natural semi-solids. The microstructure of many of these materials evolves with time, changing material properties ranging from rheology to yield strength to plasticity, and, in the case of foodstuffs, mouthfeel. One way of quantifying this microstructural evolution is time-resolved radiography [20,21] or tomography [22,23]. Soft solids found in nature include soils [24], shales [25], and magma [26], and food products such as chocolate [27]. In recent years, there has been a growing interest in the use of 4D tomography, specifically time-resolved synchrotron computed tomography (sCT), to perform non-invasive 3D structure studies of soft solids with high resolution. For example, Kareh et al. used in situ experiments to follow the evolution of semi-solid aluminum alloys under heating and loading conditions [28] whilst Karagadde et al., used in situ indentation to uncover a new fracture mechanism, the transgranular liquation cracking of semi-solid Al-Cu alloys [29].

The use of high brilliance synchrotron X-ray computed tomography (sCT), coupled with a precise cold stage, to study ice cream microstructures has recently been reported by the current authors. In this prior study, we obtained the three-dimensional quantification of microstructural changes arising from the thermal cycling at low heating and cooling rates over periods of weeks. This was undertaken to simulate storage and transport effects on the ice cream product [2]. However, the details of ice crystal development during the manufacturing stages such as those involving processing in a scraped-surface freezer and subsequent slower cooling as a block in an air blast freezer are of great interest but have yet to be studied in detail.

In the present study, we used a bespoke cold stage (capable of controlling the sample temperature precisely between 233 K and 293 K with 0.1 K accuracy) and employed high resolution (in terms of both spatial and contrast) sCT to conduct 4D (3D plus time) studies on the microstructural evolution of ice cream samples undergoing different types of thermal histories. In the first set of experiments, the ice cream samples underwent fast heating and cooling cycles to simulate aspects of manufacture and were continuously scanned, in situ and *operando*, during the process. The second set of experiments involved monitoring the microstructural evolution over long timescales and so ex situ experiments could be used to capture the time-dependent evolution. Novel image-based quantification techniques were developed to precisely and robustly evaluate the structural characteristics in the ice cream samples. Variations in the size distribution of ice crystals and unfrozen matrix were followed during the thermal-induced microstructural evolution as these are recognized as indicators for changes in texture and oral sensory perception.

2. Materials and Methods

2.1. Preparation of Ice Cream Samples

500 mL blocks of fresh ice cream containing 5% fat and 40% ice were prepared (Unilever R&D, Colworth, UK). Prior to the in situ tomographic experiments, a block of fresh ice cream was left at room temperature to melt and partially de-aerate. Kapton tubes with 3 mm inner diameters and 67 μm wall thickness (American Durafilm Co. Inc, Holliston, MA, USA) were filled with this liquid dairy mixture using a 5 mL syringe, followed by mounting them onto a specially designed cold stage which will be described in the next section.

For the ex situ samples, Kapton tubes were inserted into the central region of individual ice cream blocks and the dairy mixture was first subjected to blast freezing at 238 K and then stored at 248 K. Immediately before observation by synchrotron imaging, the Kapton tubes filled with ice cream were cut from the bulk samples on a dry ice bed (maintained at around 193 K) and then inserted into the specially designed cold stage.

2.2. Cold Stage Experimental Setup and Thermal Cycling for Ex Situ and In Situ sCT

The cold stage assembly is shown schematically in Figure 2 and described in detail in our previous paper [2]. Both the ex situ and in situ sCT experiments were conducted on beamline I13-2 of the Diamond Light Source (DLS, Harwell, UK) using a pink beam. Details of the beamline set-up are described in the next section.

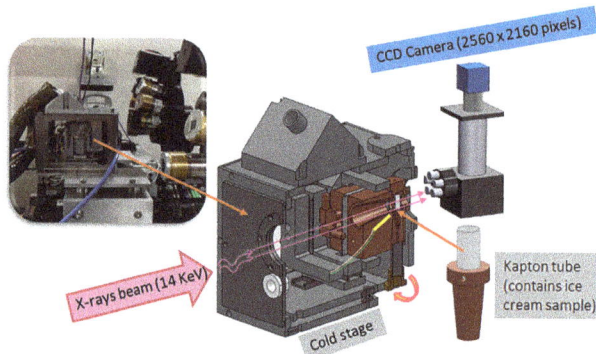

Figure 2. The experimental set-up in beamline I13, Diamond light source. The schematic shows the sectional view of the cold stage, the enlarged inset is the assembly of the ice cream sample, which consists of a 3 mm Kapton tube and the bottom copper mounts.

In order to investigate processes relevant to the manufacture of ice cream, the in situ thermal cycling experiment was performed following the thermal cycles as described. A Kapton tube containing melted ice cream mixture was firstly mounted into the cold stage at the temperature of 270 K i.e., just below the melting temperature of the ice cream formulation. It was held at 270 K for 15 min and then its temperature was rapidly reduced to 250 K with a fast rate of cooling of 5 K/min (referred to as FC) and held at this temperature for 10 min. After that, the sample was heated to 267 K, at a heating rate of 5 K/min and held there for 10 min. Subsequently, it was cooled down from this temperature to 250 K at a slower rate of 0.05 K/min (referred to as SC). The overall thermal cycle for in-situ experiments is illustrated in the temperature versus time plot of Figure 3a. The fast cooling (FC) from 270 K allows one to study the initial ice crystal growth similar to that which might occur near the wall of an SSF. Fast reheating to 267 K approximates the behavior of ice crystals when swept into the bulk of the SSF. Finally, the SC regime from 267 to 250 K is regarded as representative of cooling of a block of ice cream mix in a blast freezer.

Blocks of ex situ ice cream samples were cycled between 258 K and 268 K multiple times, as shown in Figure 3b. In each daily routine (as shown in Figure 3c), the samples were held at 258 K for 9.5 h and then the temperature went up to 268 K in 2.5 h (with a rate of 0.067 K/min). After being held at 268 K for another 9.5 h, the sample temperature cooled back to 258 K with the same thermal rate (0.067 K/min) as before. This 24-h routine repeated for 7 times over one week (C7) and 14 times over two weeks (C14) before they were scanned using synchrotron X-ray tomography. The sample without any follow-on thermal treatment was referred to as C0.

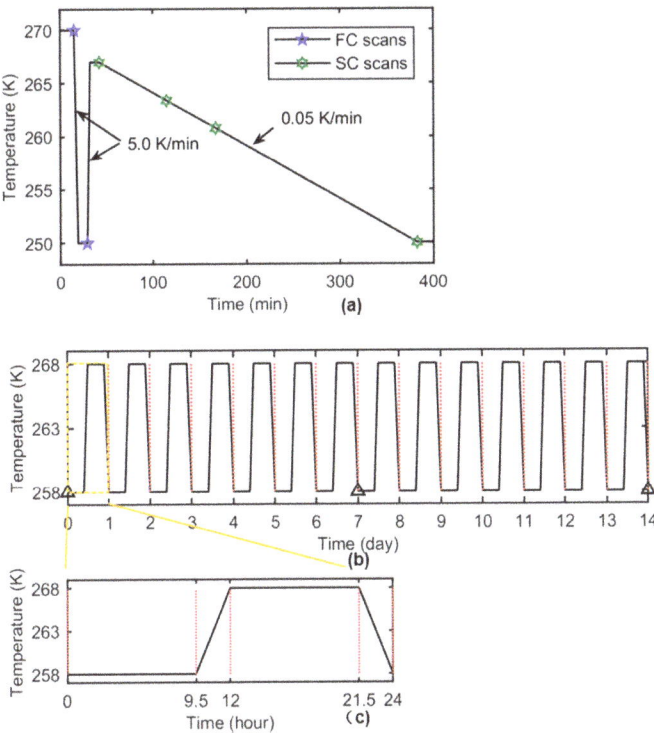

Figure 3. The thermal history of the samples during the in situ and ex situ experiments. (**a**) The overall thermal history of the in situ ice cream samples; the temperature points where the tomographic scans were reconstructed are indicated as blue pentagon and green hexagon markers. For FC, the images were reconstructed at 270 K and 250 K (blue pentagons). For SC, the reconstructed images were obtained at 266.8 K (SC0), 263.4 K (SC1), 260.8 K (SC2), and 250 K (SC3) (green hexagons). The cooling rates for FC and SC are 5.0 K/min and 0.05 K/min, respectively. (**b**) The overall thermal history of the ex situ ice cream samples, with triangular markers indicating the positions where the ice cream samples were extracted and scanned. (**c**) The zoom-in thermal profile showing a single ex situ thermal cycle.

2.3. Microstructural Characterization Using Synchrotron X-ray Computed Tomography (sCT)

Concurrently, with the thermal cycling of the sample, a series of X-ray tomographic images were acquired using sCT on I13-2 beamline at Diamond Light Source (Harwell, UK). This has a high flux undulator, producing a pink beam, with peak modes of narrow bandwidth (ca. 300 eV), with high and low bandwidth filters removing modes outside the energy ranging from 15 to 30 keV. This combined with the ca. 250 m beamline length provides excellent in-line phase contrast. A 2560 × 2160-pixel PCO Edge 5.5 CMOS camera optically coupled to a single crystal CdWO4 scintillator was used to record the projection images. The distance between the sample and the scintillator was optimized to be ~35 mm to achieve an optimum imaging quality. The final pixel size obtained was 0.81 µm for the scans.

During the in situ experiments, each tomographic run includes collecting 720 projections evenly spaced over a 180° rotation with the exposure time of 0.1 s. For FC, the scans were not recorded continuously due to the limitation of sampling rate relative to the fast cooling rate (5 K/min) and two images were acquired at the start and end points of the process, i.e., 270 K and 250 K, respectively, indicated as pentagons in Figure 3. For the same sampling limitation, there were no scans recorded during the fast heating (5 K/min). For the SC, the scans were recorded and pre-processed one by one and the 3D images were reconstructed at 266.8 K, 263.4 K, 260.8 K, and 250 K for analysis, as indicated as hexagons in Figure 3. These are referred to as SC0, SC1, SC2, and SC3 samples, respectively. For the ex situ experiments, each tomographic run includes 3601 projections over a 180° rotation with the same exposure time of 0.1 s for each projection.

2.4. Volume Data Reconstruction and Pre-Processing

The collected 2D radiographs/projections were stacked and converted into sinograms and in which any apparent continuous lines were removed by interpolation to reduce ring artifacts due to imperfections from the detector/camera. The sinograms were then used to reconstruct the volume slices using a filtered back projection (FBP) based algorithm. Because the ice cream samples were relatively low attenuating to the incident X-ray beam, the reconstructed volumes exhibited a relatively high level of noise. Therefore, the 3D volumes need to be filtered before any feature segmentation and quantification can be applied. Due to the microstructural differences in ex situ and in situ experiments, the data processing methods are slightly different.

For the ex situ experiments, a novel image processing strategy and reconstruction algorithm were used to achieve better quality images from the collected data; for details, see References [2,7]. A morphologically based method was used to quickly and robustly reduce the noise in the reconstructed volume of in situ data. First, 3D median filtering (3 × 3 × 3) was performed on all the reconstructed volumes. Then the volumes were cleared using a series of morphological operations. Both ex situ and in situ data were then binarized and labeled using global thresholding. All the volumes were subsequently checked visually, and any obvious segmentation imperfections were corrected manually using Avizo 9.4 (FEI Visualization Sciences Group, Mérignac, France).

2.5. 3D Based Quantification of Ice Crystal Dimensions

In the in situ study, an image analysis based quantification methodology of the ice crystal size in the ice cream samples was developed which is similar to the techniques for porous structure characterization for biomedical and geological samples [30]. Depending on the morphology of a porous structure, the pore size distribution can be obtained either by summarizing information of individually labeled components or estimated by measuring the variation of the interpenetrating volumes as a function of effective pore size. Here, as ice crystals appear as interconnected clusters, segmenting them into individual components is not appropriate. Therefore, we developed a method that provides an analysis similar to that of mercury intrusion porosimetry (MIP) to quantify the ice phase size distribution in the samples (see References [31,32]). In the MIP analysis, the pore size distribution within a porous sample can be determined from the cumulative volume of mercury liquid that has been forced into the pore space by externally applied high pressure. Here, we used a series of sampling spheres of varying diameter, and the size distribution in the ice crystal phase can be obtained by measuring the cumulative volume of ice crystal that can be reached by different sampling spheres.

The variation in accessible volume was used to estimate the ice crystal (or any segmented phase) size distribution, as follows:

1. A 3D distance map was first created by a Euclidean distance transform from the binarized image.
2. For the current samples, 13 sampling spheres with a diameter equally spaced between 5 µm and 40 µm were chosen to balance the quantification accuracy and computational expenses.
3. The radius of each sampling sphere (in voxels) was compared to the voxel intensities in the distance map, determining the centers of the spheres of a radius that can be completely placed within the ice phase.
4. A dilation algorithm was then applied to the voxels correlated to the sphere centers, using a spherical kernel of the same radius. The volume after dilation is the volume correlated to the sampling sphere.
5. Step 3–4 were repeated for all the sampling sphere dimensions chosen in step 2, providing a range of reachable volumes corresponding to increasing sphere diameters.
6. The volume fraction was calculated by dividing each volume by the total volume of the ice phase. Then an ice crystal size distribution can be calculated as the differential of the area under the cumulative volume percentage curve.

For the ex situ study, the reconstructed volumes from the data were cropped into a smaller volume, followed by the microstructural quantification in 3D. The three-dimensional rendering of the features, as well as the quantification of the ice crystals, was performed by manually using Avizo 9.4 (FEI Visualization Sciences Group, Mérignac, France). For more details, see References [2,7].

3. Results and Discussion

3.1. In Situ: Microstructural Evolution—2D Images

First, a tomographic scan was collected at 270 K prior to the onset of fast cooling. Except for air cells, it was not possible to resolve structural features in the material at this temperature. This indicates that it was above the freezing point of the ice cream composition under investigation and additionally significant supercooling is required before ice crystal nucleation occurs.

Figure 4a,b show representative 2D tomographic image slices recorded after the fast cooling from 270 to 250 K and Figure 4c,d show representative image slices following the fast heating at 5 K/min from 250 to 267 K and holding at that temperature for 10 min. The main features evident at low magnification in Figure 4a,c are the light grey contrast circular features which are low X-ray attenuation air cells introduced by sample handling and imaging. These are embedded in a material with a fine scale microstructure showing grey/white contrast. At higher magnification (Figure 4b,d), the details of this matrix material's microstructure are revealed. In Figure 4b, following fast cooling to 250 K, the microstructure comprises a high volume fraction of ice crystals (grey) with a lighter contrast continuous phase surrounding the crystals; the latter is the unfrozen matrix phase comprising a concentrated sucrose solution, fat globules, and solids. The ice crystals have a range of sizes and are apparently lozenge-shaped with dimensions of the order of 5 to 10 µm, with domains of aligned crystals. In Figure 4d, following fast heating from 250 K to 267 K, it is clear that the ice crystals occupy a significantly smaller volume fraction than at 250 K. Furthermore, the dimensions of the ice crystals have increased markedly following the fast heating and short, 10 min, holding time at 267 K. The coarsening could be due to factors such as Ostwald ripening and physical agglomeration during heating and holding. In Ostwald ripening, the driving force is the difference in solubility between the polydisperse particles of different curvatures. This solubility difference establishes a concentration gradient between the smaller and the larger particles, which leads to the growth of the larger particles at the expense of the smaller ones, the solute being transported through the unfrozen phase. It should also be noted that the volume fraction of air cells could potentially increase due to the photochemical cracking by the X-rays (Figure 4a,c) [33].

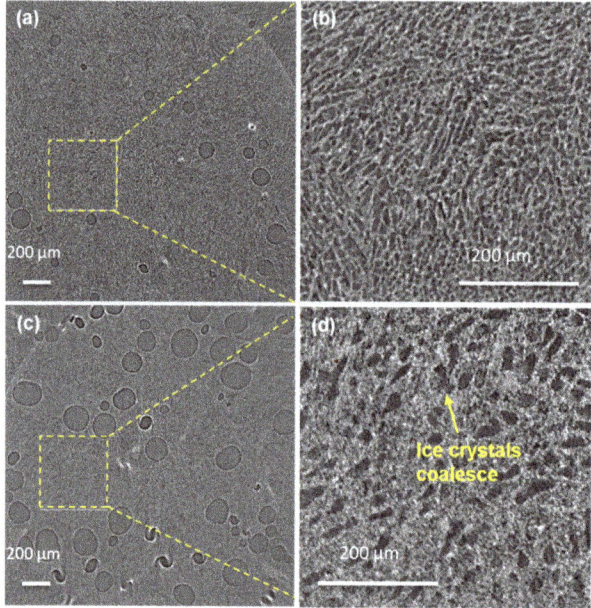

Figure 4. The reconstructed 2D tomographic slices. (**a**,**b**) Following fast cooling (FC) to 250 K. (**c**,**d**) Following heating to 266.8 K and holding for 10 min. In (**a**), the large circular dark areas with thin brighter boundaries are air cells and the matrix comprises of a dark contrast for the ice crystals and a brighter contrast for the unfrozen material. Zoom-ins of the areas highlighted by yellow squares in (**a**,**c**) are shown in (**b**,**d**) respectively.

Figure 5a–d show the 2D tomographic slices recorded at 266.8 K, 263.4 K, 260.3 K, and 250 K respectively during the slow cooling (SC) regime (0.05 K/min) following a 10 min hold at 267 K. During this regime, it is obvious that ice crystals grow in size and the morphologies became more spherical than those at 267 K supporting the proposed mechanisms of Ostwald ripening and physical agglomeration. Moreover, the ice crystal volume fraction increases significantly and the specific interface area (S_s) which is given by the interfacial area between ice (solid) and unfrozen matrix (liquid), denoted as A, divided by the total enclosed volume of ice (solid), denoted as V_s, i.e., $S_s = \frac{A}{V_s}$, decreases during slow cooling and solidification (Table 1). There is also evidence, from the morphological features, that some of the ice crystals appear to have coalesced (see coalesced structure highlighted by an arrow in Figure 5d).

Table 1. The total volume fraction and specific interface area (S_s) of ice crystals quantified by 3D tomography.

Cooling Rate	Temperature (K)	Ice Volume Fraction (%)	specific Interface Area (S_s) (mm^{-1})
FC	250.0	38	599
SC0	266.8	28	301
SC1	263.4	40	247
SC2	260.3	36	233
SC3	250.0	46	193

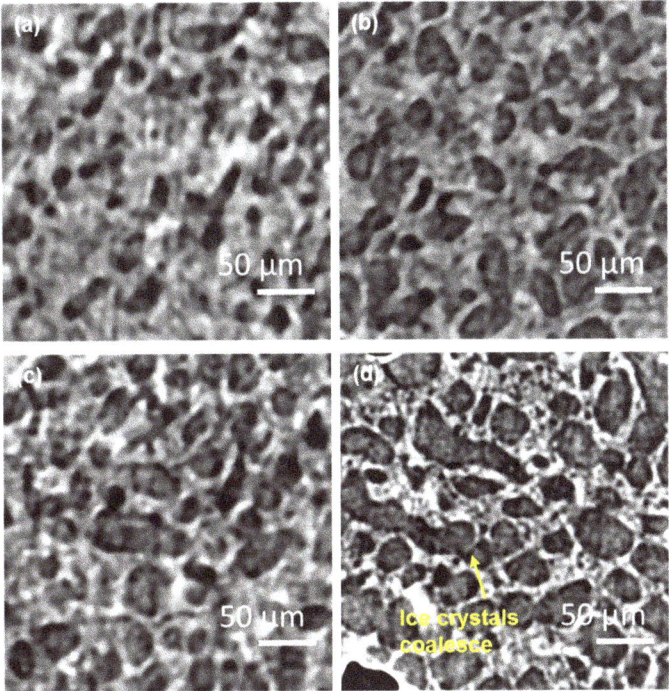

Figure 5. The 2D reconstructed tomographic slice showing the microstructural evolution of ice crystals during the slow cooling regime at the following temperatures: (**a**) 266.8 K, (**b**) 263.4 K, (**c**) 260.8 K and (**d**) 250 K. Note, for images taken at 260 K and above, coarsening occurs during image acquisition, blurring the image.

3.2. In situ: 3D Microstructural Evolution as a Function of Temperature and Time

The 3D rendering of ice crystals and un-frozen matrix from representative regions of the same size are shown in Figure 6. In agreement with the 2D observations, the ice crystals are fine after the FC stage (Figure 6a) and after the fast heating and holding (Figure 6b), the small ice crystals have transformed into much larger ones. The ice crystal size continues to increase during the slow cooling period (Figure 6b–e). The color-coded local thickness of the ice crystals in Figure 6a–e allows us to better visualize the increase of the ice crystal thickness during the experiment (see quantification in the next section). The 3D interconnected volume of the unfrozen matrix is shown in Figure 6f–j. It appears that the wall thickness of the unfrozen matrix tended to increase upon heating (from 250 K to 267 K), whilst the thickness decreases as the temperature falls and the volume fraction of the ice increases in line with Figure 1a.

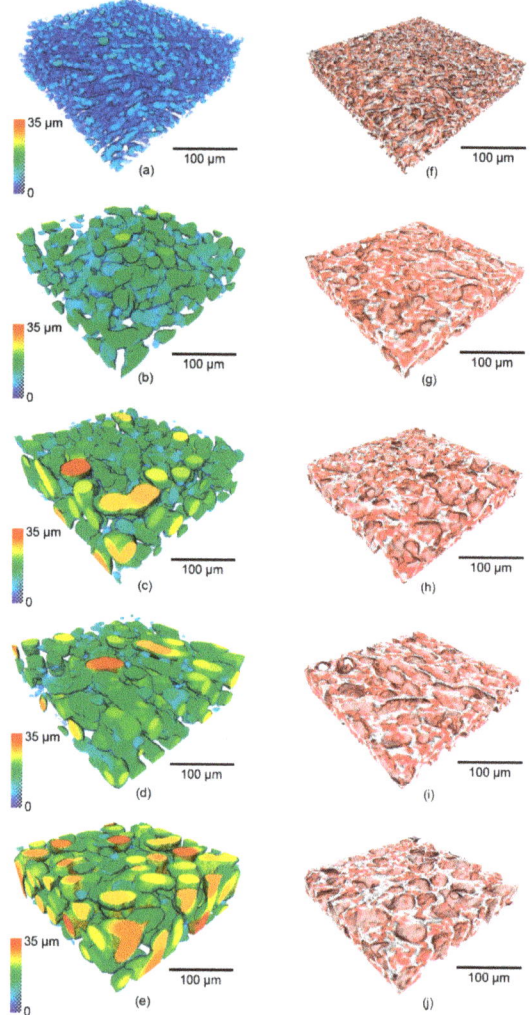

Figure 6. The 3D morphological evolution of representative volumes of (**a**–**e**) ice crystals and (**f**–**j**) the unfrozen matrix during the in situ experiment. (**a,f**) show the ice crystals and un-frozen matrix after fast cooling (FC) and holding at 250 K. During the slowing cooling (SC) period, the ice crystals, and un-frozen matrix are shown for the following temperatures: (**b,g**) 266.8 K, (**c,h**) 263.4 K, (**d,i**) 260.8 K, and (**e,j**) 250 K respectively. In (**a**–**e**), the color of the renderings correlates with the 3D local thickness of the ice crystals.

The size distributions of ice crystals and the unfrozen matrix, obtained from the 3D accessible volume method, are shown in Figure 7. As expected from the 2D tomographic slices (Figures 4 and 5) and the 3D reconstructed volume rendering (Figure 6), after the FC stage the ice crystals were very fine with a modal size of 8.2 µm and a range from around 1 to 15 µm. When the temperature was raised from 250 K to 267 K at 5 K/min and held for 10 min, the modal value of the size distribution increased to 11.2 µm. The distribution curve shifted to the right with a wider distribution ranging from 1–26 µm, which suggests the melting of smaller ice crystals and the growth of larger crystals at high temperature accompanying an overall decrease in ice volume with the increasing temperature. During slow cooling at a rate of 0.05 K/min, the modal size increased dramatically to 17.2 µm at 263.4 K (SC1) and to 20.2 µm at 260.8 K (SC2). Once the temperature had dropped below 263 K, the changes in size distribution at this cooling rate were not significant, which indicates that continued cooling below 263 K does not have a significant influence on the size of ice crystals over the timescale of the present experiments. It is presumably due to the slow molecular thermal diffusion at low temperatures, resulting in a reduced rate of coarsening.

In a complementary manner, the effects of the thermal history on the unfrozen matrix were also quantified. Figure 7b shows the size distribution of the unfrozen matrix at different temperatures, quantified using the same method as that employed for ice crystals (see Section 2.5). The unfrozen matrix is a complex 3D network that maintains the integrity of the ice cream structure. The 3D volume rendering in Figure 6f–j show that the unfrozen matrix appeared thicker between the ice crystals during the heating regime, whilst it became thinner as the sample cooled down from 267 K to 250 K, due to the melting of the ice crystals during the heating and recrystallization during cooling.

These qualitative observations are supported by the numerical data. The unfrozen matrix at 267 K (20.2 µm) after heating is much greater in thickness than that measured at 250 K (8.2 µm), which supports the idea that small crystals were melting into the unfrozen matrix as the temperature increased. During the SC regime, a trend of decreasing size of the unfrozen matrix is observed when the temperature of the sample was slowly cooled down to 250 K. This is presumably due to the growth increasing volume fraction of ice crystals, and is consistent with our quantification for the overall increased size of the ice crystal.

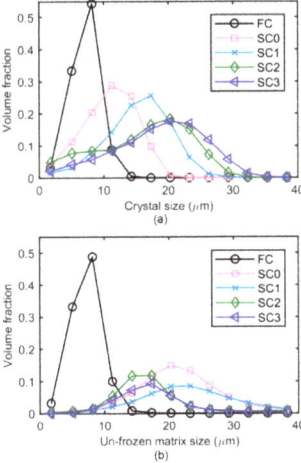

Figure 7. The size distributions of (**a**) the ice crystals and (**b**) the unfrozen matrix in the ice cream sample after fast cooling to 250 K (FC), and then after the fast reheating and holding process at 266.8 K (SC0), and during the slow cooling process at 263.4 K (SC1), 260.8 K (SC2), and 250 K (SC3), respectively.

The morphological changes of ice crystals are accompanied by changes in the total ice fraction and specific interface area (S_s) as revealed in mm^{-1} when cooled down to 250 K.

Table 1. Both the volume fraction and specific interface area (S_s) of ice crystals decreased during heating and holding, which supports the observation of the partial melting and coarsening of ice crystals. During this period, the needle-shaped ice crystals at 250 K became more spherical after the heating regime (clearly to minimize the interfacial free energy). The fraction of ice decreased significantly from 38% at 250 K to 28% 267 K, which is consistent with the melting phenomenon while heating. In the SC regime, we noted that the increase of the ice volume fraction with freezing is not completely monotonic. One possible reason for this is the relative movement of the entire volume during cooling, which might change the features that are measured.

The S_s of ice crystals first decreased dramatically from 599 to 301 mm^{-1} during the fast heating and holding period. However, the morphology of the ice crystals showed only small changes during the SC regime, as indicated by the average S_s decreasing from 301 mm^{-1} at 267 K to 247 mm^{-1} at 263 K, and then continuously decreased to 193 mm^{-1} when cooled down to 250 K.

3.3. Ex Situ: Microstructural Evolution of Ice Crystals Following Long-Term Thermal Cycling

Apart from the dynamic evolution mechanism of the individual features in ice cream which were elucidated in the in situ study, an examination of the microstructural evolution over long timescales (days/weeks) was also performed. The results are summarized in Figure 8 and reported in more detail in a previous paper [7]. After the thermal cycling from 258 to 268 K for 7 days, the size of ice crystals increased from a modal value of 45 μm in the C0 sample to a modal value of 85 μm in the C7 sample leading to coarsened ice structures and an ice cream that is likely to have a less desirable perception of texture for consumers. The size of the ice crystals was found to continuously grow even after 14-day long cycles. However, the rate of growth dramatically decreased after 7 cycles (7-day sample), with the size of ice crystals increasing only by ~10 μm from sample C7 to sample C14. It is possible that most of the surfaces of the ice crystals following 7 cycles are close to the size of the walls between the air cells. Therefore, air cells act as diffusion barriers, reducing Ostwald growth to being controlled by one-dimensional diffusion only in the plane of the wall. To demonstrate the morphological evolution in detail, a representative 3D rendering of ice crystals from the sCT data is superimposed in Figure 8.

Compared with the ex situ study, the size distribution obtained following the FC stage of the in situ work has a much smaller modal value of 8.2 μm. Such a fine scale microstructure is possible due to the fast cooling from the ice cream mix (liquid), which is different from the ex situ coarsened ice cream sample which was being "thermally abused" for a number of cycles between 258 and 268 K, i.e., the maximum temperature was well below the temperature of zero ice fraction (Figure 1a).

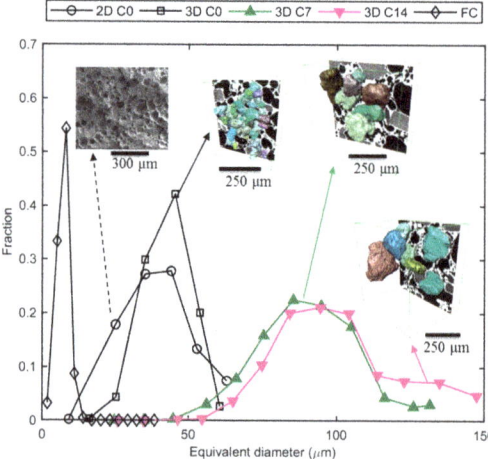

Figure 8. The size distributions of ice crystals in the ice cream samples following the FC stage from the in situ experiment, and C0, C7, C14 samples from the ex situ experiment. For the ex situ experiments, the morphology of the ice crystals is shown in the inserts as 3D color renderings of individual ice crystals. Additional data are taken from Reference [7].

4. Conclusions

In the present study, we applied in situ and *operando* time-resolved synchrotron tomography in a bespoke cold stage to quantify the fast evolution of the different microstructural phases in ice cream during processing, including ice crystals and the unfrozen matrix. To capture the long-term microstructural evolution during storage, we used an ex situ sCT methodology.

The in situ experimental results in this study reveal that the coarsening of ice crystals was due to both Ostwald ripening and physical agglomeration during heating and cooling. This change in the ice crystals size and morphology strongly influences our sensory perception of ice cream's taste. During the subsequent storage, we demonstrate that fluctuations in storage temperature can cause a partial-melting and recrystallization process, increasing the rate of coarsening. These processes were quantified, providing valuable data to both inform and validate models of the behavior of soft-solids.

Author Contributions: Conceptualization, P.D.L., E.G., P.S., G.V.D., J.B.; Methodology, E.G., J.M., P.R., D.S.E., P.D.L., P.S., J.B.; Formal Analysis, J.M., E.G.; Writing—Original Draft Preparation, J.M., E.G., P.D.L., D.G.M.; Writing—Review & Editing, All.

Funding: This work was financially supported by Unilever R&D (Colworth, UK) and by the EPSRC-UK (EP/I02249X/1, EP/J010456/1 and EP/M009688/1).

Acknowledgments: The authors acknowledge the use of the facility access in Diamond Light Source (MT12194, MT12195, MT12616 and MT17609) and Research Complex at Harwell. The authors also thank I13 staff of Diamond Light Source (especially Cipiccia) and group members for technical support.

Conflicts of Interest: The authors declare no conflict of interest.

References

1. Clarke, C. *The Science of Ice Cream*; Royal Society of Chemistry: London, UK, 2015.
2. Guo, E.; Kazantsev, D.; Mo, J.; Bent, J.; Van Dalen, G.; Schuetz, P.; Rockett, P.; StJohn, D.; Lee, P.D. Revealing the microstructural stability of a three-phase soft solid (ice cream) by 4D synchrotron X-ray tomography. *J. Food Eng.* **2018**, *237*, 204–214. [CrossRef]
3. Clarke, C. The physics of Ice Cream. *Phys. Educ.* **2003**, *38*, 248–253. [CrossRef]
4. Cook, K.L.K.; Hartel, R.W. Effect of freezing temperature and warming rate on dendrite break-up when freezing ice cream mix. *Int. Dairy J.* **2011**, *21*, 447–453. [CrossRef]
5. Chang, Y.; Hartel, R.W. Measurement of air cell distributions in dairy foams. *Int. Dairy J.* **2002**, *12*, 463–472. [CrossRef]
6. Sofjan, R.P.; Hartel, R.W. Effects of overrun on structural and physical characteristics of ice cream. *Int. Dairy J.* **2004**, *14*, 255–262. [CrossRef]
7. Guo, E.Y.; Zeng, G.; Kazantsev, D.; Rockett, P.; Bent, J.; Kirkland, M.; Van Dalen, G.; Eastwood, D.S.; StJohn, D.; Lee, P.D. Synchrotron X-ray tomographic quantification of microstructural evolution in ice cream—A multiphase soft solid. *Rsc. Adv.* **2017**, *7*, 15561–15573. [CrossRef]
8. Goff, H.D.; Verespej, E.; Smith, A.K. A study of fat and air structures in ice cream. *Int. Dairy J.* **1999**, *9*, 817–829. [CrossRef]
9. Cook, K.L.K.; Hartel, R.W. Mechanisms of Ice Crystallization in Ice Cream Production. *Compr. Rev. Food Sci. Food Saf.* **2010**, *9*, 213–222. [CrossRef]
10. Flemings, M.C.; Riek, R.; Young, K. Rheocasting. *Mater. Sci. Eng.* **1976**, *25*, 103–117. [CrossRef]
11. Van Dalen, G. A study of bubbles in foods by X-ray microtomography and image analysis. *Microsc. Anal.* **2012**, *26*, S8–S12.
12. Eisner, M.D.; Wildmoser, H.; Windhab, E.J. Air cell microstructuring in a high viscous ice cream matrix. *Colloids Surf. A: Physicochem. Eng. Asp.* **2005**, *263*, 390–399. [CrossRef]
13. Méndez-Velasco, C.; Goff, H.D. Fat structure in ice cream: A study on the types of fat interactions. *Food Hydrocoll.* **2012**, *29*, 152–159. [CrossRef]
14. Chang, Y.; Hartel, R. Stability of air cells in ice cream during hardening and storage. *J. Food Eng.* **2002**, *55*, 59–70. [CrossRef]
15. Chang, Y.; Hartel, R.W. Development of air cells in a batch ice cream freezer. *J. Food Eng.* **2002**, *55*, 71–78. [CrossRef]

16. Caillet, A.; Cogné, C.; Andrieu, J.; Laurent, P.; Rivoire, A. Characterization of ice cream structure by direct optical microscopy. Influence of freezing parameters. *LWT-Food Sci. Technol.* **2003**, *36*, 743–749. [CrossRef]
17. Cheng, J.; Ma, Y.; Li, X.; Yan, T.; Cui, J. Effects of milk protein-polysaccharide interactions on the stability of ice cream mix model systems. *Food Hydrocoll.* **2015**, *45*, 327–336. [CrossRef]
18. Pinzer, B.R.; Medebach, A.; Limbach, H.J.; Dubois, C.; Stampanoni, M.; Schneebeli, M. 3D-characterization of three-phase systems using X-ray tomography: Tracking the microstructural evolution in ice cream. *Soft Matter* **2012**, *8*, 4584–4594. [CrossRef]
19. Vicent, V.; Verboven, P.; Ndoye, F.T.; Alvarez, G.; Nicolai, B. A new method developed to characterize the 3D microstructure of frozen apple using X-ray micro-CT. *J. Food Eng.* **2017**, *212*, 154–164. [CrossRef]
20. Lee, P.; Hunt, J. Hydrogen porosity in directional solidified aluminium-copper alloys: In situ observation. *Acta Mater.* **1997**, *45*, 4155–4169. [CrossRef]
21. Leung, C.L.A.; Marussi, S.; Atwood, R.C.; Towrie, M.; Withers, P.J.; Lee, P.D. In situ X-ray imaging of defect and molten pool dynamics in laser additive manufacturing. *Nat. Commun.* **2018**, *9*, 1355. [CrossRef] [PubMed]
22. Stock, S. Recent advances in X-ray microtomography applied to materials. *Int. Mater. Rev.* **2008**, *53*, 129–181. [CrossRef]
23. Maire, E.; Withers, P.J. Quantitative X-ray tomography. *Int. Mater. Rev.* **2014**, *59*, 1–43. [CrossRef]
24. Bhreasail, Á.N.; Lee, P.; O'Sullivan, C.; Fenton, C.; Hamilton, R.; Rockett, P.; Connolley, T. In-Situ Observation of Cracks in Frozen Soil using Synchrotron Tomography. *Permafr. Periglac. Process.* **2012**, *23*, 170–176. [CrossRef]
25. Figueroa Pilz, F.; Dowey, P.J.; Fauchille, A.L.; Courtois, L.; Bay, B.; Ma, L.; Taylor, T.G.; Mecklenburgh, J.; Lee, P. Synchrotron tomographic quantification of strain and fracture during simulated thermal maturation of an organic-rich shale, UK Kimmeridge Clay. *J. Geophys. Res. Solid Earth* **2017**, *122*, 2553–2564. [CrossRef]
26. Polacci, M.; Arzilli, F.; La Spina, G.; Le Gall, N.; Cai, B.; Hartley, M.E.; Genova, D.D.; Vo, N.T.; Nonni, S.; Atwood, R.C.; et al. Crystallisation in basaltic magmas revealed via in situ 4D synchrotron X-ray microtomography. *Sci. Rep.* **2018**, *8*, 8377–8383. [CrossRef] [PubMed]
27. Reinke, S.K.; Wilde, F.; Kozhar, S.; Beckmann, F.; Vieira, J.; Heinrich, S.; Palzer, S. Synchrotron X-ray microtomography reveals interior microstructure of multicomponent food materials such as chocolate. *J. Food. Eng.* **2016**, *174*, 37–46. [CrossRef]
28. Kareh, K.M.; Lee, P.D.; Atwood, R.C.; Connolley, T.; Gourlay, C.M. Revealing the micromechanisms behind semi-solid metal deformation with time-resolved X-ray tomography. *Nat. Commun.* **2014**, *5*, 4464. [CrossRef] [PubMed]
29. Karagadde, S.; Lee, P.D.; Cai, B.; Fife, J.L.; Azeem, M.A.; Kareh, K.M.; Puncreobutr, C.; Puncreobutr, D.; Puncreobutr, T.; Puncreobutr, R.C. Transgranular liquation cracking of grains in the semi-solid state. *Nat. Commun.* **2015**, *6*, 8300. [CrossRef] [PubMed]
30. Wu, Z.Y.; Hill, R.G.; Yue, S.; Nightingale, D.; Lee, P.D.; Jones, J.R. Melt-derived bioactive glass scaffolds produced by a gel-cast foaming technique. *Acta Biomater.* **2011**, *7*, 1807–1816. [CrossRef] [PubMed]
31. Atwood, R.C.; Jones, J.R.; Lee, P.D.; Hench, L.L. Analysis of pore interconnectivity in bioactive glass foams using X-ray microtomography. *Scr. Mater.* **2004**, *51*, 1029–1033. [CrossRef]
32. Yue, S.; Lee, P.D.; Poologasundarampillai, G.; Jones, J.R. Evaluation of 3-D bioactive glass scaffolds dissolution in a perfusion flow system with X-ray microtomography. *Acta Biomater.* **2011**, *7*, 2637–2643. [CrossRef] [PubMed]
33. Mao, W.L.; Mao, H.K.; Meng, Y.; Eng, P.J.; Hu, M.Y.; Chow, P.; Cai, Y.Q.; Shu, J.; Hemley, R.J. X-ray-induced dissociation of H_2O and formation of an O-2-H-2 alloy at high pressure. *Science* **2006**, *314*, 636–638. [CrossRef] [PubMed]

© 2018 by the authors. Licensee MDPI, Basel, Switzerland. This article is an open access article distributed under the terms and conditions of the Creative Commons Attribution (CC BY) license (http://creativecommons.org/licenses/by/4.0/).

Article

In-Situ High Resolution Dynamic X-ray Microtomographic Imaging of Olive Oil Removal in Kitchen Sponges by Squeezing and Rinsing

Abhishek Shastry [1,2,*], Paolo E. Palacio-Mancheno [3], Karl Braeckman [4], Sander Vanheule [1,2], Ivan Josipovic [2], Frederic Van Assche [1,2], Eric Robles [5,*], Veerle Cnudde [2,6], Luc Van Hoorebeke [1,2] and Matthieu N. Boone [1,2,*]

1. Radiation Physics Research Group, Department of Physics and Astronomy, Ghent University, Proeftuinstraat 86/N12, B-9000 Gent, Belgium; Sander.Vanheule@UGent.be (S.V.); Frederic.Vanassche@UGent.be (F.V.A.); Luc.VanHoorebeke@UGent.be (L.V.H.)
2. Centre for X-ray Tomography, Ghent University, Proeftuinstraat 86/N12, B-9000 Gent, Belgium; Ivan.Josipovic@UGent.be (I.J.); Veerle.Cnudde@UGent.be (V.C.)
3. Procter and Gamble Corporate Functions, Surface Imaging and Microscopy Department, Mason Business Center, Mason, OH 45040, USA; palaciomancheno.pe@pg.com
4. The Procter and Gamble Company, Brussels Innovation Center, 1853 Strombeek Bever Temselaan 100, Bever, Belgium; braeckman.k@pg.com
5. The Procter and Gamble Company, Newcastle Innovation Center, Whitley Road, Longbenton, Newcastle-Upon-Tyne NE12 9TS, UK
6. Pore-Scale Processes in Geomaterials Research Group (PProGRess), Department of Geology, Ghent University, Krijgslaan 281/S8, B-9000 Gent, Belgium
* Correspondence: Abhishek.Shastry@UGent.be (A.S.); robles.es@pg.com (E.R.); Matthieu.Boone@UGent.be (M.N.B.); Tel.: +32-465539169 (A.S.); +44-7557322588 (E.R.); +32-9-264-6628 (M.N.B.)

Received: 30 July 2018; Accepted: 17 August 2018; Published: 20 August 2018

Abstract: Recent advances in high resolution X-ray tomography (μCT) technology have enabled in-situ dynamic μCT imaging (4D-μCT) of time-dependent processes inside 3D structures, non-destructively and non-invasively. This paper illustrates the application of 4D-μCT for visualizing the removal of fatty liquids from kitchen sponges made of polyurethane after rinsing (absorption), squeezing (desorption) and cleaning (adding detergents). For the first time, time-dependent imaging of this type of system was established with sufficiently large contrast gradient between water (with/without detergent) and olive oil (model fat) by the application of suitable fat-sensitive X-ray contrast agents. Thus, contrasted olive oil filled sponges were rinsed and squeezed in a unique laboratory loading device with a fluid flow channel designed to fit inside a rotating gantry-based X-ray μCT system. Results suggest the use of brominated vegetable oil as a preferred contrast agent over magnetite powder for enhancing the attenuation coefficient of olive oil in a multi fluid filled kitchen sponge. The contrast agent (brominated vegetable oil) and olive oil were mixed and subsequently added on to the sponge. There was no disintegration seen in the mixture of contrast agent and olive oil during the cleaning process by detergents. The application of contrast agents also helped in accurately tracking the movement and volume changes of soils in compressed open cell structures. With the in house-built cleaning device, it was quantified that almost 99% of cleaning was possible for contrasted olive oil (brominated vegetable oil with olive oil) dispersed in the sponge. This novel approach allowed for realistic mimicking of the cleaning process and provided closer evaluation of the effectiveness of cleaning by detergents to minimize bacterial growth.

Keywords: X-ray μCT; in-situ experiments; flow cell

1. Introduction

Sponges are ubiquitous implements used in household and industrial cleaning tasks thanks to their flexibility and absorption ability. The cleaning of sponges, i.e., removal of soil on surfaces and cleaning the sponges themselves after absorbing the soil is dependent on the mechanics/hardness of the structure (abrasion level), the interconnectedness and surface area/energy of the open cell structure (capillary forces) [1]. However, the sponge's complex porous structure may leave soil trapped in the sponge during the sorption phase. Where the soil ends up in the sponge depends on the physical properties of the soil. For example, if it is solid food particles (meat, undissolved carbohydrates, solid fat), it will most likely be deposited on the outer surface of the sponge due to filtration mechanism. However, for liquid soil (vegetable oil, dissolved proteins, dissolved starch) they can be wicked inside the porous structure due to capillary action. These residues whether on the surface or inside the pores may be used by micro-organisms (bacteria, fungi) as a source of food for their growth, giving rise to biofilm formation over time, and hence potential hygiene risks. While detergent formulations with antibacterial active agents may be used to prevent microbial growth in sponges, the use of these antibacterial agents is regulated and limited (EC regulation No 1223/2009 of the European Parliament and of the Council of 30 November 2009 on cosmetic products). Therefore, formulations that are effective in preventing bacterial growth is preferred by ensuring food soils are completely removed inside the sponge. Key to this is the mixture of ingredients in the formulation and the fluid dynamics required for removal. The fluid dynamics inside the complex deformable porous structure are however poorly understood, and therefore their effects on soil removal are not accounted for. To improve the efficiency of cleaning agents, it is important to understand how efficiently these agents are delivered throughout the complex structure of the sponge and to understand the mechanisms that govern the emulsification of the soil inside the sponge. The purpose of this paper is to develop a non-invasive imaging method that can accurately visualize the displacement and relocation of olive oil (or any other animal or plant-based oil) inside an open cell sponge in a multiphase fluid environment and to assess levels of soil removal under compressive forces. Outcomes from this paper can be used as modeling inputs for the evaluation of how sponge absorbs food soil from dishes into the sponge and how these are subsequently removed and released into the wash solution upon squeezing.

In comparison with cleaning sponges (cellulose sponges), kitchen sponges made of polyurethane are less susceptible to tearing because of its high tensile strength [2,3]. Polyurethane sponges are nontoxic and free of biocides unlike cellulose sponges which inherit chemicals to control microbial growth during polymerization process [3]. The sponges have a high water retention capability and release liquid only under pressure [4]. Nevertheless, ester, amide and urethane groups represent sites on the polymer surfaces for hydrolytic attack [5], and are subject to degradation by aqueous acids, alkalis and steam during the cleaning process. This degradation results in structural deformation of the sponge and increase in its overall surface area. The oil sorption capacity increases with the decrease in foam density over time/use and attracts bacterial adhesion. For their commercial availability and known cleaning capacity, polyurethane sponges were studied in this work over other materials.

High resolution X-ray Computed Tomography (µCT) is a non-destructive 3D imaging technique able to visualize both the external and internal structure of porous objects [6]. Throughout the last two decades, it has become an established technique in numerous research areas for visual evaluation of objects. Particularly in porous media, 3D analysis is often performed to extract quantitative parameters like density, porosity, pore size [7–10]. Also, µCT can be used as input for the generation of 3D geometries which are subsequently used for finite element (FE) [11] or fluid flow simulations [12], and pore scale modeling [13–15]. Alternatively, micro magnetic resonance imaging (MRI) can provide dynamic fluid tracking on compressed foam structures; yet, with poor spatial and temporal resolution [16]. In recent years, an increasing interest has extended X-ray imaging to the temporal dimension [17]; exploiting the non-destructive nature of µCT to assess dynamic processes of pores structures such as visualizing multiple phase flow and solute transport in real-time [14,18,19]. Accordingly, µCT was chosen as the ideal technology to track soil removal over MRI. In addition, to achieve temporal analysis of pores structures, in-situ devices have been developed for various purposes in the past including flow cells, and compression stages [20,21]; yet, the application to fluid-filled sponges with soil under dynamic compression has been poorly studied.

An attempt has been made for the imaging of olive oil and its removal from the sponge at a spatial resolution better than 30 microns. However, the possibilities of imaging olive oil in water simultaneously are limited due to the low difference in X-ray attenuating coefficients at the X-ray energies commonly used in μCT [22]. Low contrast driven by small density differences can be countered by the application of X-ray contrast agents which can bind with specificity to the olive oil constituents and enhance its X-ray attenuation against water and PU (Polyurethane). Contrast Enhanced Computed Tomography has been in place for quite some time in medical imaging [23]. However, literature on contrast agents for μCT on inert materials is limited with the research across the globe targeting mainly biological [24], biomedical [25–28] and geological samples [29,30]. On the other hand, Scanning Electron Microscopy (SEM) staining techniques for polymeric substance can be used for X-ray μCT applications. Staining agents like Osmium (Os) and Ruthenium (Ru) Tetroxide are used to target unsaturated poly-hydrocarbons (e.g., oils and waxes) but unfortunately, both staining agents are toxic and volatile. [31–33]. Similarly, phosphotungstic acid (PTA) and phosphomolybdic acid (PMA) can be used to target the conjugated unsaturated fatty oils and proteins. Nevertheless, polyurethane (PU) is not resistant to strong acids and can be hydrolysed [34]. Thus, in this work, application of contrast agents for olive oil inside a fluid filled sponge substrate was accomplished with alternative materials with different physical properties, namely magnetite powder [35] (chemical formula Fe_3O_4) and brominated vegetable oil [36]. The contrast is mainly attributed to iron's and bromine's higher attenuation coefficient values vs. water, olive oil and polyurethane. Magnetite particles are dispersed in olive oil while brominated vegetable oil is miscible with olive oil. This paper illustrates the application of these contrast agents before and after loading, for the assessment of olive oil removal/cleaning and it is believed to be the first of its kind.

In-situ experiments with X-ray Tomography requires custom build modules capable of mimicking the dynamic process under investigation while respecting the practical constraints of the measurement technique throughout the experiment [37]. In 4D-μCT, the most notable constraints are on the size and composition of the sample holder. Indeed, due to the geometrical magnification in laboratory based μCT, the diameter of the sample should be as small as possible to obtain the desired resolution in the μCT images at maximal flux efficiency (i.e., with a sufficiently small source-to-object distance). In fluid flow experiments, the sample holder completely confines the sample, and should be as transparent as possible to X-rays. Furthermore, a rotational symmetry is desired in tomographic experiments, particularly in the scope of applying corrections for region-of-interest tomography reconstruction [38]. Keeping all these into consideration the flow cell described below was developed to be able to mimic the cleaning process in a realistic way.

2. Materials and Methods

2.1. Sample Material

The experiments presented in this paper are conducted on standard kitchen sponges (non-scratch type) made from polyurethane (Spontex, Colombes, France). To improve the reproducibility of the experiments, a large number of sponges was purchased simultaneously, originating from the same production batch. To evaluate the cleaning inside the sponge, they were soiled with commercially available olive oil (Bertolli Extra Virgin, Unilever, Rotterdam, The Netherlands). The cleaning of the sponges was performed using Dreft dishwashing liquid (P&G, Cincinnati, OH, USA) and standard tap water (hard water). For the experiments, sponge samples of 3 cm diameter and 4 cm height were implemented considering the geometrical magnification of X-ray μCT and the required spatial resolution.

2.2. Scanner System

In this work, two different high-resolution CT systems were used, both custom-designed by the Ghent University Centre for X-ray Tomography (UGCT, www.ugct.ugent.be). For high quality static imaging of sponges, the HECTOR system was used. This system is based on an open-type XWT240 X-ray tube (X-ray WorX, Garbsen, Germany) and a large PerkinElmer flat-panel detector. More details on the system components can be found in [39]. The best spatial resolution for this system is approximately 4 µm. For the dynamic experiments, the Environmental Micro-CT system or EMCT [40] was used. This gantry-based system is designed specifically to conduct in-situ 4D-µCT experiments, as the source-detector system rotates around the stationary object, thus allowing for several mountings, cables and tubes into the in-situ device, i.e., the flow cell. With a L9181-02 X-ray tube (Hamamatsu Photonics, Hamamatsu, Japan) and a Xineos 1313 flat-panel detector (Teledyne DALSA, Waterloo, ON, Canada), this system is optimized for a compact design and high scanning speed. At highest speed, a full 360° rotation is performed in 12 s, with a best resolution of approximately 15 µm, partly limited by the available X-ray flux. At slower speeds, a best resolution of approximately 5 µm can be achieved. For more information about the scanning system and the relationship between scanning speed and resolution, the reader is referred to [14,40], respectively. Reconstruction of the radiographs obtained during both static and dynamics CT scans was done using Octopus Reconstruction [41] which is an in house developed software package. Octopus Analysis [10] is in house developed software and was used for 3D analysis of the reconstructed images. The 3D rendering of the sponges with the residue was made using VGStudio MAX 3.2 (Volume Graphics GmbH, Heidelberg, Germany).

2.3. Flow Cell and Its Automation

A sample holder (flow cell) with a provision of a flow channel and the capability to squeeze and flush the sample was designed and constructed in polymethymethacrylate (PMMA). The flow cell consists of a cylindrical body with a grid and a plunger. A grid is placed to hold the sample and drain the fluid out of the sponge upon squeezing. An inlet is provided through the plunger and below the grid a provision is made for the outlet. The tube dimensions are indicated in the Figure 1. The tube can be split into two parts: (1) working area of the tube and (2) region below the grid. The height of the working area is 9 cm with 3 cm inner diameter and the height of the region below the grid plate is 4 cm with 2 cm inner diameter. The height of the plunger is 12 cm, with a protrusion of 3 cm diameter and 2 cm height (Figure 1). The wall thickness of the tube is 0.5 cm and the external diameter of the tube is 4 cm. PMMA was chosen considering the mechanical strength needed for the cyclic action of the flow cell and because of its relatively low X-ray attenuation. The low X-ray attenuation of the flow cell is necessary to make sure that sufficient X-ray flux reaches the detector [42].

The flow cell can be attached to a scotch yoke mechanism connected to a stepper motor to enable automated squeezing. The supplementary parts are made from PVC and consist of a pinion wheel with a provision to attach the motor shaft, a cap to stabilize the position of the plunger and a support needed to mount the top part of the setup. The total amplitude of the plunging motion is 9 cm. Figure 2 illustrates in detail the add-on modules along with the flow cell. A NEMA23 stepper motor (RepRapWorld B.V, Nootdorp, The Netherlands) with a torque of 30.59 kg.cm is used for providing the thrust required for the reciprocating action of the plunger. The position of the inlet was altered to facilitate easy flow of water through the sponge while squeezing. The flexible hose was attached to both inlet and outlet extensions. The water flow through the flow cell is based on a gravity fed pipe flow system and no electric motor was introduced to pump in water. The outlet was connected to a suitable water basin and the water flow was controlled with a clip attached to the hose.

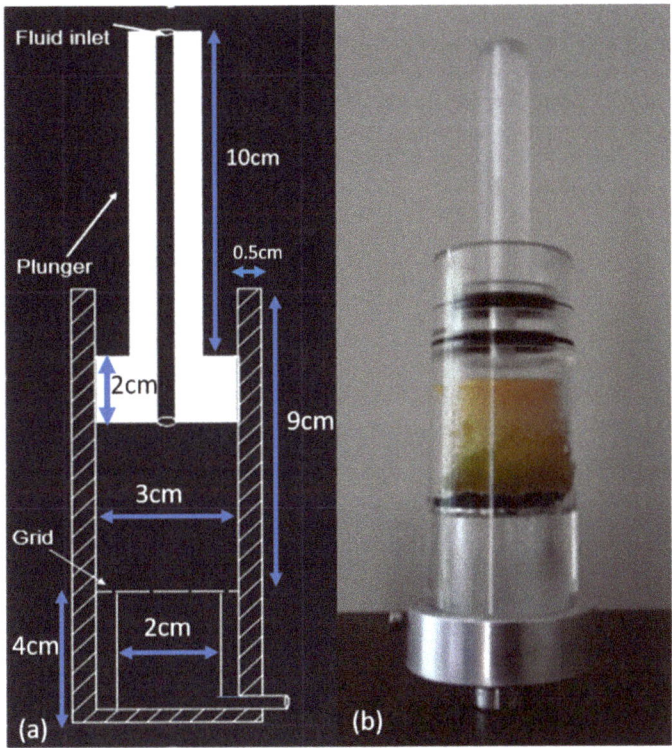

Figure 1. (**a**) 2D front view of CAD model of the sample holder (not to scale), (**b**) photo of the sample holder with sponge.

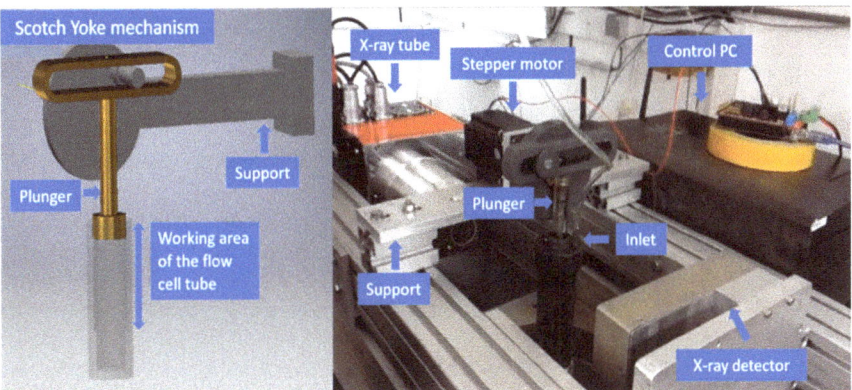

Figure 2. Picture and schematic drawing of the flow cell (Front view) with attached flow channels.

2.4. Experimental Design

Here is a brief overview of all the experiments illustrated in this article.

Preliminary studies:

- Section 2.5 describes two preliminary experiments performed to characterize the samples in more detail. In the first experiment the sponge sample was placed inside the flow cell and fluids (olive oil and water) were added on to the sponge. This system was scanned using X-ray µCT to visualize the microstructure of the sponge and to spot the fluids (water and olive oil) considering their attenuation coefficient values. In the second experiment each of the fluids (olive oil, water and detergent) were scanned separately to determine and note the difference in their attenuation coefficient values.

In-situ test:

- Section 2.7 elaborates on the third experiment where the contrasted olive oil was scanned using X-ray µCT to estimate the attenuation coefficient enhancement of olive oil due to the application of contrast agents. A fourth experiment is also described in which the developed experimental protocol was followed, aiming to demonstrate the cleaning capabilities of the custom-built flow cell and quantification of the contrasted olive oil present in the sponge before and after cleaning (i.e., removal of contrasted olive oil from the sponge) with detergents.

2.5. Sample Characterisation

To examine the imaging capability of X-ray µCT for olive oil, water and sponge, preliminary imaging of olive oil in sponges was conducted without the application of any contrast agents on the HECTOR system. For the first experiment the sponge was cut into a cylindrical cross section with a diameter of 3 cm and a height of 4 cm which was placed inside the tube of the flow cell. The flow cell with the dry sponge (sponge without olive oil and water) was scanned using X-ray µCT. Olive oil and water each 5 mL were added to the dry sponge and for 5 min the fluids were allowed to diffuse in the sponge. After this time, the wet sponge (sponge with olive oil and water) was scanned. For both these scans, 1401 projections of 1 s exposure time per projection with a voxel size of $36 \times 36 \times 36$ µm^3 were acquired over the full 360° rotation. The tube output was set at 70 kV and 30 W and the duration of the scan was around 26 min. The scans were obtained without any use of filters on the X-ray source.

In the second experiment, tubes containing pure olive oil, water and detergent outside the purview of the sponge were scanned using X-ray µCT. Scans with 2001 projections of 1 s exposure time per projection were acquired at full 360° rotation with a voxel size of $37.5 \times 37.5 \times 37.5$ µm^3 and tube output of 70 kV and 30 W. The duration of the scan was around 36 min. The scans were obtained without any use of filters on the X-ray source.

2.6. Contrast Agents

Contrast agents help to improve visualization between the targeted material (i.e., the fatty liquid) and all other constituents of the structure (sponge and detergent solution), including background. In this study, we compare two different contrast agents: magnetite powder and brominated vegetable oil. Commercially available magnetite powder (Inoxia Ltd., Surrey, UK and Natural type) with an average particle size of 40 µm was added to the olive oil at 10% wt/volume. Alternatively, bromine vegetable oil (VWR, Radnor, PA, USA) is another interesting contrast agent which was considered as suitable because it forms miscible solution with olive oil and has higher attenuation coefficient value. The concentration of brominated vegetable oil in olive oil was maintained at 10% wt/volume for the experiments.

2.7. Application of Contrast Agents on Sponges

The third experiment aimed at assessing the specificity of each of the contrast agents in olive oil and visualizes the homogeneity of the dispersion (magnetite)/solution (brominated oil). Inside two different containers of 2 cm diameter and 8.5 cm height, 0.5 g of magnetite powder and 0.5 g of brominated vegetable oil were each dispersed respectively in 5 mL of olive oil. Both solutions were observed under X-ray µCT. The tube output for these experiments remained at 80 kV and 12 W.

2401 projections with 0.1 s exposure time per projection were acquired at full 360° rotation at a voxel size of 33.5 × 33.5 × 33.5 µm³ in the scan. The duration of scan was around 5 min.

For the fourth experiment the cylindrical cross section of the sponge of 3 cm diameter and a height of 4 cm placed inside the flow cell was first introduced with 5 mL of magnetite powder dispersed olive oil (10% wt/volume) and the following experimental protocol was followed to mimic the soiling and cleaning of the sponge. Later for a new sponge sample, 5 mL of brominated vegetable oil mixed olive oil (10% wt/volume) was introduced and the experimental protocol was repeated.

One compression cycle is the movement of the plunger to its full length inside the working area of the tube and returning to its initial position. The cycle frequency was maintained at ten cycles per minute throughout the whole experiment.

Process 1: The flow cell was connected to an inlet and outlet channel for continuous flow of water through the sponge sample.

Process 2: The sponge was compressed using the plunger for one minute at 10 cycles/min to disperse the contrasted olive oil with continuous flow of water.

Process 3: 5 mL of detergent was added to the dirty sponge and squeezed for one minute at 10 cycles/min inside the flow cell without water supply.

Process 4: At Stage 1 of cleaning process, with live water feed 10 min of squeezing was performed at 10 cycles/min.

Process 5: For the Stage 2 of cleaning process another 5 mL of detergent was added onto the same sponge and again 10 min of squeezing with continuous water feed was performed at 10 cycles/min.

Process 6: In the end the sponge was removed and air dried for a day at room temperature.

At the end of each process the protocol was halted and a µCT scan was acquired using the EMCT system. The duration of the scan was around 5 min each. For these scans, 2401 projections of 0.1 s exposure time per projection with a voxel size of 33.5 × 33.5 × 33.5 µm³ were acquired over the full 360° rotation. The tube output was set to 80 kV and 12 W with no hardware filter.

2.8. Criteria for Cleaning

The process of cleaning involves squeezing and rinsing of the sponge with the addition of detergent and water to remove the dispersed contrasted olive oil. The volume of the soiled sponge was loaded in Octopus Analysis and by adjusting the threshold the contrasted olive oil present in the sponge was segmented and the volume of contrasted olive oil volume was determined. Considering the same attenuation coefficient for the contrasted olive oil, thresholding was done for the sponge after all cleaning steps. The percentage of cleaning (complementary percentage) (see further in the Section 3.3) was determined by subtracting the ratio of volume of contrasted olive oil determined after cleaning stages to the volume of contrasted olive oil before cleaning from 1 and multiplying the obtained number by 100.

Equation (1)

Percentage of cleaning for the stage one
$$= \left[1 - \frac{\text{Volume of contrasted olive oil in stage one}}{\text{Volume of contrasted olive oil before cleaning}}\right] \times 100 \quad (1)$$

3. Results

3.1. Characterization of Materials

Figure 3 illustrates the characterization of the sponge microstructure in absence of load, with and without liquids inside. In Figure 3a the pore structure of the dry sponge can be seen and Figure 3c illustrates the distribution of water and olive oil inside the sponge. With the help of 2D cross sectional images the structure and distribution of the fluids at any given layer inside the sponge can be visualized in 2D and 3D. Although the location of olive oil and water inside the sponge could be visually assessed however, in a single system it was impossible to accurately discriminate between olive oil, water and sponge material (Figure 3d).

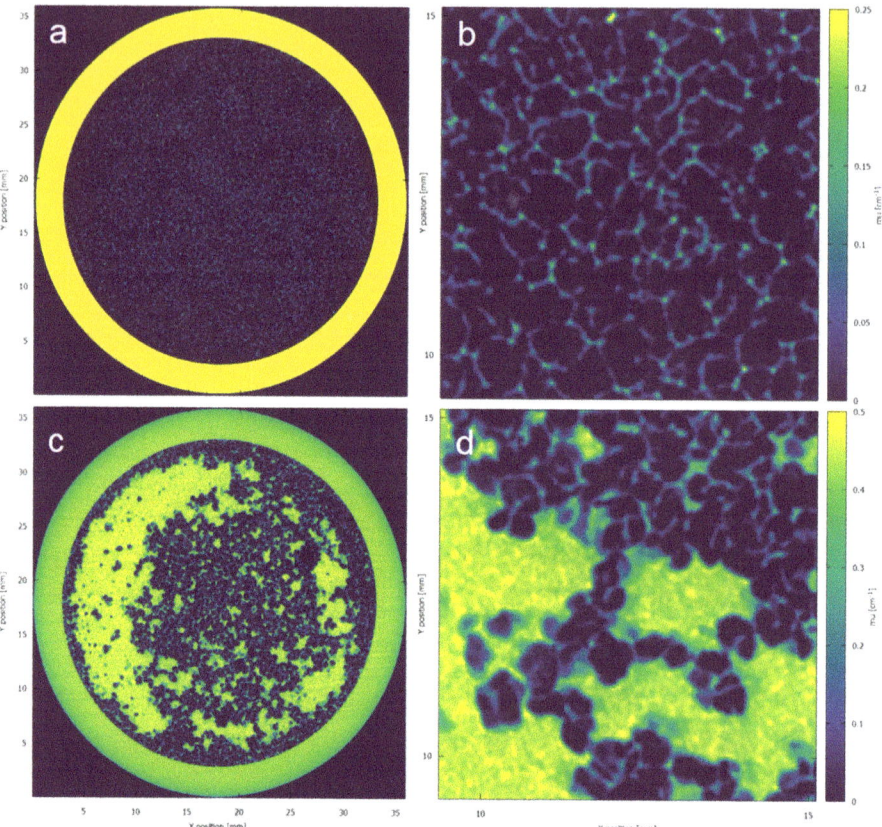

Figure 3. (a) Cross sectional image indicating the microstructure of the dry sponge. (c) Cross sectional image of sponge with water and olive oil without application of contrast agent. (b,d) Zoomed images of the result shown in (a,c), depicting the microstructure of the sponge and water and non-contrasted olive oil in the sponge, respectively. The colormap illustrates the reconstructed attenuation coefficient of different materials.

Table 1 illustrates the experimental attenuation coefficient values of each of the test fluids retrieved separately in multiple scans with same scanner settings. The images were loaded in Octopus Analysis and the average grey value was determined over a volume of interest of approximately 3140 mm^3. This grey value was later converted to a linear attenuation coefficient μ using the calibration from the reconstruction. Although these values are separated by 2σ (standard deviation), it should be noted that these are obtained in a container of pure material. In a real system such as the sponge, the features to be recognized are small and partial volume effects have a significant contribution [6]. Furthermore, these measurements are obtained from a high-quality scan, and in-situ experiments will yield higher noise levels.

Table 1. Solvents with their experimental linear X-ray attenuation coefficient (μ) and the achieved standard deviation σ (μ) indicating the close proximity in attenuation coefficient values of test fluids and attenuation coefficient enhancement of olive oil by addition of contrast agents.

Solvents	μ (cm^{-1})	σ (μ) (cm^{-1})
Olive oil	0.25	0.0158
Water	0.33	0.0261
Detergent (Dish washing liquid)	0.35	0.0297
Magnetite powder dispersed olive oil	0.45	0.0421
Brominated vegetable oil with olive oil	0.62	0.0775

3.2. Assessment of Contrast Agent Specificity

In the specificity test, both contrast agents were dispersed evenly in the solution by severe mixing. The magnetite powder solution however proved to be unstable over time due to sedimentation of the powder. The stability of the solution depends on the concentration of magnetite powder dispersed in olive oil. With lower concentrations the onset of sedimentation only happens if the solution remains stationary in its liquid state without being applied onto sponge.

The values of contrasted olive oil in Table 1 indicate that there is a considerable improvement in the attenuation coefficient of olive oil with dispersion of magnetite powder and brominated vegetable oil respectively. 10% wt/volume concentration of contrast agent dispersed in olive oil was chosen to be optimal for all the experiments considering the sedimentation property of magnetite powder in olive oil.

3.3. Experiments of Contrasted Olive Oil on Sponges: Cleaning Assessment of the Custom-Built Device and Quantification of Contrasted Olive Oil in the Sponge

Octopus Analysis software was used to calculate the volume of contrasted olive oil present in the sponge at different stages of the cleaning process and it was compared with the actual volume of contrasted olive oil added to the sponge. Although 5 mL of contrasted olive oil was added on to the sponge only 3.6 mL of magnetite powder dispersed olive oil and 1.2 mL of brominated vegetable oil with olive oil could be recorded through image segmentation (Table 2). The factors influencing this difference in value will be explained in the Discussion section.

Table 2. Volume of contrasted olive oil at different stages of cleaning indicating the removal of contrasted olive oil from the sponge. Stage 1 of cleaning process: soiled sponge thoroughly rinsed with 5 mL of detergent is squeezed for 10 min at 10 cycles/min with continuous water feed. Stage 2 of cleaning process: 5 mL of detergent is added to the same sponge and squeezed for 10 min at 10 cycles/min with continuous water feed.

Contrasted Olive Oil	Actual Volume of Contrasted Olive Oil	Uncleaned Sponge	Stage 1	Stage 2
Volume of magnetite dispersed olive oil (mL)	5	3.6	0.8	0.79
Volume of brominated vegetable oil with olive oil (mL)	5	1.2	0.08	0.01

Using Equation (1) the percentages of cleaning for the sponge were determined. In Table 3 the percentage of cleaning for the given constant volume at two cleaning stages are given. Sponge applied with brominated vegetable oil mixed olive oil showed a better cleaning percentage compared to sponge with magnetite powder dispersed olive oil and the change in the cleaning percentage between the two stages were less for the latter case.

Table 3. Percentage of cleaning for two sponges, sponge 1 (magnetite powder dispersed olive oil) and sponge 2 (brominated vegetable oil mixed olive oil).

Sponge Samples	Stage 1 of Cleaning	Stage 2 of Cleaning
Sponge 1	77.2%	77.5%
Sponge 2	92.5%	98.8%

3.4. Experiments: Dynamics of Soil Removal from Sponges under Loading

From the stack of reconstructed slices, a 3D volume of both the uncleaned and the cleaned sponges were rendered using VGStudio MAX 3.2 software. Figures 4 and 5 give the visual representation of the contrasted olive oil with respectively magnetite powder dispersed olive oil and brominated vegetable oil with olive oil in the sponge before and after two stages of cleaning along with the sponge after Process 3 (Section 2.7). The removal of contrasted olive oil in the sponge is made possible by the squeezing action of the plunger, the external compressive force overcomes the capillary forces and thereby the trapped fluid gets displaced by a non-wetting phase. Dynamic action of the plunger not only increases the pore size of the sponge but also helps in mobilizing the water droplets and detergent throughout the porous structure.

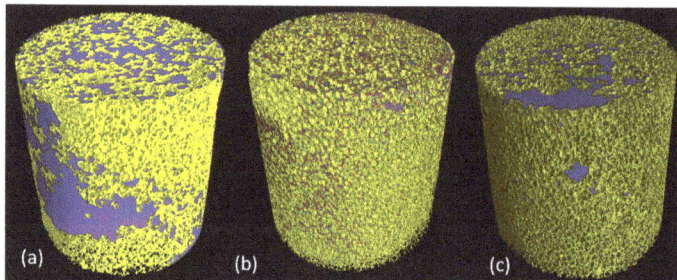

Figure 4. 3D rendering of sponge with magnetite powder dispersed olive oil (**a**) before cleaning, (**b**) intermediate Process 3 (Section 2.7) and (**c**) after two stages of cleaning. Pseudo coloration is performed based on the segmentation: blue represents magnetite powder dispersed olive oil, pale blue represents water, red represents detergent and yellow colour represents the sponge. The residue was present even after 2 stages of cleaning for the magnetite powder dispersed olive oil sponge.

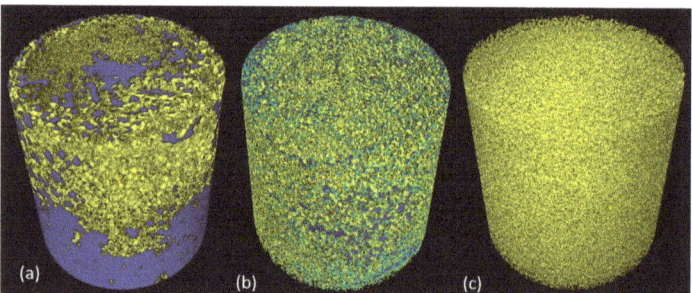

Figure 5. 3D rendering of sponge with brominated vegetable oil dispersed olive oil (**a**) before cleaning, (**b**) intermediate Process 3 (Section 2.7) and (**c**) after two stages of cleaning. Pseudo coloration is performed based on the segmentation: blue represents brominated vegetable oil with olive oil, pale blue represents water, red represents detergent and yellow colour represents the sponge. There were no traces of contrasted olive oil in the cleaned sponge.

4. Discussion

Although there is a visible difference between the olive oil and water present in the sponge, it is extremely challenging to quantitatively retrieve the interface between both liquids (Figure 3d). As shown in Table 1, the attenuation coefficient values of these fluids were in close proximity to each other.

The contrast was therefore improved by using two different contrast agents namely magnetite powder and brominated vegetable oil. The dispersion capabilities of the contrast agents in the olive oil and the influence of the contrast agent on the properties of the olive oil are of high importance and limit the practically achievable concentration. Due to the inert chemical behaviour of magnetite powder with olive oil and their difference in bulk densities, sedimentation of magnetite powder occurred at concentrations higher than 50% (wt/volume). For brominated vegetable oil, such sedimentation was not observed as the two liquids were miscible. However, with a thorough premix of the solution, the olive oil acts as a suitable carrier for magnetite powder [43] and the solution remains stable for sufficiently long time before being applied on to the sponge.

With the help of 2D cross sectional images a thin film may be seen on the walls of the sponge pores due to the adhesion of the contrasted olive oil, making it partially a closed cell structure. Also due to the uneven distribution of contrasted olive oil, the sponge tends to lose its stability and collapses towards the heavier side. This was one of the challenges that had to be faced while conducting the experimental protocol. After each step, the sponge was rearranged before the scan to have the full view of the structure.

The traces of magnetite powder dispersed olive oil were present in the sponge even after two stages of cleaning. A possible cause of this phenomenon is the magnetite powder without olive oil which tends to stick to the sponge material due to the heterogeneous solution. This separation of magnetite powder from its solution will not be helpful as the main purpose of cleaning the sponge becomes questionable. The influence of the contrast agent on the olive oil properties, particularly with respect to the interaction with the sponge material, is a very complex research question. Solving this is out of the scope of this manuscript but part of parallel research in the same research consortium. However, the use of brominated vegetable oil as contrast agent for olive oil can be justified as bromine results in a higher attenuation coefficient and the two liquids form a stable solution.

The contrasted olive oil added onto the sponge could impregnate purely by gravity and capillary forces. According to laws of capillarity, the small pores cause higher capillary pressure for the wetting phase (contrasted olive oil) to move towards non-wetting phase (air filled pores) hence making it difficult for imbibition of contrasted olive oil through the sponge. This resulted in the concentration of contrasted olive oil at the periphery (Figures 4a and 5a).

To assess the entire volume of contrasted olive oil present in the sponge, the different phases were segmented in 3D analysis software. The limitation to find the optimal threshold value together with partial volume effects were some of the reasons for the difference between the measured amount of olive oil in the uncleaned sponge and the added amount of olive oil (Table 2, 3.6 mL of magnetite powder dispersed olive oil and 1.2 mL of brominated vegetable oil mixed olive oil compared to 5 mL inserted in the system). One of the other important reasons may be the loss of contrasted olive oil through the outlet after addition, for which solutions will be sought. The amount of solution of brominated vegetable oil in olive oil left behind in the sponge was relatively less and unlike magnetite powder the dissolved brominated vegetable oil did not adhere to the sponge and therefore provides a better approach to contrasting.

As the system was not operated in vacuum, the displacement of the wetting phase by a non-wetting phase resulted in trapping of contrasted olive oil droplets inside the sponge [44]. Dynamic processes like squeezing and rinsing were necessary to mobilize and emulsify this trapped soil droplets for easy removal. Process 2 and Process 3 illustrated in Section 2.7 helped in mobilizing water-soluble particles present in the sponge and in emulsification of fatty soils respectively [45]. These processes facilitated for removal of contrasted olive oil dispersed in the sponge.

The flow cell with its cyclic action and with the help of water channels was successful in removing the contrasted olive oil present inside the sponge. Comparing the percentage of cleaning and the enhancement in attenuation coefficient of olive oil, brominated vegetable oil becomes a favorable choice as contrast agent. The 3D rendered images in Figures 4 and 5 depict the displacement and relocation of fluid inside the porous structure in relation to time and hence can be used as an input for developing a model for fatty oil absorption and removal with a detergent solution. This method can be employed to optimize a detergent formulation that can quickly wet the kitchen sponge and the oil trapped in between the pores, emulsify the fat into oil droplets and completely remove them via squeezing action.

X-ray 4D-μCT (time dependent imaging) is a suitable choice for visualizing in situ experiments as it needs no sample preparation and gives both qualitative and quantitative data. Application of suitable contrast agents have further enabled the technique to image low attenuating samples in a multiphase fluid system non-destructively.

5. Summary

A 4D-μCT approach to assess removal of fatty oil from sponge has been developed. A sample holder with a flow channel and the capability to squeeze and flush the sample was designed to follow the dynamic processes inside the sponge during oil removal. Two contrast agents were assessed to improve visualization of the fatty oil absorption, emulsification and removal. Magnetite particles provided contrast but the stability of the magnetite dispersion in olive oil and the magnetite adsorption onto the sponge inner surfaces were identified as the major problems with this approach. On the other hand, mixing brominated vegetable oil with olive oil was identified as the better approach as it forms a solution with olive oil. This fatty oil solution does not phase separate and does not adsorb strongly to the sponge inner surfaces thus behaving like olive oil. Washing the sponge with a dishwashing solution almost completely removed the oil indicating deep down cleaning and can help minimize bacterial growth inside the sponge.

Author Contributions: Conceptualization, M.N.B., V.C., E.R.; Methodology, A.S., M.N.B., E.R., K.B., P.E.P.-M., V.C., L.V.H., I.J.; Software, A.S., F.V.A., S.V.; Validation, A.S., P.E.P.-M., E.R., M.N.B.; Formal Analysis, A.S.; Investigation, A.S.; Resources, A.S., I.J., F.V.A., S.V., P.E.P.-M.; Data Curation, A.S.; Writing—Original Draft Preparation, A.S., M.N.B.; Writing—Review and Editing, all authors; Visualization, A.S., M.N.B.; Supervision, M.N.B., E.R., V.C., L.V.H.; Project Administration, E.R., M.N.B., V.C.; Funding Acquisition, E.R., V.C., M.N.B., L.V.H., K.B.

Funding: This work is funded by the European Union's Horizon 2020 research and innovation programme under grant agreement No. 722871 in the scope of the Marie Skłodowska-Curie Action ITN BioClean. The Ghent University Special Research Fund (BOF-UGent) is acknowledged for the financial support to the Centre of Expertise UGCT (BOF.EXP.2017.000007).

Acknowledgments: Philippe Van Auwegem (Ghent University, Department Physics and Astronomy) is acknowledged for his assistance in building the flow cell and Daniëlle Schram (Ghent University, Department Geology) for helping in purchase and procurement of the materials.

Conflicts of Interest: This work presents a methodology to investigate deep-down cleaning using in-situ experiments. The results of this work do not reflect any value measurement of any P&G product and as such do not pose a conflict of interest. Any dishwashing liquid can be used for this purpose.

References

1. Landrock, A.H. *Handbook of Plastic Foams: Types, Properties, Manufacture and Applications*; Elsevier: Amsterdam, The Netherlands, 1995.
2. Menges, G.; Knipschild, F. Estimation of mechanical properties for rigid polyurethane foams. *Polym. Eng. Sci.* **1975**, *15*, 623–627. [CrossRef]
3. WorldBioProducts. The Polyurethane Advantage. Available online: http://www.worldbioproducts.com/polyurethane.html (accessed on 16 August 2018).

4. Fort Richard Laboratories. White Paper on Cellulose and Polyurethane Sponges for Surface Sampling. Available online: http://www.fortrichard.com/uploads/resources/Environmental%20Monitoring/White%20Paper%20on%20Use%20of%20Cellulose%20and%20Polyurethane%20Sponge%20for%20EZ%20Reach%20Samplers%20(Updated%2010-25-11).pdf (accessed on 16 August 2018).
5. Services, T.F. Some Properties of Polyurethane Foams. Available online: https://www.technicalfoamservices.co.uk/blog/some-properties-of-polyurethane-foams/ (accessed on 16 August 2018).
6. Cnudde, V.; Boone, M.N. High-resolution X-ray computed tomography in geosciences: A review of the current technology and applications. *Earth Sci. Rev.* **2013**, *123*, 1–17. [CrossRef]
7. Maire, E.; Withers, P.J. Quantitative X-ray tomography. *Int. Mater. Rev.* **2014**, *59*, 1–43. [CrossRef]
8. Long, H.; Swennen, R.; Foubert, A.; Dierick, M.; Jacobs, P. 3D quantification of mineral components and porosity distribution in westphalian c sandstone by microfocus X-ray computed tomography. *Sediment. Geol.* **2009**, *220*, 116–125. [CrossRef]
9. Pardo-Alonso, S.; Solórzano, E.; Brabant, L.; Vanderniepen, P.; Dierick, M.; Van Hoorebeke, L. 3D analysis of the progressive modification of the cellular architecture in polyurethane nanocomposite foams via X-ray microtomography. *Eur. Polym. J.* **2013**, *49*, 999–1006. [CrossRef]
10. Brabant, L.; Vlassenbroeck, J.; De Witte, Y.; Cnudde, V.; Boone, M.N.; Dewanckele, J.; Van Hoorebeke, L. Three-dimensional analysis of high-resolution X-ray computed tomography data with morpho+. *Microsc. Microanal.* **2011**, *17*, 252–263. [CrossRef] [PubMed]
11. Koloushani, M.; Hedayati, R.; Sadighi, M.; Mohammadi-Aghdam, M. CT-based micro-mechanical approach to predict response of closed-cell porous biomaterials to low-velocity impact. *J. Imaging* **2018**, *4*, 49. [CrossRef]
12. Soltani, P.; Zarrebini, M.; Laghaei, R.; Hassanpour, A. Prediction of permeability of realistic and virtual layered nonwovens using combined application of X-ray μct and computer simulation. *Chem. Eng. Res. Des.* **2017**, *124*, 299–312. [CrossRef]
13. Chen, X.; Verma, R.; Espinoza, D.N.; Prodanović, M. Pore-scale determination of gas relative permeability in hydrate-bearing sediments using X-ray computed micro-tomography and lattice boltzmann method. *Water Resour. Res.* **2018**, *54*, 600–608. [CrossRef]
14. Bultreys, T.; Boone, M.A.; Boone, M.N.; De Schryver, T.; Masschaele, B.; Van Hoorebeke, L.; Cnudde, V. Fast laboratory-based micro-computed tomography for pore-scale research: Illustrative experiments and perspectives on the future. *Adv. Water Resour.* **2016**, *95*, 341–351. [CrossRef]
15. Bultreys, T.; Stappen, J.V.; Kock, T.D.; Boever, W.D.; Boone, M.A.; Hoorebeke, L.V.; Cnudde, V. Investigating the relative permeability behavior of microporosity-rich carbonates and tight sandstones with multiscale pore network models. *J. Geophys. Res. B Solid Earth* **2016**, *121*, 7929–7945. [CrossRef]
16. Von der Schulenburg, D.G.; Paterson-Beedle, M.; Macaskie, L.; Gladden, L.; Johns, M. Flow through an evolving porous media—Compressed foam. *J. Mater. Sci.* **2007**, *42*, 6541–6548. [CrossRef]
17. Moser, S.; Nau, S.; Salk, M.; Thoma, K. In situ flash X-ray high-speed computed tomography for the quantitative analysis of highly dynamic processes. *Meas. Sci. Technol.* **2014**, *25*, 025009. [CrossRef]
18. De Kock, T.; Boone, M.A.; De Schryver, T.; Van Stappen, J.; Derluyn, H.; Masschaele, B.; De Schutter, G.; Cnudde, V. A pore-scale study of fracture dynamics in rock using X-ray micro-CT under ambient freeze–thaw cycling. *Environ. Sci. Technol.* **2015**, *49*, 2867–2874. [CrossRef] [PubMed]
19. Van Stappen, J.F.; Meftah, R.; Boone, M.A.; Bultreys, T.; De Kock, T.; Blykers, B.K.; Senger, K.; Olaussen, S.; Cnudde, V. In situ triaxial testing to determine fracture permeability and aperture distribution for CO_2 sequestration in Svalbard, Norway. *Environ. Sci. Technol.* **2018**, *52*, 4546–4554. [CrossRef] [PubMed]
20. Boone, M.; De Kock, T.; Dewanckele, J.; Cnudde, V.; Van Loo, D.; Van de Casteele, E.; De Schutter, G.; Jacobs, P. Four-Dimensional Monitoring of Freeze-Thaw Cycles in Limestone with X-ray Computed Microtomography (micro-CT). In Proceedings of the 12th Euroseminar on Microscopy Applied on Natural Building Stones, Dortmund, Germany, 15–19 September 2009; Available online: https://biblio.ugent.be/publication/752070 (accessed on 19 December 2016).
21. De Schryver, T.; Dierick, M.; Heyndrickx, M.; Van Stappen, J.; Boone, M.A.; Van Hoorebeke, L.; Boone, M.N. Motion compensated micro-CT reconstruction for in-situ analysis of dynamic processes. *Sci. Rep.* **2018**, *8*, 7655. [CrossRef] [PubMed]
22. Brown, K.; Schlüter, S.; Sheppard, A.; Wildenschild, D. On the challenges of measuring interfacial characteristics of three-phase fluid flow with X-ray microtomography. *J. Microsc.* **2014**, *253*, 171–182. [CrossRef] [PubMed]

23. Garrett, P.R.; Meshkov, S.; Perlmutter, G. Oral contrast agents in CT of the abdomen. *Radiology* **1984**, *153*, 545–546. [CrossRef] [PubMed]
24. Descamps, E.; Sochacka, A.; De Kegel, B.; Van Loo, D.; Van Hoorebeke, L.; Adriaens, D. Soft tissue discrimination with contrast agents using micro-CT scanning. *Belg. J. Zool.* **2014**, *144*, 20–40.
25. Pauwels, E.; Van Loo, D.; Cornillie, P.; Brabant, L.; Van Hoorebeke, L. An exploratory study of contrast agents for soft tissue visualization by means of high resolution X-ray computed tomography imaging. *J. Microsc.* **2013**, *250*, 21–31. [CrossRef] [PubMed]
26. Lusic, H.; Grinstaff, M.W. X-ray-computed tomography contrast agents. *Chem. Rev.* **2012**, *113*, 1641–1666. [CrossRef] [PubMed]
27. Cormode, D.P.; Jarzyna, P.A.; Mulder, W.J.; Fayad, Z.A. Modified natural nanoparticles as contrast agents for medical imaging. *Adv. Drug Deliv. Rev.* **2010**, *62*, 329–338. [CrossRef] [PubMed]
28. Zou, J.; Hannula, M.; Misra, S.; Feng, H.; Labrador, R.H.; Aula, A.S.; Hyttinen, J.; Pyykkö, I. Micro CT visualization of silver nanoparticles in the middle and inner ear of rat and transportation pathway after transtympanic injection. *J. Nanobiotechnol.* **2015**, *13*, 5. [CrossRef] [PubMed]
29. Wellington, S.L.; Vinegar, H.J. X-ray computerized-tomography. *J. Pet. Technol.* **1987**, *39*, 885–898. [CrossRef]
30. Van Loo, D.; Bouckaert, L.; Leroux, O.; Pauwels, E.; Dierick, M.; Van Hoorebeke, L.; Cnudde, V.; De Neve, S.; Sleutel, S. Contrast agents for soil investigation with X-ray computed tomography. *Geoderma* **2014**, *213*, 485–491. [CrossRef]
31. Smith, R.W.; Bryg, V. Staining polymers for microscopical examination. *Rubber Chem. Technol.* **2006**, *79*, 520–540. [CrossRef]
32. Allan-Wojtas, P.; Hildebrand, P.; Braun, P.; Smith-King, H.; Carbyn, S.; Renderos, W. Low temperature and anhydrous electron microscopy techniques to observe the infection process of the bacterial pathogen xanthomonas fragariae on strawberry leaves. *J. Microsc.* **2010**, *239*, 249–258. [CrossRef] [PubMed]
33. Priester, J.H.; Horst, A.M.; Van De Werfhorst, L.C.; Saleta, J.L.; Mertes, L.A.; Holden, P.A. Enhanced visualization of microbial biofilms by staining and environmental scanning electron microscopy. *J. Microbiol. Methods* **2007**, *68*, 577–587. [CrossRef] [PubMed]
34. Burstein, S. Reduction of phosphomolybdic acid by compounds possessing conjugated double bonds. *Anal. Chem.* **1953**, *25*, 422–424. [CrossRef]
35. Mahdavi, M.; Ahmad, M.B.; Haron, M.J.; Namvar, F.; Nadi, B.; Rahman, M.Z.A.; Amin, J. Synthesis, surface modification and characterisation of biocompatible magnetic iron oxide nanoparticles for biomedical applications. *Molecules* **2013**, *18*, 7533–7548. [CrossRef] [PubMed]
36. Grabherr, S.; Gygax, E. X ray Contrasting Agent for Post-Mortem, Experimental and Diagnostic Angiography. European Patent EP2526973A2, 12 September 2007.
37. Buffiere, J.Y.; Maire, E.; Adrien, J.; Masse, J.P.; Boller, E. In situ experiments with X ray tomography: An attractive tool for experimental mechanics. *Exp. Mech.* **2010**, *50*, 289–305. [CrossRef]
38. Kyrieleis, A.; Titarenko, V.; Ibison, M.; Connolley, T.; Withers, P. Region-of-interest tomography using filtered backprojection: Assessing the practical limits. *J. Microsc.* **2011**, *241*, 69–82. [CrossRef] [PubMed]
39. Masschaele, B.; Dierick, M.; Loo, D.V.; Boone, M.N.; Brabant, L.; Pauwels, E.; Cnudde, V.; Hoorebeke, L.V. Hector: A 240 kv micro-ct setup optimized for research. *J. Phys. Conf. Ser.* **2013**, *463*, 012012. [CrossRef]
40. Dierick, M.; Van Loo, D.; Masschaele, B.; Van den Bulcke, J.; Van Acker, J.; Cnudde, V.; Van Hoorebeke, L. Recent micro-ct scanner developments at ugct. *Nucl. Instrum. Methods Phys. Res. B* **2014**, *324*, 35–40. [CrossRef]
41. Vlassenbroeck, J.; Dierick, M.; Masschaele, B.; Cnudde, V.; Van Hoorebeke, L.; Jacobs, P. Software tools for quantification of X-ray microtomography. *Nucl. Instrum. Methods Phys. Res. A* **2007**, *580*, 442–445. [CrossRef]
42. Boone, M.; Bultreys, T.; Masschaele, B.; Van Loo, D.; Van Hoorebeke, L.; Cnudde, V. In-Situ, Real Time Micro-CT Imaging of Pore Scale Processes, the Next Frontier for Laboratory Based Micro-CT Scanning. In Proceedings of the 30th International Symposium of the Society of Core Analysts, Snowmass, CO, USA, 21–26 August 2016; Available online: https://biblio.ugent.be/publication/8130807 (accessed on 21 December 2016).
43. Taufiq, A.; Saputro, R.; Hidayat, N.; Hidayat, A.; Mufti, N.; Diantoro, M.; Patriati, A.; Putra, E.; Nur, H. Fabrication of magnetite nanoparticles dispersed in olive oil and their structural and magnetic investigations. In *Proceedings of the IOP Conference Series: Materials Science and Engineering, 2017*; IOP Publishing: Bristol, UK, 2017; p. 012008.

44. Moura, M.; Fiorentino, E.A.; Måløy, K.J.; Schäfer, G.; Toussaint, R. Impact of sample geometry on the measurement of pressure—Saturation curves: Experiments and simulations. *Water Resour. Res.* **2015**, *51*, 8900–8926. [CrossRef]
45. Rapp, B.E. *Microfluidics: Modeling, Mechanics and Mathematics*; William Andrew: Norwich, NY, USA, 2016.

© 2018 by the authors. Licensee MDPI, Basel, Switzerland. This article is an open access article distributed under the terms and conditions of the Creative Commons Attribution (CC BY) license (http://creativecommons.org/licenses/by/4.0/).

Article

Preservation of Bone Tissue Integrity with Temperature Control for In Situ SR-MicroCT Experiments

Marta Peña Fernández [1], Enrico Dall'Ara [2], Alexander P. Kao [1], Andrew J. Bodey [3], Aikaterina Karali [1], Gordon W. Blunn [4], Asa H. Barber [1,5] and Gianluca Tozzi [1,*]

1. Zeiss Global Centre, School of Mechanical and Design Engineering, University of Portsmouth, PO1 3DJ, Portsmouth, UK
2. Department of Oncology and Metabolism and INSIGNEO Institute for in Silico Medicine, University of Sheffield, S1 3DJ, Sheffield, UK
3. Diamond Light Source, Oxfordshire, OX11 0DE, UK
4. School of Pharmacy and Biomedical Sciences, University of Portsmouth, PO1 2DT, Portsmouth, UK
5. School of Engineering, London South Bank University, SE1 0AA, London, UK
* Correspondence: gianluca.tozzi@port.ac.uk; Tel.: +44-(0)-23-9284-2514

Received: 28 September 2018; Accepted: 30 October 2018; Published: 1 November 2018

Abstract: Digital volume correlation (DVC), combined with in situ synchrotron microcomputed tomography (SR-microCT) mechanics, allows for 3D full-field strain measurement in bone at the tissue level. However, long exposures to SR radiation are known to induce bone damage, and reliable experimental protocols able to preserve tissue properties are still lacking. This study aims to propose a proof-of-concept methodology to retain bone tissue integrity, based on residual strain determination using DVC, by decreasing the environmental temperature during in situ SR-microCT testing. Compact and trabecular bone specimens underwent five consecutive full tomographic data collections either at room temperature or 0 °C. Lowering the temperature seemed to reduce microdamage in trabecular bone but had minimal effect on compact bone. A consistent temperature gradient was measured at each exposure period, and its prolonged effect over time may induce localised collagen denaturation and subsequent damage. DVC provided useful information on irradiation-induced microcrack initiation and propagation. Future work is necessary to apply these findings to in situ SR-microCT mechanical tests, and to establish protocols aiming to minimise the SR irradiation-induced damage of bone.

Keywords: bone; X-ray radiation; tissue damage; SR-microCT; digital volume correlation; temperature control

1. Introduction

Bone is a highly heterogenous, anisotropic and hierarchical material that is organised at various levels to optimise its mechanical competence [1]. Thus, it is essential to understand the mechanics of its different components and the structural relationships between them at the different dimensional scales [2–4]. This is of fundamental importance since many musculoskeletal pathologies, such as osteoporosis, are associated with alterations in bone quality at the micro- and nanoscale [5]. Therefore, novel techniques aim at characterising the deformation mechanisms of bone in a three-dimensional (3D) manner, from apparent to tissue level, and establishing their links with bone structure [6–8].

To date, the only experimental method that allows for 3D strain measurements within the bone structure is digital volume correlation (DVC) in combination with in situ microcomputed tomography (microCT) testing [9–11]. DVC has been widely used in bone mechanics to investigate full-field displacement and strain in cortical [12] and trabecular [13,14] bone at different dimensional scales and loading conditions, providing a unique insight to the 3D deformation of such complex material. Nevertheless, in order to characterise bone failure mechanisms at the tissue level, high-resolution microCT is needed [11,15,16]. High-energy synchrotron radiation (SR) microCT has proven to provide fast high-quality image acquisition of bone microstructure with high spatial resolution (~1 µm), and together with in situ mechanical studies, it has allowed for a detailed coupling between 3D bone microstructure and deformation [6,17,18]. Furthermore, recent studies have combined in situ SR-microCT mechanics with DVC to investigate the internal strain and microdamage evaluation of cortical bone [12], trabecular bone [14] and bone-biomaterial systems [19], enhancing the understanding of bone failure at the microscale.

However, it is known that high exposures to SR X-ray radiation lead to a deterioration in the mechanical properties of bone as a consequence of collagen matrix degradation [20,21]. Similarly, ionising radiation, such as gamma rays, commonly used to sterilise bone allografts [22], and X-rays, negatively affects the mechanical and biological properties of the tissue by the degradation of the collagen present in the bone matrix [20,23–27]. Specifically, radiation produces reactive free radicals by the radiolysis of water molecules, which splits the polypeptides chains of the collagen and induces cross-linking reactions, causing collagen denaturation [28–30]. In clinical practice, the adverse effects of gamma radiation during sterilization have been successfully reduced by irradiating the bone while frozen [31,32]. Lowering the temperature is beneficial, as it reduces the mobility of free radicals and, therefore, their ability to interact with collagen molecules [33,34]. Particularly, Hamer et al. [31] observed that cortical bone irradiated at low temperatures (−78 °C) was less brittle and had less collagen damage when compared to the bone irradiated at room temperature. Additionally, Cornu et al. [32] showed that ultimate strength, stiffness and work to failure were not reduced significantly on trabecular bone irradiated under dry ice. In the field of high-resolution X-ray imaging of biological samples, protection against radiation damage is also essential to preserve their integrity. Cryofixation methods have been demonstrated to protect biological samples from visible structural damage and have enabled cryo-soft X-ray tomography (cryo-SXT) to become the only imaging modality able to provide nanoscale 3D information of whole cells in a near-native state [35–37]. However, soft X-rays (~0.1–1 keV) are not able to penetrate bone tissue, nor can they be accommodated for in situ mechanics protocols. Furthermore, cryotechniques involve freeze-drying of the specimens at −150 °C and have been shown to induce microdamage and significantly reduce torsional strength, compressive yield stress and compressive modulus of cortical bone [32,38–40]. Hence, low temperatures positively influence bone preservation during irradiation. However, mechanical testing of bone in such conditions, below the freezing temperature of water, cannot be conducted, as the mechanical properties of bone would be affected. In fact, due to the large water content of bone, ice crystals may cause structural damage to the tissue [23].

Therefore, it is essential to define some guidelines in order to preserve bone tissue integrity and mechanics during in situ SR-microCT experiments. Very recently, DVC applied to SR-microCT images of trabecular bone was used to investigate the influence of SR irradiation-induced microdamage on the bone's apparent mechanics [14]. Microcracks were detected in the bone tissue after long exposures to SR radiation, despite the apparent elastic properties remaining unaltered. Also, high local strain levels were observed that corresponded to the microdamaged areas. However, reducing the total exposure to SR X-ray radiation was able to preserve bone integrity and plasticity. The results of that study [14] provided important information on bone degradation and residual strain accumulation resulting from SR X-ray exposure, but the study had some limitations. Firstly, bone specimens were subjected to cyclic mechanical loading during SR-microCT imaging; thus, the full-field strain measurements were not entirely due to SR irradiation but also to the mechanics. In fact, DVC results showed that even at reduced exposures to SR radiation, there were some regions of high strain concentration, which may have been induced by the mechanical load and further enhanced by the irradiation. Secondly, reducing the total exposure by decreasing the exposure time per projection during SR-microCT acquisition notably decreased image quality and, consequently, DVC performance. Hence, further evaluation and optimisation of the imaging setup is needed in order to preserve bone integrity while maximising image quality for reliable DVC-computed full-field measurement within the bone tissue.

In this context, there is a clear need to define experimental protocols for in situ SR-microCT mechanics able to preserve bone tissue integrity against SR X-ray radiation-induced damage, exploiting the research conducted in different fields. The aim of this study is, therefore, to propose a novel proof-of-concept methodology to retain bone tissue integrity, based on residual strain determination via DVC, by decreasing the environmental temperature during SR-microCT testing.

2. Materials and Methods

2.1. Specimen Preparation

Samples were obtained from a fresh bovine femur. A section (20 mm in thickness) was cut with a hacksaw from the proximal diaphysis of the femur and a diamond-coated core drill was used to extract 4 mm cylindrical compact (n = 2) and 6 mm trabecular (n = 2) bone specimens under constant water irrigation. The ends of the cores were trimmed to achieve a 12 mm length for the compact and a 16 mm length for the trabecular bone specimens. Brass endcaps were used to embed the ends of the specimens (~2 mm), ensuring perpendicularity between the bone cores and the endcap bases. Samples were kept frozen at −20 °C and thawed for approximately 2 h in saline solution at room temperature before imaging.

2.2. SR-MicroCT Imaging

SR-microCT was performed at the Diamond-Manchester Imaging Branchline I13-2 (Figure 1a) of Diamond Light Source (DLS), Oxfordshire, UK. A partially coherent polychromatic 'pink' beam (5–35 keV) of parallel geometry was generated by an undulator from an electron storage ring of 3.0 GeV. The undulator gap was set to 5 mm for data collection and, to limit bone damage, 11 mm for low-dose alignment. The beam was reflected from the platinum stripe of a grazing-incidence focusing mirror and high-pass filtered with 1.4 mm pyrolytic graphite, 3.2 mm aluminium and 50 μm steel. The propagation (sample-to-scintillator) distance was approximately 40 mm. Images were recorded by a sCMOS (2560 × 2160 pixels) pco.edge 5.5 (PCO AG, Kelheim, Germany) detector which was coupled to a 500 μm-thick $CdWO_4$ scintillator and a visual light microscope with a 4× objective lens, providing a total magnification of 8×. This resulted in an effective voxel size of 0.81 μm and a field of view of 2.1 × 1.8 mm². A total of 1801 projection images were collected over 180° of continuous rotation ('fly-scan'), with an exposure time of 512 ms per projection (11 ms overhead per exposure), adopting the imaging conditions reported in [14]. The total scanning time was approximately 15 min. The projection images were flat-field- and dark-field-corrected prior to image reconstruction using SAVU [41], which incorporated ring artefact suppression and optical distortion correction [42]. Each specimen underwent five full consecutive tomographic data collections.

2.3. In Situ Testing and Temperature Control

Specimens were placed within an in situ testing device (CT5000-TEC, Deben, Bury Saint Edmunds, UK) and kept in saline solution during image acquisition (Figure 1a). The device is equipped with a 5 kN load cell, Peltier heated and cooled jaws with a temperature range from −20 °C to +160 °C and an environmental chamber. A small preload (2–5 N) was first applied to ensure good end-contact and avoid motion artefacts during tomographic acquisition, after which the actuator was stopped, and the jaws' positions held throughout the test. Bone specimens (N = 1 compact and N = 1 trabecular) were imaged at room temperature ($T_{room} \approx 23$ °C) and at ~0 °C (N = 1 compact and N = 1 trabecular) by cooling and keeping the Peltier jaws at the target temperature. A thermocouple (Type K, RS Pro, RS Components, Corby, UK) was also attached to the surface of the bone samples and was used during the in situ test to monitor the temperature directly at the tissue during image acquisition and between tomographies. Temperature measurements and recordings were processed with a thermocouple data logger (USB TC-08, Pico Technology, St Neots, UK). For reliable temperature measurements, the thermocouple was calibrated prior to the experiment.

2.4. Image Post-Processing

Five datasets were obtained for each specimen and further processed using Fiji platform [43]. After image reconstruction, each 3D dataset consisted of 2000 images (2400 × 2400 pixels) with 32-bit grey-levels. Images were converted to 8-bit greyscale and cropped to parallelepipeds (volume of interest (VOI)) with a cross-section of 1400 × 1400 pixels (1.134 × 1.134 mm^2) and a height equal to 1800 pixels (1.46 mm) in the centre of the scanned volume (Figure 1b,c). Noise in the images was reduced by applying a nonlocal means filter [44], where the variance of the noise was automatically estimated for each dataset [45]. The five consecutive scans per specimen were first rigidly registered using the first acquired dataset as a reference. The 3D rigid registration was based on sum of squares differences as a similarity measurement between the reference and each target image. Finally, the filtered VOIs were masked by setting to zero-intensity the non-bony voxels (i.e., Haversian and Volkmann's canals in compact bone and bone marrow space in trabecular bone). A binary image (value of one for bone voxel and zero elsewhere) was first created using Otsu's threshold algorithm followed by a despeckling filter to remove 3D regions less than three voxels in volume both in white and black areas, which are mainly related to nonfiltered noise. Additionally, isolated pixels were removed, and small holes were filled by using a series of morphological operations as described in [16]. The quality of the binary images was checked by visual inspection. Masked images, with the original greyscale value in the bony voxels and zero elsewhere, were obtained by multiplying the filtered image with the final binary image (Figure 1d,e).

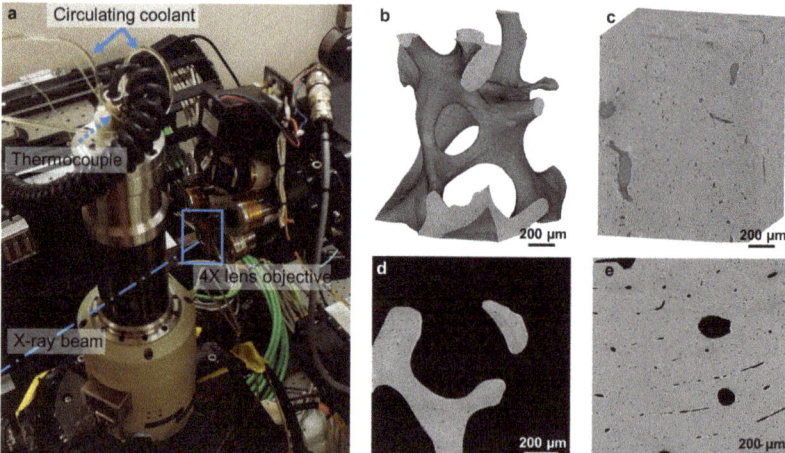

Figure 1. (**a**) Experimental setup at I13-2 beamline. The direction of the beam is indicated by the dashed-dotted line. Specimens were scanned within a loading device using a 4× lens objective. The temperature in the device was controlled with a circulating coolant and monitored on the tissue via an additional thermocouple attached to the surface of the specimens. SR-microCT reconstructed volume of interest (VOI) (1.13 × 1.13 × 1.46 mm^3) analysed for (**b**) trabecular and (**c**) compact bones with an effective voxel size of 0.81 µm. Two-dimensional cross-section through the middle of the VOI after masking the bone marrow (**d**) from the trabecular bone and the Haversian and Volkmann's canals (**e**) from the compact bone.

2.5. Digital Volume Correlation

Digital volume correlation (DaVis v10.0, LaVision, Göttingen, Germany) was carried out to evaluate the residual strain in the bone tissue due to progressive damage induced by X-ray exposure to SR radiation during SR-microCT at different temperatures. DaVis software is based on a local approach of correlation, which has been widely used in bone mechanics [13,14,46]. Details on the operating principles of the software are reported elsewhere [16,47]. DVC was applied to the masked images to avoid large strain artefacts in regions with insufficient greyscale pattern (i.e., bone marrow) [16]. A different multi-pass scheme was used for the DVC computation on compact and trabecular specimens after an evaluation of the baseline strains in the first two consecutive tomograms for the four specimens, obtained in a nominal 'zero-strain' state, where the irradiation-induced damage was considered minimal (Supplementary Materials). A final subvolume of 32 voxels, reached via successive (predictor) passes using subvolumes of 112, 56, 48 and 40 voxels, was used for the compact bone, whereas, for the trabecular bone, a final subvolume of 64 voxels, reached via successive passes of 112, 88, 80 and 72 voxels, was adopted. Given the voxel size of the SR-microCT images, the final DVC measurement spatial resolution was 25.9 µm for compact and 51.8 µm for trabecular bones. Additionally, in both cases, subvolumes with a correlation coefficient below 0.6 were removed from the resultant displacement vectors to avoid artefacts due to poor correlation. The different processing schemes for both bone typologies mainly depended on the higher number of features (i.e., osteocyte lacunae) available in the compact bone specimens compared to the trabecular ones, which allowed a smaller subvolume size to be used for the former [11]. To evaluate the 3D full-field residual strain distribution in the bone tissue over time in relation to the damage induced by continuous X-ray exposure to SR radiation, DVC was performed by registering the reference image (first acquired tomogram) with each of the remaining tomograms. First (ε_{p1}) and third (ε_{p3}) principal strains and maximum shear (γ_{max}) strain were computed within the bone volume after a bicubic interpolation of the measured strain. Furthermore, in order to couple the initiation and propagation of microcracks in the tissue with the displacement and first principal strain directions, dedicated MATLAB (v2018a, MathWorks, Natick, MA, USA) scripts were developed. The MATLAB scripts allow for the representation of any set of orthogonal slices within the volume and for the computation of the displacement and first principal strain values and their corresponding direction for each subvolume.

3. Results

3.1. In Situ Testing and Temperature Control

Temperature readings from the thermocouple attached to the surface of the bone specimens suggested a consistent temperature gradient ($\Delta T = 0.4$ °C) at each exposure period (Figure 2a) corresponding to the opening (rise in temperature) and closing (drop in temperature) of the X-ray shutter. Small fluctuations in the temperature were recorded once the X-ray shutter was open, as they are more evident during tomographic acquisition compared to the steady position. However, those fluctuations were far less important than the temperate gradients recorded between consecutive tomographies. The stress-relaxation curves recorded during in situ testing showed that the X-ray beam significantly influenced the relaxation behaviour of the trabecular bone specimen at room temperature (Figure 2b). A consistent increase in the force was recorded after the start of each tomographic acquisition. This trend was not observed for the compact bone specimen.

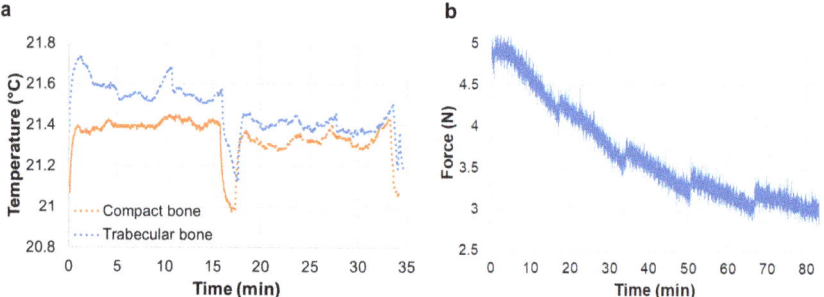

Figure 2. (a) Temperature readings measured using a thermocouple attached to the compact and trabecular bone surface at room temperature. The solid line corresponds to thermocouple readings during ~15 min with the X-ray shutter opened and the thermocouple in the beam path. Dotted lines represent thermocouple readings during tomographic acquisition. The sudden drop and consequent rise in temperature coincide with the closing and opening of the X-ray shutter. (b) Force readings in trabecular bone specimen at room temperature during five consecutive tomograms. An increase in the force was observed and corresponded with the opening of the X-ray shutter.

3.2. Compact Bone

No damage was visually detected in the compact bone specimens after five tomograms, either at room temperature or 0 °C. The residual ε_{p1} distribution (Figure 3) did not show any notable changes in the tissue after the acquisition of two (Figure 3a) and five (Figure 3b) tomograms, with some localised areas of higher residual strain in the specimen imaged at room temperature. The strain histograms (Figure 3c) showed peak values below 1000 µε for both specimens, and no clear trends were observed between exposure to SR radiation and peak strain values. However, histograms showed tails with higher strains after five tomograms at room temperature compared to 0 °C. Similar findings were observed for the residual ε_{p3} and γ_{max} (Figure S2), suggesting a strain redistribution between consecutive tomographies, which did not cause important damage overall. Highly strained regions of the specimen tested at room temperature were localised around the Haversian and Volkmann's canals (Figure 4). Residual strain in a region of approximately 20 µm surrounding the canals was compared to the strain in the internal bone matrix volume. Particularly, the cumulative histograms of γ_{max} (Figure 4e) after two and five acquired tomograms showed slightly higher strains around the canals for the same bone volume percentage.

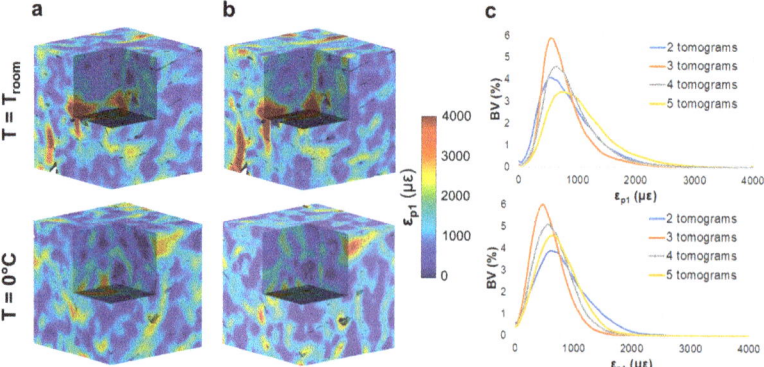

Figure 3. Three-dimensional first principal strain (ε_{p1}) distribution in compact bone tissue imaged at room temperature (top) and 0 °C (bottom) after two (**a**) and five (**b**) acquired tomograms. A representative cube (~1 mm^3) in the centre of the analysed VOI is represented. Histograms of the residual strain distribution (**c**) in the tissue are shown for all the acquired tomograms.

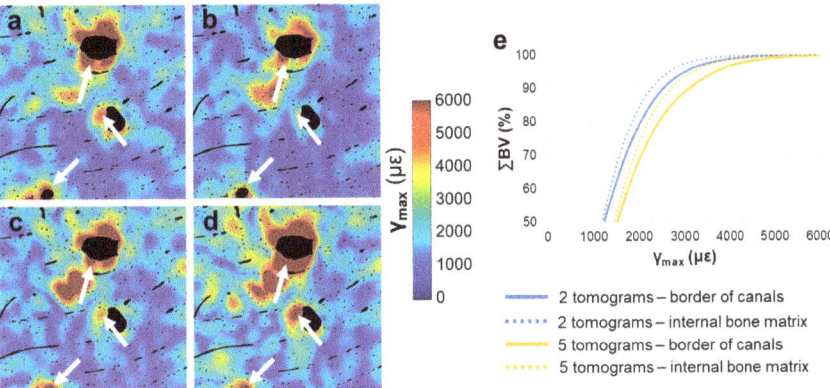

Figure 4. Maximum shear strain (γ_{max}) distribution in compact bone tissue imaged at room temperature. Cross-sections in 2D are shown after (**a**) two, (**b**) three, (**c**) four and (**d**) five acquired tomograms. Arrows indicate highly strained regions. A cumulative histogram of the residual strain (**e**) in the tissue voxels around the canals (solid lines) and the remaining bone matrix (dotted lines) is shown for after two and five tomograms.

3.3. Trabecular Bone

A visual inspection of the reconstructed images showed the presence of several microcracks after five tomograms, corresponding to ~80 min of total exposure to SR X-ray radiation, in the trabecular bone specimen at room temperature. However, decreasing the temperature to 0 °C facilitated tissue preservation, as microdamage was not observed. Furthermore, the high levels of residual strain measured with DVC correlated well with the microdamage visible from the images. The histograms of residual strain distributions (Figure 5) after each tomogram highlighted the differences between the two trabecular bone specimens. On one hand, the specimen imaged at room temperature showed a consistent increase in residual strain when increasing the exposure to X-ray radiation (Figure 5a–c). This trend was clearly observed in ε_{p1} (Figure 5a), for which strain peak values increased from ~1500 to ~3000 µε after two and five consecutive scans, respectively. ε_{p3} (Figure 5b) peak values were found to be below −1500 µε, whereas peak γ_{max} (Figure 5c) ranged from ~2000 µε to ~3500 µε after two and five tomograms, respectively. The residual strain accumulation was less evident for the trabecular bone specimen maintained at 0 °C (Figure 5d–f). In fact, peak strain values remained below ±1000 µε for ε_{p1} (Figure 5d) and ε_{p3} (Figure 5 e), respectively, and below 2000 µε for γ_{max} (Figure 5f) after five tomograms. The 3D full-field strain distribution in the trabecular bone (Figure 6) was accumulated in the tissue after each tomogram. In particular, for the specimen at room temperature (Figure 6, top), it could be seen that ε_{p1} was increasing after each tomography, and regions of high residual strains after two full tomographies (Figure 6a, top) were progressively enlarged, reaching strain values of over 4000 µε after five tomograms (Figure 6d, top). This strain accumulation was less pronounced in the specimen at 0 °C (Figure 6, bottom), although some areas of high strain concentration were observed after each tomogram. Furthermore, some strain redistributions could be seen after three (Figure 6b, bottom) and four (Figure 6c, bottom) full tomographies.

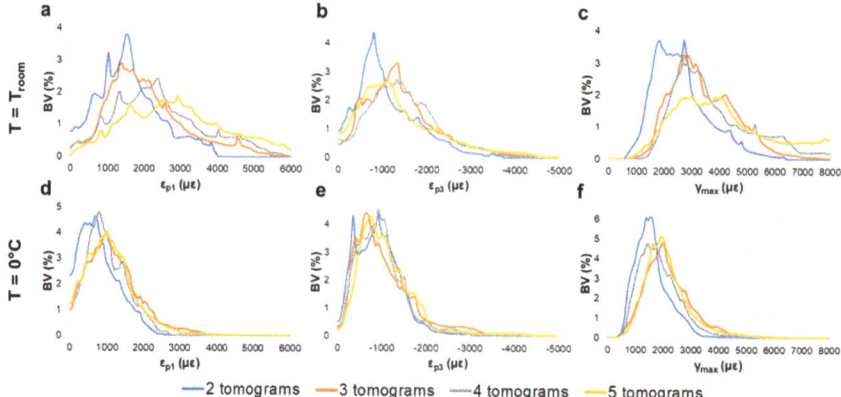

Figure 5. Histograms of the residual strain distribution in trabecular bone tissue imaged at room temperature (top) and 0 °C (bottom). (**a,d**) First principal strains (ε_{p1}), (**b,e**) third principal strains (ε_{p3}) and (**c,f**) maximum shear strains (γ_{max}) after each acquired tomogram are shown.

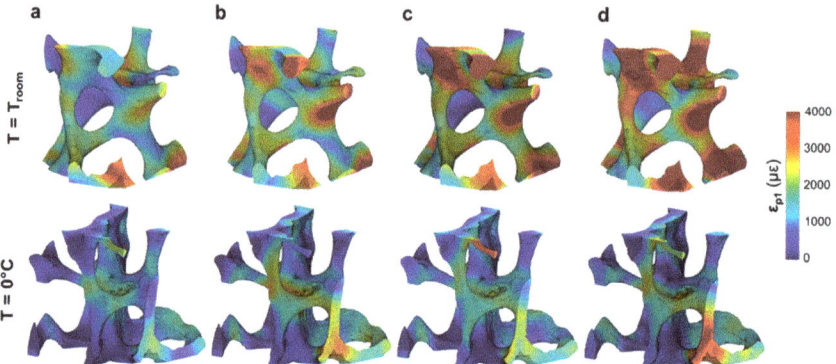

Figure 6. Three-dimensional full-field first principal strain (ε_{p1}) distribution in trabecular bone tissue imaged at room temperature (top) and 0 °C (bottom) after the acquisition of (**a**) two, (**b**) three, (**c**) four and (**d**) five consecutive tomograms. A representative cube (~1 mm^3) in the centre of the analysed VOI is represented.

3.4. Tracking of Crack Formation

Microcracks were clearly visible in the trabecular bone specimen imaged at room temperature after five tomograms (Figure 7a,b). A region inside a trabecula (Figure 7b) was tracked during the in situ test to couple the residual strain accumulation with the crack formation. The displacement field around the damaged region (Figure 7c–f) suggested a relative motion between regions at both sides of the cracks since the earliest stages, before cracking was visible (Figure 7c–e). In fact, low displacements were found on one side, and those were mainly directed toward the positive z-direction, whereas, in the neighbouring side, displacements were progressively increased and reoriented toward the negative z-direction. After cracking (Figure 7f), displacements further increased around the crack, and a pronounced reorientation of their direction was observed. A deeper look at the displacement in the orthogonal planes (Figure 8), before and after crack formation, evidenced the discontinuities in the displacement field in proximity to the crack. Particularly, before crack formation (Figure 8a), displacement showed a high misorientation in the XY and XZ planes. After cracking (Figure 8b), the displacement field at one end of the crack was found perpendicular to the crack direction (XY plane), whereas it seemed aligned with the crack on the other end, which may indicate the further propagation direction. Both ε_{p1} and γ_{max} showed a progressive increase in the microcracked region, reaching values above 4000 µε for ε_{p1} (Figure 9b) and approximately 5000 µε for γ_{max} (Figure 10b) in the damaged area. In general, tensile strains were the most correlated to microdamage detection. In fact, the directions of ε_{p1} (Figure 9) suggested a combination of tensile and shear modes of crack formation. In addition, the principal directions before cracking seemed to be highly disordered throughout the analysed volume. In particular, the highlighted vectors before cracking (Figure 9a) exhibited a very abrupt change in orientation, whereas the same areas after cracking (Figure 9b) were considerably aligned with the microcrack. γ_{max} (Figure 10) increased after crack formation, and discontinuities at both sides of the crack were observed (Figure 10b). Moreover, higher shear strain levels were found at one side of the crack (XY plane), which also corresponded to principal strains and displacements perpendicular to the crack, thus possibly suggesting the direction of crack propagation.

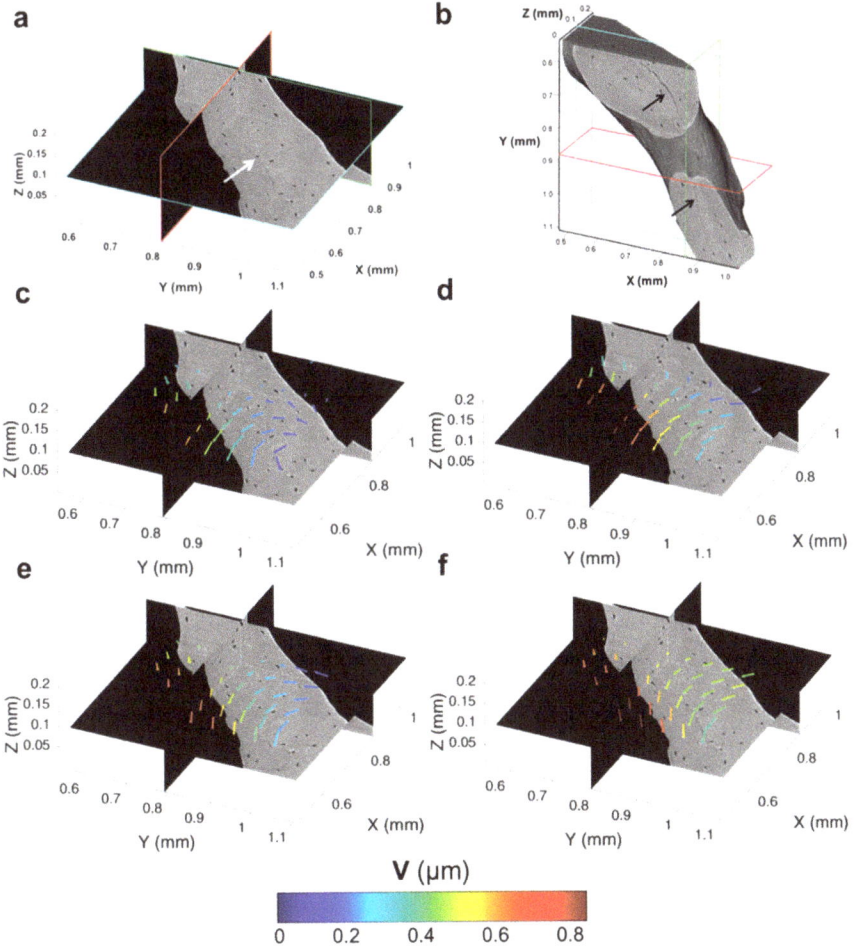

Figure 7. Microcrack tracking in trabecular bone tissue imaged at room temperature. (**a**) Representative orthoslices and (**b**) 3D representation of the trabecular bone region tracked over time. Arrows indicate the microcracks visible in the tissue. (**c**–**f**) Digital volume correlation (DVC)-computed displacement field (V) in each subvolume on the analysed region of interest around a microcrack at different time points corresponding to the acquisition of (**c**) two, (**d**) three, (**e**) four and (**f**) five tomograms. Vector lengths are identical, and the colour code refers to the V magnitude in micrometres.

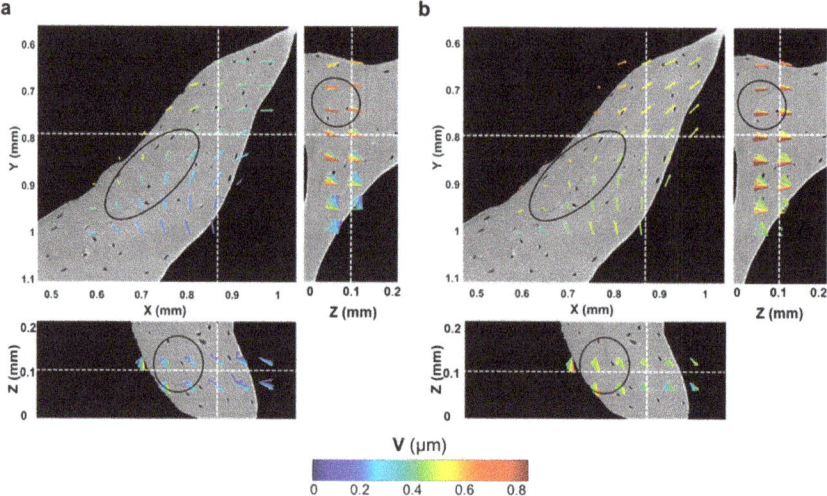

Figure 8. DVC-computed displacement field through the region of interest analysed around the microcracked area (**a**) before (fourth tomogram) and (**b**) after cracking was visible (fifth tomogram). Oval regions highlight damaged areas of bone tissue. Vector lengths are identical, and the colour code refers to the displacement vector length (V) in micrometres.

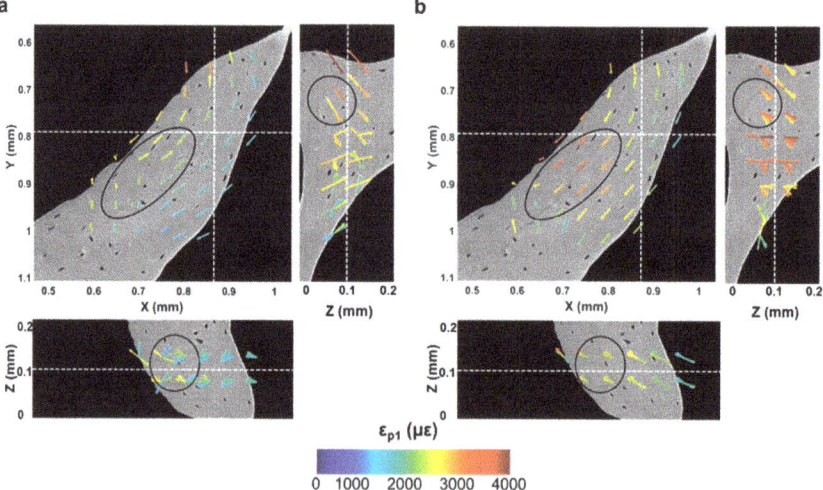

Figure 9. DVC-computed first principal strain (ε_{p1}) through the region of interest analysed around the microcracked area (**a**) before (fourth tomogram) and (**b**) after cracking was visible (fifth tomogram). Vectors indicate first principal strain directions in each subvolume. Oval regions highlight damaged areas of bone tissue, which correspond to high orientation changes in the principal strain direction before and after cracking. Vector lengths are identical, and the colour code refer to the $\varepsilon p1$ magnitude.

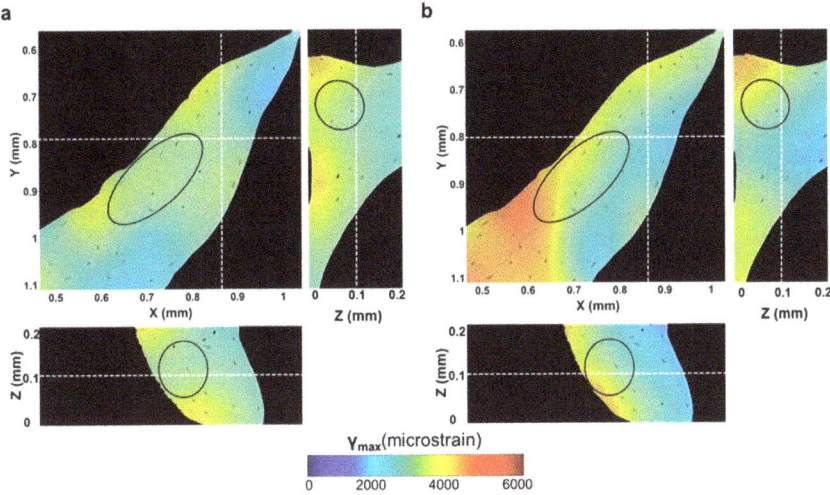

Figure 10. DVC-computed maximum shear strain (γ_{max}) through the region of interest analysed around the microcracked area (**a**) before (fourth tomogram) and (**b**) after cracking was visible (fifth tomogram). Oval regions highlight damaged areas of bone tissue, which correspond to an increase in shear strain values before and after cracking. High discontinuities in shear strains were identified in the damage region (**b**), which may suggest the direction of crack propagation.

4. Discussion

The proof-of-concept experiment reported herein enabled important understanding of the SR X-ray radiation-induced damage to the integrity of bone tissue. The residual strain accumulation caused by SR X-ray radiation was quantified for the first time using DVC applied to in situ SR-microCT images, and the effect of the environmental temperature on the SR irradiation-induced damage in bone tissue was addressed. It is known that irradiation has a deleterious effect on the structural and mechanical properties of bone as a result of collagen matrix degradation due to the formation of collagen cross-links and eventual rupture of the collagen fibres [20,27]. Several studies have addressed the effect of high-energy SR X-ray radiation on the mechanical properties of bone [20,21,26,48], and safe dose values (35 kGrays) were defined to preserve bone mechanics [20]. However, during in situ SR-microCT studies, a reduction of the dose is related to a reduction in the total exposure to SR radiation and, therefore, the signal-to-noise ratio of the acquired tomograms, with a consequent reduction in image quality and increased DVC errors [14]. Therefore, new protocols need to be defined in order to preserve bone tissue while maintaining good image quality. Furthermore, whether bone integrity can be preserved by controlling the temperature during in situ SR experiments still remains unexplored.

The overall change in temperature during image acquisition was minimal ($\Delta T = 0.4\,°C$) (Figure 2a) and in line with previous reports on SR beam heating [49,50]. Wallander and Wallentin [51] showed that X-ray-induced heating can lead to significant temperature increase (i.e., nanowire at 8 °C above room temperature) at typical synchrotron beamline fluxes. As a strategy for reducing the X-ray-induced heating, it was suggested to improve the heat transfer of the target material to the surroundings, for example, by immersing the samples in liquid [51]. However, it still remains unclear whether that thermal gradient in a very short period of time (opening/closing of the beam shutter) may induce collagen degradation. As specimens were held between the loading stage platens during the in situ test, the effect of the X-ray beam on the stress-relaxation behaviour of the specimens could be observed (Figure 2b), similar to the data reported in [52]. With only a fixed preload applied, an increase in the load was identified for the trabecular bone specimen at each cyclic period that corresponded with the opening of the X-ray shutter. Both trabecular and compact bone exhibit a highly viscoelastic behaviour;

however, this is more evident for trabecular bone due to the large content of bone marrow in its cavities. Thus, the loadcell of the loading stage was not accurate enough to capture any changes in the stress-relaxation behaviour for the compact bone specimen. Heat causes a transformation of the collagen molecule, known as the collagen shrinkage phenomenon [53], whereby the collagen molecule develops a contractile force that is held constant [54,55] at a given temperature (shrinkage temperature). This shrinkage behaviour is related to the cross-links in the collagen and its stability [53]. Even though the specimens in the current study were kept at a constant temperature (~23 °C), the beam-induced temperature rise of 0.4 °C may contribute to the activation of a similar contractile force, which is a clear indicator of the harmful effects of the SR irradiation on bone tissue.

The results obtained from the current study have shown that reducing the temperature to 0 °C notably reduced the irradiation-induced microdamage and residual strain in trabecular bone specimens (Figure 6). However, minimal effect was observed for compact bone (Figure 3). Nguyen et al. [30] reported that the mechanical properties of compact bone were decreased by a lower dose than that affecting trabecular bone. However, it has been shown here (Figure 3) that the structural integrity of compact bone tissue was not compromised, as microcracks were not detected as in the trabecular bone tissue. In any case, specimens were not mechanically tested; thus, whether the regions of high strain concentration found in compact bone (Figure 3) influence the mechanical properties is still unknown. Furthermore, Peña Fernández et al. [14] showed that the presence of microcracks was not always related to changes in the apparent elastic properties of the irradiated bone.

Although the overall residual strain in compact bone imaged at room temperature was low, with peak strain values below 1000 µε for ε_{p1} (Figure 3), some highly strained regions were identified in close proximity to Haversian and Volkmann's canals (Figure 4). Canals and osteocyte lacunae are known to act as stress concentrating features in specimens subjected to mechanical load [6,12]; however, the effect of irradiation on these specific sites has never been considered. Haversian canals contain unbound water [56], and as ionising radiation produces the release of free radicals via radiolysis of water molecules [29], it is expected that a larger number of free radicals, which could interact with the collagen and induce cross-linking reactions, are found in proximity to the canals due to the higher water content.

Lowering the environmental temperature to 0 °C had a positive effect on the DVC-measured residual strain in trabecular bone, which showed a peak principal strain value below 1000 µε (Figure 5); furthermore, no microdamage was visually detected on the reconstructed tomograms. These results are consistent with medical studies on the effect of gamma irradiation, where it was shown that irradiating bone specimens while frozen did not affect the mechanical properties of bone [31,34]. In fact, decreasing the temperature reduces the mobility of the water, and, therefore, decreases the mobility of highly reactive oxygen free radicals produced by high-energy X-ray radiation. Impairing that mobility protects the collagen by reducing cross-linking reactions within its molecules [57,58]. The effect of freezing on the mechanical properties of bone has been previously studied [59–62] and no statistical differences were found after freezing, nor after several freeze-thaw cycles [63,64]. It should be noted that, during the proposed experiment, specimens were immersed in saline solution at 0 °C, and ice crystals, which may cause structural damage to the tissue [63], were not observed at any stage of the experiment.

The irradiation-induced damage in the trabecular bone imaged at room temperature resulted in microcracks that were visible in the tissue even if the specimen was not subjected to any mechanical load. At the nanoscale, SR irradiation-induced free radical attack of the collagen network results in a cross-linking reaction that degrades the structural integrity of the collagen fibres [20,29,30]. Previous studies using atomic force microscopy have shown that crack formation and bone fracture occur between the mineralised collagen fibrils. Fantner et al. [65] proposed that the mineralised collagen fibres are held together by a nonfibrillar organic matrix that acts as a glue. The glue resists the separation of the mineralised collagen fibrils, avoiding the formation of cracks, when a load is applied to the bone. During the formation of microcracks, work that stretches the glue molecules would be required to separate the mineralised collagen fibrils. Irradiation may affect that mechanism by damaging the sacrificial bonds, as a result of the observed shrinkage behaviour, which could lead to the rupture of those bonds after prolonged exposure to irradiation and consequent microcrack formation. At the macroscale, DVC-computed displacements (Figure 7) suggest a vortex motion around the microcracked region, which results in a shrinkage process of the material and the formation of a microcrack that follows an unusual pattern in fracture mechanics. The denaturation of the collagen may not be homogeneous throughout the bone tissue; therefore, crack propagation would follow the degeneration process of the collagen.

DVC was successfully used to understand crack formation and propagation in bone. Christen et al. [12] investigated the initiation and propagation of microcracks in cortical bone using DVC; however, full-field displacements and strains were only evaluated in terms of magnitude, but the directions were not explored. Additionally, specimens were pre-cracked before mechanical testing; thus, crack initiation and propagation was expected around the notch region. In this study, microcracks were not induced by mechanical loading, but by SR irradiation instead. Discontinuities in the displacement field (Figure 8a) corresponded to high-orientation changes in the strain field (Figure 9a) that could indicate crack formation. Furthermore, perpendicularity of displacement (Figure 8b) and principal strains (Figure 9b) to the crack might be related to a crack propagation front. Similar crack formation mechanisms were observed in clay deformation using digital image correlation (DIC) following desiccation [66,67]. Like the results herein reported, in opening mode, the direction of the crack was perpendicular to that of ε_{p1}, whereas, for cracks in mixed opening-sliding mode, ε_{p1} was found parallel to the direction of the crack (Figure 9b). Those studies [66,67] concluded that cracks formed a network which is found after thermal shocks, and the authors emphasized the need to develop a multiscale approach to better understand crack formation and propagation. Similar to those findings, irradiation-induced microcracks need to be further investigated at different dimensional levels to properly understand the formation mechanisms.

This study has some limitations. First, only one specimen per bone type was tested at each temperature, and the mechanical properties of the bone specimens were not evaluated after irradiation. Residual strain maps suggested that a decrease in the temperature had a beneficial effect on preserving bone integrity and mechanics, but specimens were maintained far below physiological conditions (~37 °C); thus, it could be argued that the mechanical properties of bone tissue could have been altered. Further analyses are needed to properly assess the effect of the environmental temperature during in situ SR-microCT experiments, translating the findings of the proposed methodology to in situ SR-microCT bone mechanics. Moreover, a combination of techniques at different dimensional scales would enhance the knowledge of the irradiation-induced damage in bone tissue.

5. Conclusions

The 3D full-field residual strain distribution of compact and trabecular bone subjected to high-energy SR irradiation was computed using DVC applied to SR-microCT images acquired at different temperatures. Lowering the temperature during irradiation to only 0 °C had a positive effect on trabecular bone tissue, which—unlike such bone imaged at room temperature—did not present visible microcracks, and residual strain values were not increased with further radiation. However, a minimal effect was observed in compact bone. A shrinkage behaviour induced by both the beam-induced temperature and high-energy irradiation may well be the source of the irradiation-induced damage and microcracks in bone tissue. DVC applied to high-resolution SR-microCT images has proven to be a useful tool for understanding crack formation and propagation in bone tissue. Further work is needed to clearly establish protocols for the application of SR-microCT to the in situ mechanics of bone and potentially extend the knowledge to other biological tissues in order to minimise SR irradiation-induced damage.

Supplementary Materials: The following are available online at http://www.mdpi.com/1996-1944/11/11/2155/s1, Figure S1: Relationship between (**a**) MAER and (**b**) SDER with the sub-volume size for the four bone specimens, Figure S2: Histograms of the residual strain distribution in compact bone tissue imaged at room temperature (top) and 0 °C (bottom). (**a**) Third principal strains (ε_{p3}) and (**b**) maximum shear strains (γ_{max}) after each acquired tomogram are shown, Table S1: Random errors for the three displacement components for compact and trabecular bone specimens. Median values of the two specimens per group are shown.

Author Contributions: Conceptualization, M.P.F., E.D., G.T.; Methodology, M.P.F., E.D., G.T.; Validation, M.P.F., G.T.; Formal Analysis, M.P.F., A.J.B.; Investigation, M.P.F., A.P.K., A.J.B., A.K.; Resources, G.W.B., G.T.; Data Curation, M.P.F.; Writing-Original Draft Preparation, M.P.F.; Writing-Review & Editing, M.P.F., E.D., A.P.K., A.J.B., G.T.; Visualization, M.P.F.; Supervision, G.T.; Project Administration, G.T.; Funding Acquisition, A.H.B., G.W.B., G.T.

Funding: The project was partially funded by the Engineering and Physical Sciences Research Council (EPSRC, Frontier Grant Multisim, EP/K03877X/1).

Acknowledgments: The authors would like to thank Diamond Light Source (UK) for time at the Diamond-Manchester Imaging Branchline I13-2 and its associated Data Beamline [68] under proposal MT16497, and the Zeiss Global Centre (University of Portsmouth) for post-processing. We further acknowledge Dave Hollis (LaVision Ltd.) for assistance with DaVis software, Kazimir Wanelik for help during the experiment and image reconstruction at Diamond Light Source, and Kamel Madi for fruitful discussions during the experiment.

Conflicts of Interest: The authors declare no conflict of interest.

References

1. Rho, J.Y.; Kuhn-Spearing, L.; Zioupos, P. Mechanical properties and the hierarchical structure of bone. *Med. Eng. Phys.* **1998**, *20*, 92–102. [CrossRef]
2. Fratzl, P.; Gupta, H.S.; Paschalis, E.P.; Roschger, P. Structure and mechanical quality of the collagen–mineral nano-composite in bone. *J. Mater. Chem.* **2004**, *14*, 2115–2123. [CrossRef]
3. Currey, J.D. How well are bones designed to resist fracture? *J. Bone Miner. Res.* **2003**, *18*, 591–598. [CrossRef] [PubMed]
4. Weiner, S.; Traub, W.; Wagner, H.D. Lamellar bone: Structure–function relations. *J. Struct. Biol.* **1999**, *126*, 241–255. [CrossRef] [PubMed]
5. de Bakker, C.M.J.; Tseng, W.-J.; Li, Y.; Zhao, H.; Liu, X.S. Clinical Evaluation of Bone Strength and Fracture Risk. *Curr. Osteoporos. Rep.* **2017**, *15*, 32–42. [CrossRef] [PubMed]
6. Voide, R.; Schneider, P.; Stauber, M.; Wyss, P.; Stampanoni, M.; Sennhauser, U.; van Lenthe, G.H.; Müller, R. Time-lapsed assessment of microcrack initiation and propagation in murine cortical bone at submicrometer resolution. *Bone* **2009**, *45*, 164–173. [CrossRef] [PubMed]
7. Wolfram, U.; Wilke, H.J.; Zysset, P.K. Damage accumulation in vertebral trabecular bone depends on loading mode and direction. *J. Biomech.* **2011**, *44*, 1164–1169. [CrossRef] [PubMed]
8. Li, S.; Demirci, E.; Silberschmidt, V.V. Variability and anisotropy of mechanical behavior of cortical bone in tension and compression. *J. Mech. Behav. Biomed. Mater.* **2013**, *21*, 109–120. [CrossRef] [PubMed]
9. Bay, B.K.; Smith, T.S.; Fyhrie, D.P.; Saad, M. Digital volume correlation: Three-dimensional strain mapping using X-ray tomography. *Exp. Mech.* **1999**, *39*, 217–226. [CrossRef]
10. Grassi, L.; Isaksson, H. Extracting accurate strain measurements in bone mechanics: A critical review of current methods. *J. Mech. Behav. Biomed. Mater.* **2015**, *50*, 43–54. [CrossRef] [PubMed]
11. Dall'Ara, E.; Peña-Fernández, M.; Palanca, M.; Giorgi, M.; Cristofolini, L.; Tozzi, G. Precision of DVC approaches for strain analysis in bone imaged with µCT at different dimensional levels. *Front. Mater.* **2017**, *4*, 31. [CrossRef]
12. Christen, D.; Levchuk, A.; Schori, S.; Schneider, P.; Boyd, S.K.; Müller, R. Deformable image registration and 3D strain mapping for the quantitative assessment of cortical bone microdamage. *J. Mech. Behav. Biomed. Mater.* **2012**, *8*, 184–193. [CrossRef] [PubMed]
13. Gillard, F.; Boardman, R.; Mavrogordato, M.; Hollis, D.; Sinclair, I.; Pierron, F.; Browne, M. The application of digital volume correlation (DVC) to study the microstructural behaviour of trabecular bone during compression. *J. Mech. Behav. Biomed. Mater.* **2014**, *29*, 480–499. [CrossRef] [PubMed]

14. Peña Fernández, M.; Cipiccia, S.; Bodey, A.J.; Parwani, R.; Dall'Ara, E.; Blunn, G.; Pani, M.; Barber, A.H.; Tozzi, G. Effect of SR-microCT exposure time on the mechanical integrity of trabecular bone using in situ mechanical testing and digital volume correlation. *J. Mech. Behav. Biomed. Mater.* **2018**, *88*, 109–119. [CrossRef] [PubMed]
15. Palanca, M.; Bodey, A.J.; Giorgi, M.; Viceconti, M.; Lacroix, D.; Cristofolini, L.; Dall'Ara, E. Local displacement and strain uncertainties in different bone types by digital volume correlation of synchrotron microtomograms. *J. Biomech.* **2017**, *58*, 27–36. [CrossRef] [PubMed]
16. Peña Fernández, M.; Barber, A.H.; Blunn, G.W.; Tozzi, G. Optimisation of digital volume correlation computation in SR-microCT images of trabecular bone and bone-biomaterial systems. *J. Microsc.* **2018**, *00*, 1–16. [CrossRef]
17. Thurner, P.J.; Wyss, P.; Voide, R.; Stauber, M.; Stampanoni, M.; Sennhauser, U.; Müller, R. Time-lapsed investigation of three-dimensional failure and damage accumulation in trabecular bone using synchrotron light. *Bone* **2006**, *39*, 289–299. [CrossRef] [PubMed]
18. Larrue, A.; Rattner, A.; Laroche, N.; Vico, L.; Peyrin, F. Feasibility of micro-crack detection in human trabecular bone images from 3D synchrotron microtomography. *Annu. Int. Conf. IEEE Eng. Med. Biol. Proc.* **2007**, 3918–3921. [CrossRef]
19. Tozzi, G.; Peña Fernández, M.; Parwani, R.; Bodey, A.J.; Dall'Ara, E.; Blunn, G.W.; Barber, A.H. Micromechanics and DVC of bone-biomaterial systems produced by osteorgenerative biomaterials in vivo. In Proceedings of the 23rd Congress of the European Society of Biomechanics (ESB 2017), Seville, Spain, 2–5 July 2017.
20. Barth, H.D.; Launey, M.E.; MacDowell, A.A.; Ager, J.W.; Ritchie, R.O. On the effect of X-ray irradiation on the deformation and fracture behavior of human cortical bone. *Bone* **2010**, *46*, 1475–1485. [CrossRef] [PubMed]
21. Barth, H.D.; Zimmermann, E.A.; Schaible, E.; Tang, S.Y.; Alliston, T.; Ritchie, R.O. Characterization of the effects of x-ray irradiation on the hierarchical structure and mechanical properties of human cortical bone. *Biomaterials* **2011**, *32*, 8892–8904. [CrossRef] [PubMed]
22. Singh, R.; Singh, D.; Singh, A. Radiation sterilization of tissue allografts: A review. *World J. Radiol.* **2016**, *8*, 355. [CrossRef] [PubMed]
23. Hamer, A.J.; Strachan, J.R.; Black, M.M.; Ibbotson, C.J.; Stockley, I.; Elson, R.A. Biomechanical Properties of Cortical Allograft Bone Using a New Method of Bone Strength Measurement: A Comparison of Fresh, Fresh-Frozen and Irradiated Bone. *J Bone Jt. Surg. Br* **1996**, *78*, 363–368. [CrossRef]
24. Currey, J.D.; Foreman, J.; Laketić, I.; Mitchell, J.; Pegg, D.E.; Reilly, G.C. Effects of ionizing radiation on the mechanical properties of human bone. *J. Orthop. Res.* **1997**, *15*, 111–117. [CrossRef] [PubMed]
25. Vastel, L.; Meunier, A.; Siney, H.; Sedel, L.; Courpied, J.P. Effect of different sterilization processing methods on the mechanical properties of human cancellous bone allografts. *Biomaterials* **2004**, *25*, 2105–2110. [CrossRef] [PubMed]
26. Singhal, A.; Deymier-Black, A.C.; Almer, J.D.; Dunand, D.C. Effect of high-energy X-ray doses on bone elastic properties and residual strains. *J. Mech. Behav. Biomed. Mater.* **2011**, *4*, 1774–1786. [CrossRef] [PubMed]
27. Flanagan, C.D.; Unal, M.; Akkus, O.; Rimnac, C.M. Raman spectral markers of collagen denaturation and hydration in human cortical bone tissue are affected by radiation sterilization and high cycle fatigue damage. *J. Mech. Behav. Biomed. Mater.* **2017**, *75*, 314–321. [CrossRef] [PubMed]
28. Gouk, S.-S.; Kocherginsky, N.M.; Kostetski, Y.Y.; Moser, M.O.; Yang, P.; Lim, T.-M.; Sun, W.Q.; Moser, H.O.; Yang, P.; Lim, T.-M.; et al. Synchrotron radiation-induced formation and reaction of free radicals in the human acellular dermal matrix. *Radiat. Res.* **2005**, *163*, 535–543. [CrossRef] [PubMed]
29. Akkus, O.; Belaney, R.M.; Das, P. Free radical scavenging alleviates the biomechanical impairment of gamma radiation sterilized bone tissue. *J. Orthop. Res.* **2005**, *23*, 838–845. [CrossRef] [PubMed]
30. Nguyen, H.; Morgan, D.A.F.; Forwood, M.R. Sterilization of allograft bone: Effects of gamma irradiation on allograft biology and biomechanics. *Cell Tissue Bank.* **2007**, *8*, 93–105. [CrossRef] [PubMed]
31. Hamer, A.J.; Stockley, I.; Elson, R.A. Changes in allograft bone irradiated at different temperatures. *J. Bone Joint Surg. Br.* **1999**, *81*, 342–344. [CrossRef] [PubMed]
32. Cornu, O.; Boquet, J.; Nonclercq, O.; Docquier, P.L.; Van Tomme, J.; Delloye, C.; Banse, X. Synergetic effect of freeze-drying and gamma irradiation on the mechanical properties of human cancellous bone. *Cell Tissue Bank.* **2011**, *12*, 281–288. [CrossRef] [PubMed]

33. Hiemstra, H.; Tersmette, M.; Vos, A.H.V.; Over, J.; Berkel, M.P.; Bree, H. Inactivation of human immunodeficiency virus by gamma radiation and its effect on plasma and coagulation factors. *Transfusion* **1991**, *31*, 32–39. [CrossRef] [PubMed]
34. Grieb, T.A.; Forng, R.Y.; Stafford, R.E.; Lin, J.; Almeida, J.; Bogdansky, S.; Ronholdt, C.; Drohan, W.N.; Burgess, W.H. Effective use of optimized, high-dose (50 kGy) gamma irradiation for pathogen inactivation of human bone allografts. *Biomaterials* **2005**, *26*, 2033–2042. [CrossRef] [PubMed]
35. Carzaniga, R.; Domart, M.C.; Collinson, L.M.; Duke, E. Cryo-soft X-ray tomography: A journey into the world of the native-state cell. *Protoplasma* **2014**, *251*, 449–458. [CrossRef] [PubMed]
36. Duke, E.; Dent, K.; Razi, M.; Collinson, L.M. Biological applications of cryo-soft X-ray tomography. *J. Microsc.* **2014**, *255*, 65–70. [CrossRef] [PubMed]
37. Carzaniga, R.; Domart, M.-C.; Duke, E.; Collinson, L.M. Correlative Cryo-Fluorescence and Cryo-Soft X-Ray Tomography of Adherent Cells at European Synchrotrons. *Methods Cell Biol.* **2014**, *124*, 151–178. [CrossRef] [PubMed]
38. Voggenreiter, G.; Ascherl, R.; Blumel, G.; Schmit-Neuerburg, K.P. Effects of preservation and sterilization on cortical bone grafts. *Arch. Orthop. Trauma Surg.* **1994**, *113*, 294–296. [CrossRef] [PubMed]
39. Cornu, O.; Banse, X.; Docquier, P.L.; Luyckx, S.; Delloye, C. Effect of freeze-drying and gamma irradiation on the mechanical properties of human cancellous bone. *J. Orthop. Res.* **2005**, *18*, 426–431. [CrossRef] [PubMed]
40. Yin, L.; Venkatesan, S.; Webb, D.; Kalyanasundaram, S.; Qin, Q.H. Effect of cryo-induced microcracks on microindentation of hydrated cortical bone tissue. *Mater. Charact.* **2009**, *60*, 783–791. [CrossRef]
41. Atwood, R.C.; Bodey, A.J.; Price, S.W.T.; Basham, M.; Drakopoulos, M. A high-throughput system for high-quality tomographic reconstruction of large datasets at Diamond Light Source. *Philos. Trans. R. Soc. A Math. Phys. Eng. Sci.* **2015**, *373*, 20140398. [CrossRef] [PubMed]
42. Vo, N.T.; Atwood, R.C.; Drakopoulos, M. Radial lens distortion correction with sub-pixel accuracy for X-ray micro-tomography. *Opt. Express* **2015**, *23*, 32859–32868. [CrossRef] [PubMed]
43. Schindelin, J.; Arganda-Carreras, I.; Frise, E.; Kaynig, V.; Longair, M.; Pietzsch, T.; Preibisch, S.; Rueden, C.; Saalfeld, S.; Schmid, B.; et al. Fiji: An open-source platform for biological-image analysis. *Nat. Methods* **2012**, *9*, 676–682. [CrossRef] [PubMed]
44. Buades, A.; Coll, B.; Morel, J.-M. Non-Local Means Denoising. *Image Process. Line* **2011**, *1*, 490–530. [CrossRef]
45. Immerkær, J. Fast noise variance estimation. *Comput. Vis. Image Underst.* **1996**, *64*, 300–302. [CrossRef]
46. Tozzi, G.; Danesi, V.; Palanca, M.; Cristofolini, L. Elastic Full-Field Strain Analysis and Microdamage Progression in the Vertebral Body from Digital Volume Correlation. *Strain* **2016**, *52*, 446–455. [CrossRef]
47. Palanca, M.; Tozzi, G.; Cristofolini, L.; Viceconti, M.; Dall'Ara, E. 3D Local Measurements of Bone Strain and Displacement: Comparison of Three Digital Volume Correlation Approaches. *J. Biomech. Eng.* **2015**, *137*, 1–14. [CrossRef] [PubMed]
48. Deymier-Black, A.C.; Singhal, A.; Almer, J.D.; Dunand, D.C. Effect of X-ray irradiation on the elastic strain evolution in the mineral phase of bovine bone under creep and load-free conditions. *Acta Biomater.* **2013**, *9*, 5305–5312. [CrossRef] [PubMed]
49. Witala, M.; Han, J.; Menzel, A.; Nygård, K. In situ small-angle X-ray scattering characterization of X-ray-induced local heating. *J. Appl. Crystallogr.* **2014**, *47*, 2078–2080. [CrossRef]
50. Bras, W.; Stanley, H. Unexpected effects in non crystalline materials exposed to X-ray radiation. *J. Non-Cryst. Solids* **2016**, *451*, 153–160. [CrossRef]
51. Wallander, H.; Wallentin, J. Simulated sample heating from a nanofocused X-ray beam. *J. Synchrotron Radiat.* **2017**, *24*, 925–933. [CrossRef] [PubMed]
52. Wolfram, U.; Schwiedrzik, J.; Bürki, A.; Rack, A.; Olivier, C.; Peyrin, F.; Best, J.; Michler, J.; Zysset, P.K. Microcrack evolution in microindentation of ovine cortical bone investigated with lapsed SRmicroCT. In Proceedings of the 23rd Congress of the European Society of Biomechanics (ESB 2017), Seville, Spain, 2–5 July 2017.
53. Zioupos, P.; Currey, J.D.; Hamer, A.J. The role of collagen in the declining mechanical properties of aging human cortical bone. *J. Biomed. Mater. Res.* **1999**, *45*, 108–116. [CrossRef]
54. Mitchell, T.W.; Rigby, B.J. In vivo and in vitro aging of collagen examined using an isometric melting technique. *Biochim. Biophys. Acta* **1975**, *393*, 531–541. [CrossRef]
55. Lee, J.M.; Pereira, C.A.; Abdulla, D.; Naimark, W.A.; Crawford, I. A multi-sample denaturation temperature tester for collagenous biomaterials. *Med. Eng. Phys.* **1995**, *17*, 115–121. [CrossRef]

56. Unal, M.; Creecy, A.; Nyman, J.S. The Role of Matrix Composition in the Mechanical Behavior of Bone. *Curr. Osteoporos. Rep.* **2018**, *16*, 205–215. [CrossRef] [PubMed]
57. Ginoza, W. Radiosensitive Molecular Weight of Single-Stranded Virus Nucleic Acids. *Nature* **1963**, *199*, 453. [CrossRef] [PubMed]
58. Kempner, E.S.; Haigler, H.T. The influence of low temperature on the radiation sensitivity of enzymes. *J. Biol. Chem.* **1982**, *257*, 13297–13299. [PubMed]
59. Panjabi, M.M.; Krag, M.; Summers, D.; Videman, T. Biomechanical time-tolerance of fresh cadaveric human spine specimens. *J. Orthop. Res.* **1985**, *3*, 292–300. [CrossRef] [PubMed]
60. Linde, F.; Sørensen, H.C.F. The effect of different storage methods on the mechanical properties of trabecular bone. *J. Biomech.* **1993**, *26*, 1249–1252. [CrossRef]
61. Kang, Q.; An, Y.H.; Friedman, R.J. Effects of multiple freezing-thawing cycles on ultimate indentation load and stiffness of bovine cancellous bone. *Am. J. Vet. Res.* **1997**, *58*, 1171–1173. [PubMed]
62. Borchers, R.E.; Gibson, L.J.; Burchardt, H.; Hayes, W.C. Effects of selected thermal variables on the mechanical properties of trabecular bone. *Biomaterials* **1995**, *16*, 545–551. [CrossRef]
63. Lee, W.; Jasiuk, I. Effects of freeze-thaw and micro-computed tomography irradiation on structure-property relations of porcine trabecular bone. *J. Biomech.* **2014**, *47*, 1495–1498. [CrossRef] [PubMed]
64. Mazurkiewicz, A. The effect of trabecular bone storage method on its elastic properties. *Acta Bioeng. Biomech.* **2018**, *20*, 21–27. [CrossRef] [PubMed]
65. Fantner, G.E.; Hassenkam, T.; Kindt, J.H.; Weaver, J.C.; Birkedal, H.; Pechenik, L.; Cutroni, J.A.; Cidade, G.A.G.; Stucky, G.D.; Morse, D.E.; et al. Sacrificial bonds and hidden length dissipate energy as mineralized fibrils separate during bone fracture. *Nat. Mater.* **2005**, *4*, 612–616. [CrossRef] [PubMed]
66. Hedan, S.; Fauchille, A.L.; Valle, V.; Cabrera, J.; Cosenza, P. One-year monitoring of desiccation cracks in Tournemire argillite using digital image correlation. *Int. J. Rock Mech. Min. Sci.* **2014**, *68*, 22–35. [CrossRef]
67. Wei, X.; Hattab, M.; Bompard, P.; Fleureau, J.-M. Highlighting some mechanisms of crack formation and propagation in clays on drying path. *Géotechnique* **2016**, *66*, 287–300. [CrossRef]
68. Bodey, A.J.; Rau, C. Launch of the I13-2 data beamline at the Diamond Light Source synchrotron. *J. Phys. Conf. Ser.* **2017**, *849*, 012038. [CrossRef]

© 2018 by the authors. Licensee MDPI, Basel, Switzerland. This article is an open access article distributed under the terms and conditions of the Creative Commons Attribution (CC BY) license (http://creativecommons.org/licenses/by/4.0/).

Article

Dynamic Tomographic Reconstruction of Deforming Volumes

Clément Jailin * and Stéphane Roux

LMT (ENS Paris-Saclay/CNRS/University Paris-Saclay), 61 avenue du Président Wilson, F-94235 Cachan, France; stephane.roux@ens-paris-saclay.fr
* Correspondence: clement.jailin@ens-paris-saclay.fr

Received: 17 July 2018; Accepted: 6 August 2018; Published: 9 August 2018

Abstract: The motion of a sample while being scanned in a tomograph prevents its proper volume reconstruction. In the present study, a procedure is proposed that aims at estimating both the kinematics of the sample and its standard 3D imaging from a standard acquisition protocol (no more projection than for a rigid specimen). The proposed procedure is a staggered two-step algorithm where the volume is first reconstructed using a "Dynamic Reconstruction" technique, a variant of Algebraic Reconstruction Technique (ART) compensating for a "frozen" determination of the motion, followed by a Projection-based Digital Volume Correlation (P-DVC) algorithm that estimates the space/time displacement field, with a "frozen" microstructure and shape of the sample. Additionally, this procedure is combined with a multi-scale approach that is essential for a proper separation between motion and microstructure. A proof-of-concept of the validity and performance of this approach is proposed based on two virtual examples. The studied cases involve a small number of projections, large strains, up to 25%, and noise.

Keywords: tomographic reconstruction; dynamic tomography; motion compensation; projection-based digital volume correlation

1. Introduction

Tomography is a non-destructive imaging technique that enables the visualization of the bulk of the observed specimen. Tomography is now widely used in many fields (e.g., medical imaging for diagnostic [1], biology [2], material science [3], etc.), performed with various waves (e.g., X-ray, neutron, electron, terahertz, optics, ultrasound, etc.) depending on the experiment and material absorption and or scattering. Different instruments have been developed with different flux, space and time resolutions (e.g., for X-rays medical scanners, synchrotron, lab-CT, etc.) giving access to a wide range of imaging devices and performances.

To image the 3D structure, the specimen rotates over 180° or 360° with respect the source-detector pair and at a series of distributed angles radiographs are acquired. Radiographs are transformed with dark-fields and white-fields, to extract the relative beam absorption, transformed with a logarithm (Beer-Lambert law) or more sophisticated treatments for beam hardening, in order to obtain so-called projections of the local coefficient of absorption of the sample. The collection of projections at all angles constitutes a so-called *sinogram*. Then, from the sinogram, reconstruction algorithms [4] have been developed to reconstruct the 3D imaged volume. This technique relies on the strict satisfaction of conditions, in particular concerning the geometry of the set-up and the motion of the sample as a rigid rotation with the prescribed axis and angles.

The required time for a full 3D scan varies depending on the flux (and exposition time), type of camera and rotation speed of the device. Since the beginning of the development of these techniques, the time required to acquire a tomographic scan has constantly decreased [5]. Recent papers have reported on ultra-fast tomographies, at up to 20 Hz in synchrotron beamlines, that allow extremely fast processes to be captured [6,7].

Motion of the sample during the scan is one of the main issue of tomography that leads to poor quality, blurry volumes [8]. This is the case for medical imaging (as the patient or imaged organ may move), in vivo measurements [9], for electron tomography [10] (because of the extremely small scales of observation, one cannot guarantee a fixed rotation axis at nanometer accuracy) for usually minute to hour long acquisitions, fast mechanical behavior or continuous in situ experiments [6]. Wrong or imprecise estimates of the calibration parameters (that may even vary along the scan) can also be seen as motions in the sinogram space and have the same deleterious consequences for the volume reconstruction.

Sophisticated methods have been developed to avoid or limit motion perturbations especially for periodic motion, for instance using a trigger for acquisition of radiographs based on a specific signal to captures always the same phase as can be done for cardiac or respiratory motion in medical imaging [11–13].

Many works have been devoted to correcting imperfect acquisitions as a post-processing treatment. For automatic (re)calibration, *online* methods, based on the motion of the sample itself during the scanning process, have been applied as a post-process after reconstruction to evaluate a corrected set of calibration parameters [14]. When dealing with electron tomography, (TEM or STEM), the voxel scale makes this problem quite limiting. The identified motion of the specimen is often regularized as being composed of rigid body motions [15,16]. However, in addition to accounting for the slight deviation of the rigid body motion of the sample from the ideal perfect rotation, motion description can be enriched by taking into account more precisely the physics of the electron trajectory in inhomogeneous magnetic fields leading to distortions [17], or sample warping due to irradiation [18,19] for electron tomography. These kinematic degrees of freedom have to be inferred at each projection, and for this fiducial markers (such as gold nanoparticles) are used. In all those cases, deformations can be treated as a slight perturbation, with strains of order of a few 10^{-3} at most. Similarly, in Optical Projection Tomography, OPT, Zhu et al. [20] face similar reconstruction artifacts due to motion for in vivo imaging. Motion is here regularized in time as a polynomial series, and the coefficient describing motions—essentially rigid body motions—are determined from robust quantities (geometric moments) that can be computed over the entire region of interest.

Very early, corrections were also applied in the sinogram space ([21]), with affine transforms [22]. Projection-based measurement methods (e.g., Projection-based Digital Volume Correlation (P-DVC) [23], 3D–2D registration [24,25]) have been developed to correct for rigid body motions (due to a rigid patient motion or variation of calibration parameters) from the radiograph data directly.

Yet, a deforming body with a significant strain and variation in time is a much more demanding case. Projection-based Digital Volume Correlation (P-DVC) has been shown to address part of the problem with complex 4D—3D space + time—kinematic identification [26–28]. First if the reference 3D geometry is well known, the displacement field can be evaluated on the fly as the sample is being deformed. This method requires a high quality reference volume and a series of deformed projections. A single projection per motion state is required to capture the full 4D (space-time) kinematics. Alternatively, imperfect acquisition conditions (but no sample strain) can also be corrected using a similar technique, without a pre-determined 3D reference geometry [23], considering that the deforming projection stack is the one used for the reconstruction.

Similar developments have been carried out very early in the context of medical imaging where periodic motion is frequent (heart beat, breathing). In particular, Refs. [29–33] have proposed to determine the motion of the sample from projection data. Small amplitude displacement fields with a periodic modulation in time were considered and identified using highly regularized kinematic models.

However, very often, a reference reconstructed volume is known, and is used as a prior for determining the motion [29–31,33,34]. This is often the case for radio-therapy treatment where the key issue is to irradiate the targeted region, in spite of a spurious motion, and hence the goal is to identify the displacement field in 3D, and a fast determination is more valuable than a very precise one.

In a similar spirit, [31,35,36] do not consider a reference to be known but rather use a phase signal (say from an electrocardiogram) to extract from a long sinogram projections coming from a similar phase of the motion, and reconstruct a low quality volume for a series of phase. Registration of the reconstructed volumes [35,36] allows the displacement field to be estimated and interpolated for the entire range of accessed phase. Then, back-correcting for this motion a deformed reconstruction grid is obtained [37] on which the projection data can be backprojected using a classical FBP/FDK algorithms [4]. In this way, each ray follows the deformed sample at each projection angle. The obtained volume has a better quality than the initial one (more details and sharper edges). Katsevich [38] proposed a mathematical study of the generalized inverse Radon transform, using a modified filtered backprojection, showing convergence in appropriate space. Further mathematical consideration lead Hahn [39] to focus on smooth boundaries of subdomains in the volume as the latters produce singularities in projections (in sinogram space, a diverging density appears along tangent planes) that can be tracked in time easily. In the following, inverse Radon transforms and filtered backprojection will not be considered, although they constitute an attractive alternative to the modified algebraic reconstruction algorithm used hereafter.

Dynamic tomography methods based on multiple volume acquisitions have been recently developed. Ruhlandt et al. [40] proposed an approach without prior knowledge of a phase for each angle, nor of a reference volume, developed for phase contrast imaging at a synchrotron facility. The displacement field that animates a volume at time t is measured from the analysis of the motion-blurred reconstructed volume at time $t-1$ and $t+1$, then interpolated linearly. A full 4D space-time 'movie' of the phenomenon could be obtained. This method however requires the use of many acquired 3D volumes for the displacement field measurement, thus a high dose. The measured displacement has a relatively small amplitude compared to the volume texture characteristic scale. A criterion based on the image reconstruction quality is not easy to set and the quality has to be appreciated visually. A similar recent technique [41] deals with the correction of a volume using Digital Volume Correlation and an extended Simultaneous Algebraic Reconstruction Technique (SART) algorithm. To be able to correct a single rotation volume, the authors sub-sampled the acquired projections in 2 sub-acquisitions from which the motion is evaluated and further involved in the reconstruction strategy. This technique is however not suited to large and irregular displacements. In [42], the volume sub-sampling is performed more easily because of an especially designed sampling acquisition strategy (that cannot be adapted to any tomography). One displacement field, constant in time, is estimated for each successive pair of reconstructed volumes and is used to correct the reconstruction procedure. In this latter reference, although the tackled displacements and deformations are important during the entire test, the incremental displacement between all acquired volumes is small.

In most of these studies, the displacement and strain fields between scans was relatively small (strain of approximately 1% and uniformly distributed), and often the time (or phase) is believed to be known.

The present study proposes a strategy to reconstruct both the reference geometry and its large motion from a single sinogram. No periodic signal is used to constrain the kinematics. The recorded projections are the data that drive the measurement of the kinematic field, as is proposed in P-DVC. This however requires a "model," here a reconstructed 3D volume, to be known in order to measure the displacement field. It is proposed here to "learn" this model from the projection data itself using a multiscale approach.

The standard reconstruction methods are briefly presented in Section 2, so that the introduction of motion can be cast in a similar framework. Section 3 details the joint determination of the reconstructed image and the motion experienced during the scan. The latter algorithm makes use of ideas comparable to those of P-DVC for the motion, and Algebraic Reconstruction Techniques

(ART) for the microstructure and exploits a multiscale approach to disentangle microstructure and motion from the sinogram. Two virtual test cases of moving samples validate the procedure (Section 4). The first example is performed with the Shepp-Logan phantom with large deformation up to 20%. The second example is a checkerboard with a more complex temporal pulsating motion.

2. Motionless X-ray Tomography

X-ray tomographic reconstruction is based on the relative beam intensity attenuation for each discrete detector position $r = [r, z]$ (where z is parallel to the specimen rotation axis, and r is perpendicular to it) and rotation angle. For simplicity, and because the present paper is a proof of concept, the displacement field is assumed to lie in a plane perpendicular to the rotation axis, so that each slice z remains independent from its neighbors, and the problem turns two dimensional. Hence, only one line of the detector is considered, for a unique value of z (omitted from now on).

Let us briefly recall the principle of tomography for a parallel beam: a projection $p(r, \theta)$ is defined as the line integral of $f(x)$ along a direction e_θ, or

$$p(r, \theta) = \int_{\mathcal{D}(r,\theta)} f(x) \, dx \tag{1}$$

where $\mathcal{D}(r, \theta)$ is the line parallel to e_θ hitting the detector plane at position r. Different projection and interpolation algorithms exist. In the following procedure, the Matlab function radon.m is used.

Tomography consists of recording a set of N_θ projections $p(r, \theta)$ for a collection of angles $\theta(t)$ as the sample is rotated over a complete (or half) rotation about a fixed axis parallel to the detector plane. For a still sample, and a continuous rotation, $p(r, \theta(t))$, written $p(r, \theta)$, is the Radon transform of $f(x)$, $p(r, \theta) = \mathcal{R}[f(x)]$ and hence the $f(x)$ can be computed from an inverse Radon transform, $f(x) = \mathcal{R}^{-1}[p(r, \theta)]$. Let us introduce the indicator function $\mathcal{I}_\mathcal{E}(x)$ of the domain $\mathcal{E}$ within which the volume is to be reconstructed. The ray length in $\mathcal{E}$ for a specific detector position r and rotation angle θ, is simply $L(r, \theta) = \mathcal{R}_\theta[\mathcal{I}_\mathcal{E}(x)]$. It is useful to introduce the backprojection operator, $\mathcal{B}_\theta$, which to each point x of the line $\mathcal{D}(r, \theta)$ within $\mathcal{E}$, gives a value $1/L(r, \theta)$. Thus for any $p(r, \theta)$, $\mathcal{R}_\theta[\mathcal{B}_\theta[p(r, \theta)]] = p(r, \theta)$.

Tomography is now a very mature field and numerous powerful algorithms have been devised in order to deal with a discrete set of angles, with fan-bean or cone-beam projections [43], with laminography [44], etc.

However, $f(x)$ is always assumed to stand for a rigid and still object (independent of time or rotation angle). From the collection of acquired projections, different algorithms exist to reconstruct the 3D volume [4] and fall into two categories: Fourier-domain algorithms and algebraic algorithms.

Fourier space reconstructions

With Filtered Back-Projection (FBP), each projection, $p(r, \theta)$ is first "filtered" with a ramp, or Ram-Lak filter, eventually windowed. Ignoring such windowing, in Fourier space, $\mathcal{F}[p(r, \theta)](k, \theta)$ is multiplied by $|k|$, inverse Fourier transformed, and then back-projected in real space, thereby producing a field $g_\theta(x)$ that is invariant along the direction e_θ. These fields $g_\theta(x)$ are simply summed over all visited angles θ, producing the sought initial image, $f(x)$

$$f(x) = \sum_{\theta=1}^{N_\theta} g_\theta(x) \tag{2}$$

Iterative reconstructions

Other reconstruction methods have received much attention, namely iterative algebraic approaches which tolerate deviations from the ideal conditions of the previous Fourier space reconstruction such as for instance having access to a continuous range of angles, covering the entire half (or full) rotation. Those methods exploit the linear structure of the problem to solve, but for computational efficiency, they avoid the writing of the linear system. They are based on the minimization over volumes, $\psi(x)$, of the functional, $\Gamma_{ART}[\psi]$, equal to the quadratic norm of the difference $\rho(t,\theta)$ between the acquired projections and the projected reconstructed volume

$$\Gamma_{ART}[\psi] = \sum_{r,\theta} \|\rho(t,\theta)\|^2 \tag{3}$$
$$\rho(t,\theta) = p(r,\theta) - \mathcal{R}_\theta[\psi(x)]$$

then

$$f = \underset{\psi}{\text{Argmin}}\, \Gamma_{ART}[\psi] \tag{4}$$

Additional prior information may easily be added to this functional through regularization, in order to compensate limited angle range for projections, or coarse sampling for example. This generally leads to better quality reconstructions than FBP algorithms at the expense of a higher computational cost.

To solve this huge linear inverse problem, ART algorithms essentially consist of iterative updates of the volume. Successively visiting each angle, the projection of the volume is compared with the acquired one. The difference is back projected and used to correct the volume (sometimes multiplied by a damping coefficient, not considered in our case). Faster convergence rate is observed when angles are not sampled in consecutive order but rather with a large difference between successive angles. This can be achieved for instance with a permutation of the angle order. A convergence criterion on the functional value can be used to stop the number of iterations ($\Gamma_{ART}[f] < \epsilon$), with ϵ, a threshold value with respect to noise and artifact acquisition. Generally a few iterations (N_{ART}) are required for convergence. The algorithm for this method is detailed in Algorithm 1.

Algorithm 1: Standard algebraic reconstruction procedure, $f \leftarrow \text{ART}(p)$.

$n \leftarrow 1$; ▶ Initialization
$f^{(n)} \leftarrow 0$; ▶ Initialization
$\rho(r,\theta) \leftarrow p(r,\theta)$; ▶ Initialization
Choose a permutation, π, over N_θ indices;
while $\|\rho(r,\theta)\| > \epsilon$ **do**
  for $k \leftarrow 1$ **to** N_θ **do**
    $m \leftarrow \pi(k)$;
    $\rho(r,\theta_m) \leftarrow p(r,\theta_m) - \mathcal{R}_{\theta_m}[f^{(n)}(x)]$;
    $\Delta f^{(n+1)}(x) \leftarrow \mathcal{B}_{\theta_m}[\rho(r,\theta_m)]$;
    $f^{(n+1)}(x) \leftarrow f^{(n)}(x) + \Delta f^{(n+1)}(x)$;
    Implement additional constraints on $f^{(n+1)}$ (e.g., positivity);
    $n \leftarrow n+1$;
  end
end

During the reconstruction procedure, additional information, defined as constraints, can be added. Those regularizations allow the reconstruction of high quality volume with a few or missing angles, noisy projections and artifacts etc. This may come from prior knowledge of the different phases of the sample (as DART algorithms proposed by [45], reconstruction with binary images [46], Total Variation [47]), dictionary learning [48], etc. However, because those regularizations are independent from the following proposed reconstruction with motion compensation, it is not considered hereafter apart from the positivity constraint $f(x) \leftarrow \max(\psi(x), 0)$.

3. Data Driven Reconstruction of Non-Rigid Samples

It is proposed to study a specimen that moves during the acquisition with a space/time displacement field $u(x,t)$ such that, at any time, the sample is expressed with respect to a reference state $f(x + u(x,t))$.

For a still object f and p are bijectively related to each other through the inverse Radon transform. The introduction of motion causes a non-trivial nullspace and thus the loss of bijectivity. The reconstruction of the volume from the previously introduced algorithms (i.e., ignoring motion) leads to a low quality, blurry, volume.

It is to be noted that the FBP reconstruction procedure has been extended to take motion into account in [35,36,40] The driving idea is to apply the back-projection step on the currently deformed geometry of the to-be-reconstructed sample, or equivalently to transport the back-projection onto the initial geometry, unwarping the motion, so that the X-ray beam would then follow non-straight paths. In Ref. [40], the motion is estimated from the registration of two reconstructions of the volume at different instants of time and linear interpolation.

$$f(x) = \sum_{t=1}^{N_\theta} g_{\theta(t)}(x - u(x,t)) \tag{5}$$

Because this approach requires different volumes to estimate the displacement field, it is not suited when the motion is fast and when only a single scan can be acquired. Moreover, it is difficult to estimate a quality criterion but visual on the reconstructed volumes thus on the measured kinematics.

A recently developed Digital Volume Correlation (DVC) procedure called Projection-based DVC [26] allows to identify the 4D [49] (space-time) displacement field of sample from an initially reconstructed volume and its moving projections. An extension of this method has been applied to an *online* calibration (i.e., calibrated from the sample motion during the scanning process) of the tomograph [23]. An initial (blurry) reconstruction was performed from a set of initial parameters. The comparison between the projection of the blurry sample and the acquired projections is, in addition to the acquisition noise and artifacts, the signature the erroneous projection geometry parameters that can be identified and corrected. The sample could be re-positioned for each angle by a rigid body motion. Because the motion was simple and of low amplitude, the correction could be applied on the sinogram itself leading to very significant improvement on the quality of the reconstruction. However, more complex displacements, or larger amplitudes (involving larger displacement variations perpendicular to the ray) would render the corrections on the projection inaccessible.

It is proposed to introduce here a new two-step algorithm based on ART reconstruction on the one hand and P-DVC on the other hand to identify both a complex and large displacement field and volume texture with a single scan performed on a moving and deforming object. The ART functional is naturally extended to account for the motion as

$$\Gamma_{\text{motion-ART}}[\psi, v] = \sum_{r,t} \|\mathcal{R}_{\theta(t)}[\psi(x + v(x,t))] - p(r,t)\|^2 \tag{6}$$

where the summation over time extends over the N_t acquired projections (and not just a full rotation) then

$$(f, u) = \underset{\psi, v}{\text{Argmin}} \, \Gamma_{\text{motion-ART}}[\psi(x), v(x,t)] \tag{7}$$

The updating procedure (indexed by l) is split into two parts that are repeated alternatively:

- A volume reconstruction from an iterative *dynamic* ART algorithm assuming a known motion (described in Section 3.1);
- An update of the motion from P-DVC with a given reconstructed sample (described in Section 3.2).

However, as such, this procedure does not tolerate large displacement amplitudes. To increase the robustness and fast convergence, a multi-scale approach is coupled to the previous two-step procedure, resolving first the large scale features of both microstructure and motion, and progressively enriching the description with finer details. The complete multi-scale procedure is described in Section 3.3.

3.1. Dynamic Reconstruction

The dynamic reconstruction used in this article is an extension of the standard ART algorithm, and will follow the same structure as Algorithm 1. Considering the inner "for" loop, at time t (and angle $\theta(t)$), the volume is warped with the measured displacement field

$$\tilde{f}^{(n-1)}(x,t) = f^{(n-1)}(x + u(x,t)) \tag{8}$$

(initially $u(x,t) = 0$). The computed projection of $\tilde{f}^{(n-1)}$ along $\theta(t)$ is compared with the recorded projection and the residual (i.e., their difference)

$$\rho^{(n)}(r,t) = p(r,t) - \mathcal{R}_{\theta(t)}\tilde{f}^{(n-1)}(x,t) \tag{9}$$

is normalized and back-projected $\Delta \tilde{f}^{(n)} = \mathcal{B}_{\theta(t)}[\rho^{(n)}(r,t)]$.

Finally the correction term is unwarped to the frame of the undeformed state, $\widehat{\Delta f}^{(n)}(x) = \Delta \tilde{f}^{(n)}(x - u)$ so that it matches the reference configuration and it is added to the volume, $f^{(n)} = f^{(n-1)} + \widehat{\Delta f}^{(n)}$. Let us emphasize that theoretically, $\widehat{\Delta f}^{(n)}$ should have been defined implicitly as obeying $\widehat{\Delta f}^{(n)}(x + u) = \Delta \tilde{f}^{(n)}(x)$. The two expressions are equivalent only for small strains and rotations, otherwise the unwarping should involve the Eulerian rather than the Lagrangian displacement, and one can be computed from the other. Let us also note that for not too large strains and rotations, ignoring the difference between Eulerian and Lagrangian displacements simply slows down the convergence, but the final solution is not affected. In the present case, the choice was made to use the Eulerian registration to achieve the convergence for engineering strains as large as 20%. A convergence criterion has to be chosen as in the ART procedure. Nevertheless, the criterion based on the functional value cannot be used in this case as the reconstruction is unperfect. A convergence criterion based on the variation of the functional or a maximum number of iteration N_{DynART} can be set. The procedure is described in Algorithm 2.

Algorithm 2: Proposed motion-corrected algebraic reconstruction procedure, $f \leftarrow \text{DynART}(p, u)$.

$n \leftarrow 1$;   ▶ Initialization
$f^{(1)} \leftarrow 0$;   ▶ Initialization
Choose a permutation, π, over N_t acquisition times;
for $i \leftarrow 1$ **to** N_{DynART} **do**
$\quad$ **for** $t \leftarrow 1$ **to** N_t **do**
$\quad\quad \tau \leftarrow \pi(t)$;
$\quad\quad \tilde{f}^{(n)}(x, \tau) = f^{(n)}(x + u(x, \tau))$;   ▶ warp
$\quad\quad \rho(r, \tau) \leftarrow p(r, \tau) - \mathcal{R}_{\theta(\tau)}[\tilde{f}^{(n)}(x, \tau)]$;
$\quad\quad \Delta \tilde{f}^{(n+1)}(x) = \mathcal{B}_{\theta(\tau)}[\rho(r, \tau)]$;
$\quad\quad \widehat{\Delta f}^{(n+1)}(x) = \Delta \tilde{f}^{(n+1)}(x - u(x, \tau))$;   ▶ unwarp
$\quad\quad f^{(n+1)}(x) = f^{(n)}(x) + \widehat{\Delta f}^{(n+1)}(x)$;
$\quad\quad$ Implement additional constraints on $f^{(n+1)}$ (e.g., positivity);
$\quad\quad n \leftarrow n + 1$;
$\quad$ **end**
end

As earlier mentioned, additional priors can be added in this procedure at the end of the inner "for" loop. In the following, only a positivity constraint for f is added at each iteration.

3.2. Motion Identification

The full procedure is a staggered two-step process where alternatively the volume is reconstructed from a frozen displacement, and the motion is identified from a frozen estimate of the microstructure. The second step is described now.

At step l, the reconstructed volume, $f_l(x)$, although imperfect, is now considered as reliable. The projected residual fields $\rho_l(r,t)$ (computed at the end of the previous procedure when the volume is no more updated) contains patterns that are the signature of an incomplete motion correction. For the identification of the displacement field, the functional for a given f can be linearized around the previously identified displacement field $u_l = u_{l-1} + \delta u$

$$\delta u = \underset{\delta v}{\mathrm{Argmin}} \sum_{r,t} \|\mathcal{R}_{\theta(t)}[\nabla \tilde{f}_l(x,t)\delta v(x,t)] - \rho_l(r,t)\|^2 \tag{10}$$

For a better conditioning, the space and time dependencies of motion may be regularized, either using "weak regularization", with a penalty on spatial or temporal rapid variation of the displacement field to be added to the above cost function, or reverting to "strong regularization" by choosing a parametrization space composed of smooth functions of space and time. At this regularization step, any additional information pertaining to the experiment (e.g., synchronous measurements from sensors of different modalities such as force, pressure or temperature measurements, cardiac phase etc.) can be incorporated in the kinematic model through functional dependencies on such parameters. Qualitative features may also be incorporated, for instance, the sudden occurrence of a crack, may be accounted for by allowing a temporal discontinuity in concerned degrees of freedom for the kinematics.

The chosen reduced basis is composed respectively of N_τ time functions, $\varphi_i(t)$, and N_s vector spatial shape functions $\Phi_j(x)$ such as

$$u(x,t) = \sum_{i=1}^{N_\tau} \sum_{j=1}^{N_s} \alpha_{ij} \varphi_i(t) \Phi_j(x) \tag{11}$$

with α_{ij} the time and space amplitudes that weight the basis functions. Setting $\varphi_i(0) = 0$, the reference state is at initial time or angle $\theta = 0$, $u(x,0) = 0$.

The minimization of the functional $\Gamma_{\text{motion-ART}}$, Equation (6), with respect to the displacement parameters $\delta \alpha$ is performed using Newton's descent method. This procedure requires the computation of the advected image gradient and Hessian of $\Gamma_{\text{motion-ART}}$. They are built from the projected sensitivities

$$S_{ij}(r,t) = \frac{\partial \mathcal{R}_{\theta(t)} \tilde{f}(x,t)}{\partial \alpha_{ij}} = \varphi_i(t) \mathcal{R}_{\theta(t)}[\Phi_j(x) \nabla \tilde{f}(x,t)] \tag{12}$$

Numerically, the sensitivities are computed from finite differences. The Hessian matrix and second member built from those sensitivities is

$$H_{ijkl} = \sum_{r,t} S_{ij}(r,t) S_{kl}(r,t) \tag{13}$$

$$b_{ij} = \sum_{r,t} \rho(r,t) S_{ij}(r,t) \tag{14}$$

thus the vector of displacement amplitude correction $\delta \alpha$ is the solution of the linear system

$$[H]\delta \alpha = b \tag{15}$$

from which the displacement is updated. This procedure is repeated until the projection residual is no longer decreasing. Algorithm 3 summarizes the determination of the displacement field. In the following test cases, a single iteration in this algorithm is performed before updating the volume.

Algorithm 3: Displacement identification procedure, $u \leftarrow \text{PDVC}(f_\lambda, p, u_0)$.

$u(x,t) \leftarrow u_0(x,t)$; ▶ Displacement initialization
$p_{\text{old}}(r,t) \leftarrow p(r,t) - \mathcal{R}_{\theta(t)}[f_\lambda(x,t)]$; ▶ Initialization
$\text{Progress} \leftarrow 1 + \epsilon$; ▶ Initialization to force the first loop
while $\text{Progress} > \epsilon$ **do**
$\quad \tilde{f}(x,t) \leftarrow f_\lambda(x + u(x,t))$; ▶ Volume advection
$\quad \rho(r,t) \leftarrow p(r,t) - \mathcal{R}_{\theta(t)}[\tilde{f}(x,t)]$; ▶ Projection residual
$\quad \text{Progress} \leftarrow \|\rho(r,t) - \rho_{\text{old}}(r,t)\|$;
$\quad$ **for** $i \leftarrow 1$ **to** N_t **do**
$\quad\quad$ **for** $j \leftarrow 1$ **to** N_s **do**
$\quad\quad\quad \chi_j(x,t) \leftarrow \mathcal{R}_{\theta(t)}[\Phi_j(x) \nabla \tilde{f}(x,t)]$;
$\quad\quad\quad S_{ij}(r,t) \leftarrow \varphi_i(t) \chi_j(x,t)$; ▶ Projected sensitivities
$\quad\quad$ **end**
$\quad$ **end**
$\quad H \leftarrow \sum_{r,t} S \otimes S$; ▶ Hessian
$\quad b \leftarrow \sum_{r,t} \rho S$; ▶ Second member
$\quad \delta\alpha \leftarrow H^{-1} b$;
$\quad u(x,t) \leftarrow u(x,t) + \delta\alpha \varphi(t) \Phi(x)$; ▶ Motion identification
$\quad \rho_{\text{old}}(r,t) \leftarrow \rho(r,t)$;
end

3.3. Multi-Scale Approach

If displacement magnitude is bounded by a length scale λ, one expects that the reconstruction is fair at a scale larger than λ. Hence, if the original image is convoluted with a Gaussian of width λ, it should well match its sinogram. One convenient property of the projection is that the projection of the convoluted image is the convolution of the original projection with a Gaussian of the same width. However, because of motion, this matching is not perfect but just fair. It means that one may estimate a better match by treating the deformation as a *slight* perturbation.

More precisely, the recorded projections are convoluted by the Gaussian of width λ,

$$\check{p}_\lambda(r, \theta) = \sum_{r'} G_\lambda(r') p(r - r', \theta) \tag{16}$$

where, $G_\lambda(r) = 1/(2\pi\lambda^2) \exp(-|r|^2/(2\lambda^2))$.

Using the progressively identified displacement field, a more accurate determination of f can be achieved using the above described reconstruction. Because a large part of the displacement is expected to be captured in u, the idea is to repeat the above procedure but with a smaller Gaussian filter, namely cutting down λ by a factor of two. Thus at each iteration, the displacement correction being smaller and smaller, convergence to the actual displacement field is expected. A convergence criterion is chosen on the norm of the residual variation or on the norm of the displacement correction.

The summary of the complete procedure is described in Algorithm 4.

Algorithm 4: Complete dynamic tomography procedure, $(f, u) \leftarrow \text{DynTomo}(p)$

$u(x, t) \leftarrow 0$; ▶ Displacement initialization
for $i \leftarrow N_{scale}$ **to** 0 **by** -1 **do**
    $\lambda \leftarrow 2^i$;
    Compute $\breve{p}_\lambda$; ▶ Gaussian filtering and downsampling
    $\rho \leftarrow \breve{p}_\lambda$;
    Residual $\leftarrow 1 + \epsilon$; ▶ Initialization to force the first loop
    while Residual $> \epsilon$ **do**
        $f(x) \leftarrow \text{DynART}(\breve{p}_\lambda, u)$; ▶ Algorithm 2
        $u \leftarrow \text{PDVC}(f, \breve{p}_\lambda, u)$; ▶ Algorithm 3
        $\rho_{old} \leftarrow \rho$;
        Compute $\rho(r, t)$; ▶ Projection residual
        Residual $\leftarrow \|\rho - \rho_{old}\|$;
    end
end

4. Test Case

Two numerical test cases are proposed to validate the procedure. To build the input data, two geometries are chosen, and two kinematics (one per case) deformed and projected at all considered angle. The obtained projections are then corrupted by a white Gaussian noise (standard deviation of 1% of the gray level dynamic of the projections and are used as the virtual experimental inputs for our procedure.

Both examples are carried out on 512 × 512 pixel images. The beam is parallel, and only $N_\theta = 300$ projections are acquired over a single 360° rotation.

- The first application corresponds to a moving Shepp-Logan phantom with large displacement magnitude (up to 37 pixels) and large engineering strains (27%). Large strains are chosen here in order to highlight the robustness of the proposed procedure as compared with previously studied examples where strains were about 1% [40].
- The second test is performed on a checkerboard with smaller displacements but a more complex time evolution composed of two separated modes: a steady drift superimposed to a high frequency pulsating motion.

In both test cases, the displacement bases chosen for the inverse problem were similar to the ones used for performing the direct problem, so that no additional model error (apart from noise) is introduced. The space functions $\Phi(x)$ are composed of four C4 mesh elements (4-node square elements with bilinear interpolations). The space basis N_s is hence composed of 18 degrees of freedom.

4.1. Shepp-Logan Phantom Case

In this test case, the Shepp-Logan phantom is used and deformed up to 27%. For this test case, a single time evolution (linear drift in time) is applied. The imposed displacement field can be written

$$v(x, \theta) = \theta / N_\theta \cdot V(x) \tag{17}$$

The nine nodal displacements are given in Table 1 in x and y directions.

The reference and deformed phantoms are shown Figure 1. The maximum displacement amplitude is 37 pixels. The first reconstruction of the image (standard ART procedure), presented Figure 2a is very blurry. Some parts of the phantom are split in two. The initial projected residual fields are very high and stresses that the reconstruction is not properly performed.

Table 1. Applied nodal displacements for $V(x)$ in pixels, in x (left) and y (right).

$y\backslash x$	66	256	446
446	15	0	4
256	19	−4	19
66	26	10	15

$y\backslash x$	66	256	446
446	4	0	1
256	26	30	0
66	0	24	34

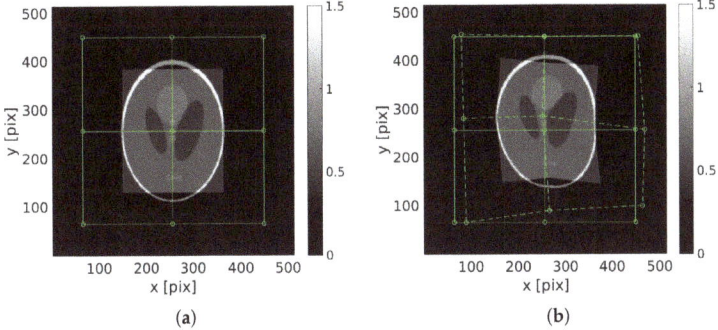

Figure 1. (a) Reference image and the 9-node mesh the node of which are subjected to a random displacement, assumed to be linear in time; (b) deformed phantom at final time N_θ.

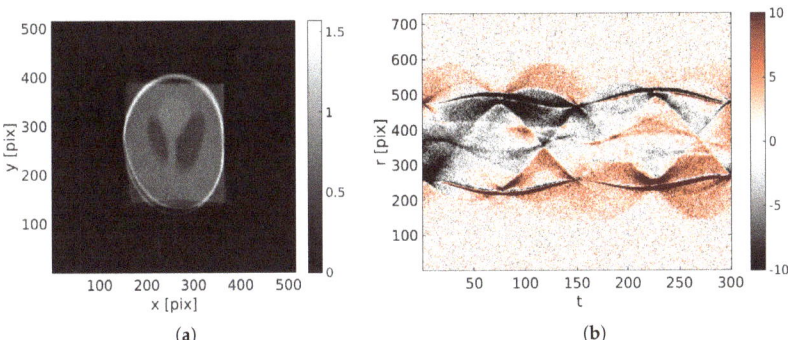

Figure 2. (a) Initial reconstruction with $u(x,t) = 0$; (b) initial projected residual fields $\rho(r,t)$. Please note that the color amplitude that is saturated in this image has been selected to be the same with the corrected residuals shown further down (Figure 3).

Before using the proposed procedure, the multi-scale procedure presented Section 3.3 is applied to the projections to willingly blur the reconstruction. After 60 iterations (i.e., volume updates), the displacement field has converged. The corrected reconstructed volume is presented Figure 3. The edges are sharp and the gray level amplitudes are correct. The projected residual fields (true metric of our procedure) is mostly composed of the white Gaussian noise meaning that the proposed procedure has been successful.

The displacement error computed on the nodal values displays a standard deviation of 3.10 pixel. This result validates the procedure.

As a last validation of the phantom reconstruction quality, the reconstruction is compared to the reference volume f. It is shown in Figure 4 that the reconstructed shape and positioning is very good. The final difference displays a "ghost" of the phantom that points out a small intensity error that does not appear in the residual fields.

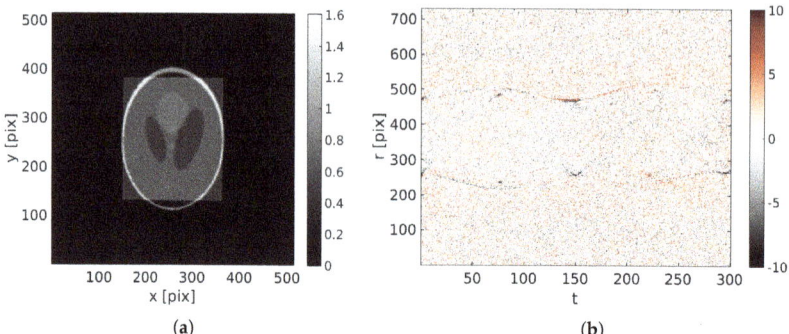

Figure 3. (a) Reconstructed image with the identified displacement field; (b) final projected residual fields $\rho(r,t)$.

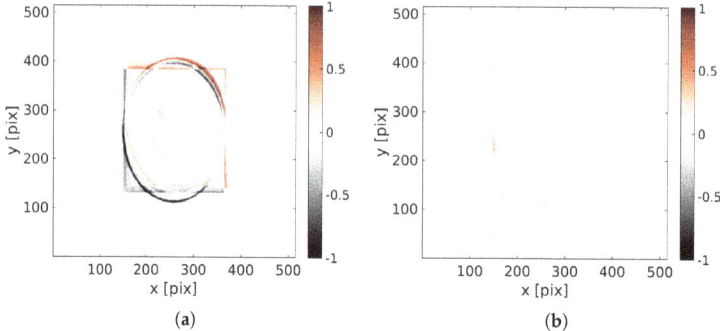

Figure 4. Difference between the reference volume and the (a) initial (i.e., ART(p)) and (b) final ones (i.e., DynTomo(p)).

4.2. Pulsating Checkerboard Case

This second test case is here based on a checkerboard composed of 8×8 squares of 35×35 pixels each. This square shaped pattern is chosen to exhibit reconstruction errors very clearly since sharp and straight boundaries are very easily detected, and hence the visual perception is a very severe test.

In this example, the imposed (supposed unknown) displacement field is composed of the sum of two parts:

- A pulsating motion: Temporally, a shifted cosine function $(1 - \cos(...))$ (obeying the constraint of being null at time 0) evolution with a non-integer number of periods to avoid symmetry (here 2.35 periods during the full-rotation scan). Spatially, the displacement field is a centered dilatation/contraction;
- A linear drift in time for all nodes with random directions and amplitudes.

The applied displacement field can be written

$$u(x,\theta) = (1 - \cos(2.35 \cdot 2\pi \cdot \theta/N_\theta)) \cdot V_1(x) + \theta/N_\theta \cdot V_2(x) \tag{18}$$

with the nodal values presented in Tables 2 and 3.

Table 2. Applied nodal displacements for $V_1(x)$ in pixels, in x (left) and y (right).

y\x	50	256	462
462	28	0	−28
256	0	0	0
50	28	0	−28

y\x	50	256	462
462	−28	0	−28
256	0	0	0
50	28	0	28

Table 3. Applied nodal displacements for $V_2(x)$ in pixels, in x (left) and y (right).

y\x	50	256	462
462	-13	22	−17
256	−22	34	17
50	17	-30	22

y\x	50	256	462
462	17	−22	9
256	22	22	−17
50	−22	17	0

The nodal displacement vectors $V_1(x)$ and $V_2(x)$ are shown in Figure 5a. The reference image, the deformed one at the end of the scan and the chosen C4 mesh are shown in Figure 5. The maximum strain is about 25%.

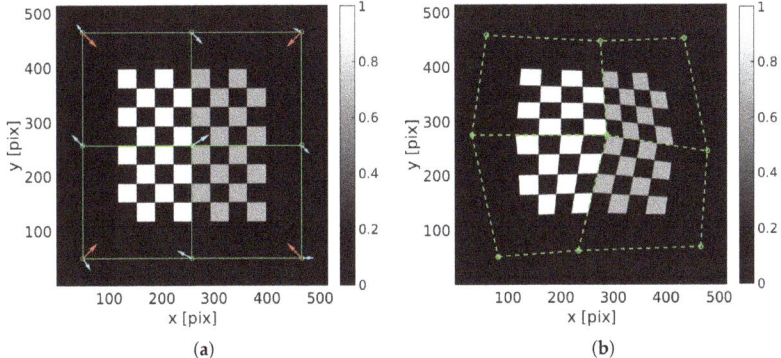

Figure 5. (**a**) Reference image (unknown) and applied nodal displacement field $V_1(x)$ in red and $V_2(x)$ in light blue; (**b**) deformed checkerboard at time N_t.

Because of the large motion amplitude, the initial reconstruction (i.e., obtained from a standard ART procedure for which $u(x, \theta) = 0$) is fuzzy and its quality is very poor as can be judged from Figure 6a. The projection of this blurred volume is compared with the initial projection to generate the initial projected residual fields $\rho(r, t)$ (see Figure 6b).

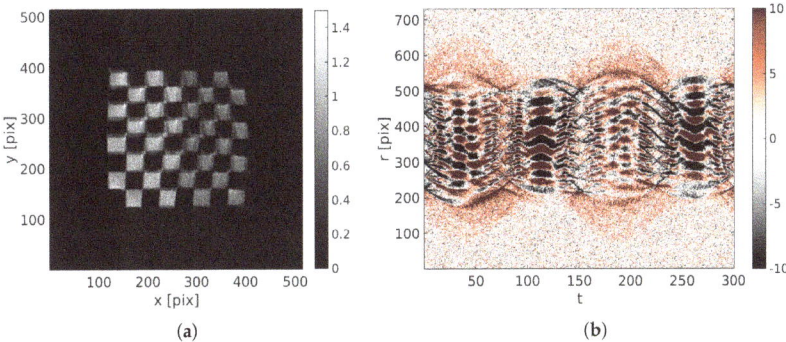

Figure 6. (**a**) Initial reconstruction with $u(x, t) = 0$; (**b**) initial projected residual fields $\rho(r, t)$.

After 60 iterations, (i.e., 60 updates of the reconstruction) — performed in approximately 2 h—the 38 degrees of freedom that drive the displacement field (18 spatial times 2 temporal degrees of freedom) have converged to a steady value. A small standard deviation of the displacement field error with respect to the prescribed displacement of less than 1.2 pixel remains at the end. Considering the large imposed motion amplitude, the estimated kinematics is deemed quite satisfactory.

The final reconstruction and projected residuals are shown in Figure 7. The reconstruction has sharp edges and its constituting squares have been correctly reconstructed. Zooms in the initial and corrected specimen are shown in Figure 8. The projected residual field, where all features of the initial sinogram have been completely erased, and only white Gaussian noise remains, means that the reconstruction has been quite successful.

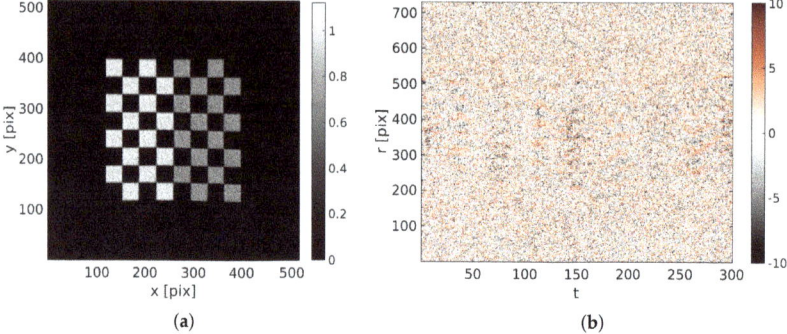

Figure 7. (a) Reconstructed image with the identified displacement field; (b) final projected residual fields $\rho(r,t)$.

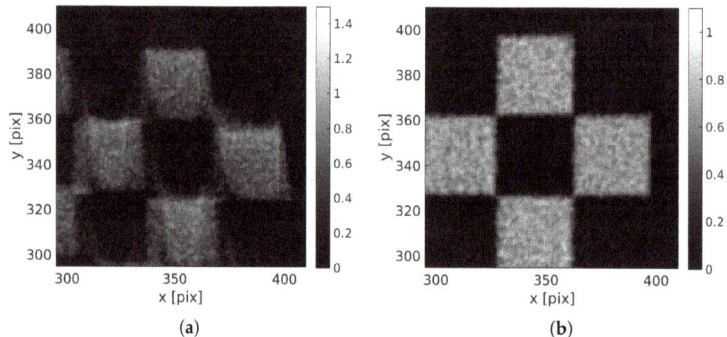

Figure 8. Zoom in the reconstructed volume (a) with a standard non-corrected volume and (b) with the proposed procedure.

To correctly appreciate the quality of the achieved volume, a difference with the initial perfect one is shown Figure 9. This difference highlights a perfect positioning of the reconstruction, and only slight discrepancies of the gray level intensity on the bright squares are visible.

The full procedure (i.e., 60 complete iterations for 512 × 512 pixel images) is performed in approximately 1 h. What takes time (and iterations) is the computation of the sensitivities that requires the deformed volumes over the entire range of time. However, it is worth mentioning that the code has not been optimized, since only a proof of concept was aimed at.

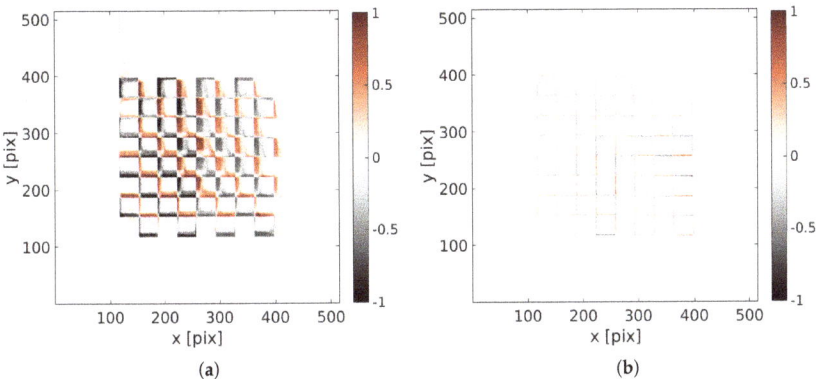

Figure 9. Difference between the initial and perfect image and (**a**) the initial reconstruction (i.e., ART(p)) and (**b**) the achieved volume (i.e., DynTomo(p)). A good positioning is reached at the end.

5. Discussion and Conclusions

An innovative algorithm is presented to perform simultaneously a dynamic reconstruction of a moving sample with the identification of the full 2D space and time displacement field. The method is derived from Algebraic Reconstruction Techniques coupled with Projection based Digital Volume Correlation. The iterative algorithm is based on two steps:

- For a given displacement field, a dynamic algebraic reconstruction algorithm is proposed. Each iteration of the procedure consists in comparing the acquired projection with the projected warped volume (deformed with current displacement field). The projected residual is backprojected, unwarped to match the reference space and added to the volume;
- For a given reference volume, a P-DVC analysis allows the displacement field to be identified. The projection of the (unperfect) warped volume is compared with the acquired projections. The residual can be read as motion using the computed sensitivity fields. An update of the displacement field is then performed.

A multiscale procedure has been proposed as an essential ingredient to properly correct large displacements. The acquired projections are first convoluted with a Gaussian kernel of large width (low pass filter) to increase its correlation length and capture large corrections from the linearized P-DVC functional. The Gaussian filter is then progressively reduced, following the residual norm evolution, to identify finer details.

The post treatment procedure, that exploits the same data as a standard acquisition (same number of projections and standard projection operator), has been tested with two challenging numerical examples (with large displacements and strains). The first is a Shepp Logan phantom with large displacement fields (up to 1/4 of the phantom length). The second is a checkerboard with a pulsating motion in time. Both examples are corrupted by a white Gaussian noise that probes the robustness with respect to the acquisition noise. The two applications show a nearly perfect identification of the displacement field and dynamic reconstruction. Performed with a parallel projection algorithm for simplicity, the exact same method can be applied with any projection model.

The proposed dynamic reconstruction algorithm has been devised as an extension of the ART algorithms. It is convenient with those approaches to include in the process an *a priori* knowledge of the scanned specimen (assumption on the gray levels, its variations, the number of phases, its sizes, etc.). Many different regularization have been proposed in the literature that enable to obtain high quality reconstructions, with less artifacts, from less projections or missing angles, etc. Because those regularizations are independent of the current algorithms, it was chosen not to implement them and focus on the proposed method performances without any 'additional help'.

Nevertheless, they are fully compatible with the proposed approach and can be implemented in a transparent fashion. When aiming to perform ultra-fast acquisitions with a few angles, they would certainly be very precious to accelerate convergence, and improve reconstruction quality.

In the proposed examples, the optical flow was kept constant. Some applications may require to include a gray level variation model. A perspective of this work could be the scan of in situ mechanical test with high strains, the identified deformation could be used to correct for absorption evolution of the material considering a constant beam intensity.

The proposed procedure shows performances that can be beneficial to numerous fields. The clear reconstruction of the moving sample allows for qualitative and quantitative analyses:

- Combined with Digital Volume Correlation [50] between well reconstructed volumes;
- Combined with image segmentation for diagnosis from radiology;
- Combined with ultra-fast tomography acquisition as recently available from some synchrotron beam-lines [6,51].

This is key for data assimilation [52] and model identification and validation in material science [53] with CT-scan as already developed with MRI [54].

Author Contributions: Methodology, C.J. and S.R.; Programming, C.J.; Data analysis, C.J.; Writing, C.J. and S.R.; Project Administration, S.R.;

Funding: This research was funded by the French "Agence Nationale de la Recherche", program "Investissements d'avenir", grant number ANR-10-EQPX-37 MATMECA.

Conflicts of Interest: The authors declare no conflict of interest.

References

1. Wang, G.; Yu, H.; De Man, B. An outlook on X-ray CT research and development. *Med. Phys.* **2008**, *35*, 1051–1064. [CrossRef] [PubMed]
2. Fogelqvist, E.; Kördel, M.; Carannante, V.; Önfelt, B.; Hertz, H.M. Laboratory cryo X-ray microscopy for 3D cell imaging. *Sci. Rep.* **2017**, *7*, 13433. [CrossRef] [PubMed]
3. Baruchel, J.; Buffière, J.; Maire, E.; Merle, P.; Peix, G. *X-ray Tomography in Material Sciences*; Hermès Science: Paris, France, 2000.
4. Kak, A.; Slaney, M. *Principles of Computerized Tomographic Imaging*; IEEE Press: Piscataway, NJ, USA, 1988.
5. Salvo, L.; Suéry, M.; Marmottant, A.; Limodin, N.; Bernard, D. 3D imaging in material science: Application of X-ray tomography. *C. R. Phys.* **2010**, *11*, 641–649. [CrossRef]
6. Maire, E.; Le Bourlot, C.; Adrien, J.; Mortensen, A.; Mokso, R. 20 Hz X-ray tomography during an in situ tensile test. *Int. J. Fract.* **2016**, *200*, 3–12. [CrossRef]
7. Dobson, K.; Coban, S.B.; McDonald, S.A.; Walsh, J.; Atwood, R.C.; Withers, P.J. 4-D imaging of sub-second dynamics in pore-scale processes using real-time synchrotron X-ray tomography. *Solid Earth* **2016**, *7*, 1059–1073. [CrossRef]
8. Milanfar, P. A model of the effect of image motion in the Radon transform domain. *IEEE Trans. Image Process.* **1999**, *8*, 1276–1281. [CrossRef] [PubMed]
9. Walker, S.; Schwyn, D.A.; Mokso, R.; Wicklein, M.; Müller, T.; Doube, M.; Stampanoni, M.; Krapp, H.; Taylor, G. In vivo time-resolved microtomography reveals the mechanics of the blowfly flight motor. *PLoS Biol.* **2014**, *12*, e1001823. [CrossRef] [PubMed]
10. Ercius, P.; Alaidi, O.; Rames, M.; Ren, G. Electron Tomography: A Three-Dimensional Analytic Tool for Hard and Soft Materials Research. *Adv. Mater.* **2015**, *27*, 5638–5663. [CrossRef] [PubMed]
11. Bonnet, S.; Koenig, A.; Roux, S.; Hugonnard, P.; Guillemaud, R.; Grangeat, P. Dynamic X-ray computed tomography. *Proc. IEEE* **2003**, *91*, 1574–1587. [CrossRef]
12. Mahesh, M.; Cody, D.D. Physics of cardiac imaging with multiple-row detector CT. *Radiographics* **2007**, *27*, 1495–1509. [CrossRef] [PubMed]
13. Flohr, T.; Ohnesorge, B. Cardiac Gating. In *Integrated Cardiothoracic Imaging with MDCT*; Rémy-Jardin, M., Rémy, J., Eds.; Springer: Berlin/Heidelberg, Germany, 2009; pp. 23–36.

14. Ferrucci, M.; Leach, R.K.; Giusca, C.; Carmignato, S.; Dewulf, W. Towards geometrical calibration of X-ray computed tomography systems—A review. *Meas. Sci. Technol.* **2015**, *26*, 092003. [CrossRef]
15. Kim, J.; Nuyts, J.; Kyme, A.; Kuncic, Z.; Fulton, R. A rigid motion correction method for helical computed tomography (CT). *Phys. Med. Biol.* **2015**, *60*, 2047. [CrossRef] [PubMed]
16. Sun, T.; Kim, J.; Fulton, R.; Nuyts, J. An iterative projection-based motion estimation and compensation scheme for head X-ray CT. *Med. Phys.* **2016**, *43*, 5705–5716. [CrossRef] [PubMed]
17. Lawrence, A.; Bouwer, J.C.; Perkins, G.; Ellisman, M.H. Transform-based backprojection for volume reconstruction of large format electron microscope tilt series. *J. Struct. Biol.* **2006**, *154*, 144–167. [CrossRef] [PubMed]
18. Printemps, T.; Bernier, N.; Bleuet, P.; Mula, G.; Hervé, L. Non-rigid alignment in electron tomography in materials science. *J. Microsc.* **2016**, *263*, 312–319. [CrossRef] [PubMed]
19. Fernandez, J.J.; Li, S.; Bharat, T.A.; Agard, D.A. Cryo-tomography tilt-series alignment with consideration of the beam-induced sample motion. *J. Struct. Biol.* **2018**, *202*, 200–209. [CrossRef] [PubMed]
20. Zhu, S.; Dong, D.; Birk, U.J.; Rieckher, M.; Tavernarakis, N.; Qu, X.; Liang, J.; Tian, J.; Ripoll, J. Automated motion correction for in vivo optical projection tomography. *IEEE Trans. Med. Imaging* **2012**, *31*, 1358–1371. [PubMed]
21. Lu, W.; Mackie, T. Tomographic motion detection and correction directly in sinogram space. *Phys. Med. Biol.* **2002**, *47*, 1267. [CrossRef] [PubMed]
22. Roux, S.; Desbat, L.; Koenig, A.; Grangeat, P. Exact reconstruction in 2D dynamic CT: Compensation of time-dependent affine deformations. *Phys. Med. Biol.* **2004**, *49*, 2169. [CrossRef] [PubMed]
23. Jailin, C.; Buljac, A.; Bouterf, A.; Poncelet, M.; Hild, F.; Roux, S. Self-calibration for lab-μCT using space-time regularized projection-based DVC and model reduction. *Meas. Sci. Technol.* **2018**, *29*, 024003. [CrossRef]
24. Otake, Y.; Schafer, S.; Stayman, J.; Zbijewski, W.; Kleinszig, G.; Graumann, R.; Khanna, A.; Siewerdsen, J. Automatic localization of vertebral levels in X-ray fluoroscopy using 3D-2D registration: A tool to reduce wrong-site surgery. *Phys. Med. Biol.* **2012**, *57*, 5485. [CrossRef] [PubMed]
25. Ouadah, S.; Stayman, J.; Gang, G.; Ehtiati, T.; Siewerdsen, J. Self-calibration of cone-beam CT geometry using 3D–2D image registration. *Phys. Med. Biol.* **2016**, *61*, 2613. [CrossRef] [PubMed]
26. Leclerc, H.; Roux, S.; Hild, F. Projection savings in CT-based digital volume correlation. *Exp. Mech.* **2015**, *55*, 275–287. [CrossRef]
27. Taillandier-Thomas, T.; Roux, S.; Hild, F. A soft route toward 4D tomography. *Phys. Rev. Lett.* **2016**, *117*, 025501, doi:10.1103/PhysRevLett.117.025501. [CrossRef] [PubMed]
28. Jailin, C.; Bouterf, A.; Poncelet, M.; Roux, S. In situ μ CT-scan Mechanical Tests: Fast 4D Mechanical Identification. *Exp. Mech.* **2017**, *57*, 1327–1340. [CrossRef]
29. Zeng, R.; Fessler, J.; Balter, J. Respiratory motion estimation from slowly rotating X-ray projections: Theory and simulation. *Med. Phys.* **2005**, *32*, 984–991. [CrossRef] [PubMed]
30. Zeng, R.; Fessler, J.; Balter, J. Estimating 3-D respiratory motion from orbiting views by tomographic image registration. *IEEE Trans. Med. Imaging* **2007**, *26*, 153–163. [CrossRef] [PubMed]
31. Li, T.; Koong, A.; Xing, L. Enhanced 4D cone-beam CT with inter-phase motion model. *Med. Phys.* **2007**, *34*, 3688–3695. [CrossRef] [PubMed]
32. Delmon, V.; Vandemeulebroucke, J.; Pinho, R.; Oliva, M.V.; Sarrut, D.; Rit, S. In-room breathing motion estimation from limited projection views using a sliding deformation model. *J. Phys. Conf. Ser.* **2014**, *489*, 012026. [CrossRef]
33. Suzuki, Y.; Fung, G.; Shen, Z.; Otake, Y.; Lee, O.; Ciuffo, L.; Ashikaga, H.; Sato, Y.; Taguchi, K. Projection-based motion estimation for cardiac functional analysis with high temporal resolution: A proof-of-concept study with digital phantom experiment. In Proceedings of the Medical Imaging 2017: Physics of Medical Imaging. International Society for Optics and Photonics, Orlando, FL, USA, 11–16 February 2017; Volume 10132.
34. Li, T.; Schreibmann, E.; Yang, Y.; Xing, L. Motion correction for improved target localization with on-board cone-beam computed tomography. *Phys. Med. Biol.* **2006**, *51*, 253. [CrossRef] [PubMed]
35. Prummer, M.; Wigstrom, L.; Hornegger, J.; Boese, J.; Lauritsch, G.; Strobel, N.; Fahrig, R. Cardiac C-arm CT: Efficient motion correction for 4D-FBP. In Proceedings of the 2006 IEEE Nuclear Science Symposium Conference Record, San Diego, CA, USA, 29 October–1 November 2006; Volume 4, pp. 2620–2628.

36. Prummer, M.; Hornegger, J.; Lauritsch, G.; Wigstrom, L.; Girard-Hughes, E.; Fahrig, R. Cardiac C-arm CT: A unified framework for motion estimation and dynamic CT. *IEEE Trans. Med. Imaging* **2009**, *28*, 1836–1849. [CrossRef] [PubMed]
37. Rit, S.; Sarrut, D.; Desbat, L. Comparison of analytic and algebraic methods for motion-compensated cone-beam CT reconstruction of the thorax. *IEEE Trans. Med. Imaging* **2009**, *28*, 1513–1525. [CrossRef] [PubMed]
38. Katsevich, A. An accurate approximate algorithm for motion compensation in two-dimensional tomography. *Inverse Prob.* **2010**, *26*, 065007. [CrossRef]
39. Hahn, B.N. Motion Estimation and Compensation Strategies in Dynamic Computerized Tomography. *Sens. Imaging* **2017**, *18*, 10. [CrossRef]
40. Ruhlandt, A.; Töpperwien, M.; Krenkel, M.; Mokso, R.; Salditt, T. Four dimensional material movies: High speed phase-contrast tomography by backprojection along dynamically curved paths. *Sci. Rep.* **2017**, *7*, 6487. [CrossRef] [PubMed]
41. De Schryver, T.; Dierick, M.; Heyndrickx, M.; Van Stappen, J.; Boone, M.A.; Van Hoorebeke, L.; Boone, M.N. Motion compensated micro-CT reconstruction for in-situ analysis of dynamic processes. *Sci. Rep.* **2018**, *8*, 7655. [CrossRef] [PubMed]
42. Guangming, Z.; Ramzi, I.; Ran, T.; Gilles, L.; Peter, W.; Wolfgang, H. Space-time Tomography for Continuously Deforming Objects. *ACM Trans. Gr.* **2018**, *37*. [CrossRef]
43. Feldkamp, L.; Davis, L.; Kress, J. Practical cone-beam algorithm. *J. Opt. Soc. Am. A* **1984**, *1*, 612–619. [CrossRef]
44. Helfen, L.; Baumbach, T.; Mikulik, P.; Kiel, D.; Pernot, P.; Cloetens, P.; Baruchel, J. High-resolution three-dimensional imaging of flat objects by synchrotron-radiation computed laminography. *Appl. Phys. Lett.* **2005**, *86*, 071915. [CrossRef]
45. Batenburg, K.; Sijbers, J. DART: A practical reconstruction algorithm for discrete tomography. *IEEE Trans. Image Process.* **2011**, *20*, 2542–2553. [CrossRef] [PubMed]
46. Roux, S.; Leclerc, H.; Hild, F. Efficient binary tomographic reconstruction. *J. Math. Imaging Vis.* **2014**, *49*, 335–351. [CrossRef]
47. Candès, E.; Romberg, J.; Tao, T. Robust uncertainty principles: Exact signal reconstruction from highly incomplete frequency information. *IEEE Trans. Inf. Theory* **2006**, *52*, 489–509. [CrossRef]
48. Xu, Q.; Yu, H.; Mou, X.; Zhang, L.; Hsieh, J.; Wang, G. Low-dose X-ray CT reconstruction via dictionary learning. *IEEE Trans. Med. Imaging* **2012**, *31*, 1682–1697. [PubMed]
49. Jailin, C.; Bujac, A.; Hild, F.; Roux, S. Fast 4D tensile test monitored via X-CT: Single projection based Digital Volume Correlation dedicated to slender samples. *J. Strain Anal. Eng. Des.* **2018**, in press.
50. Buljac, A.; Jailin, C.; Mendoza, A.; Neggers, J.; Taillandier-Thomas, T.; Bouterf, A.; Smaniotto, B.; Hild, F.; Roux, S. Digital Volume Correlation: Review of Progress and Challenges. *Exp. Mech.* **2018**, *58*, 661–708. [CrossRef]
51. Maire, E.; Withers, P.J. Quantitative X-ray tomography. *Int. Mater. Rev.* **2014**, *59*, 1–43. [CrossRef]
52. Moireau, P.; Bertoglio, C.; Xiao, N.; Figueroa, C.A.; Taylor, C.; Chapelle, D.; Gerbeau, J.F. Sequential identification of boundary support parameters in a fluid-structure vascular model using patient image data. *Biomech. Model. Mechanobiol.* **2013**, *12*, 475–496. [CrossRef] [PubMed]
53. Grédiac, M.; Hild, F.; Eds. *Full-Field Measurements and Identification in Solid Mechanics*; John Wiley & Sons: Hoboken, NJ, USA, 2012.
54. Voit, D.; Zhang, S.; Unterberg-Buchwald, C.; Sohns, J.M.; Lotz, J.; Frahm, J. Real-time cardiovascular magnetic resonance at 1.5 T using balanced SSFP and 40 ms resolution. *J. Cardiovasc. Magn. Reson.* **2013**, *15*, 79. [CrossRef] [PubMed]

© 2018 by the authors. Licensee MDPI, Basel, Switzerland. This article is an open access article distributed under the terms and conditions of the Creative Commons Attribution (CC BY) license (http://creativecommons.org/licenses/by/4.0/).

Article

Incipient Bulk Polycrystal Plasticity Observed by Synchrotron In-Situ Topotomography

Henry Proudhon [1,*], Nicolas Guéninchault [1,†], Samuel Forest [1] and Wolfgang Ludwig [2,3]

1. MINES ParisTech, PSL Research University, MAT—Centre des Matériaux, CNRS UMR 7633, BP 87 91003 Evry, France; ngueninchault@xnovotech.com (N.G.); samuel.forest@mines-paristech.fr (S.F.)
2. ESRF, The European Synchrotron, CS 40220, 38043 Grenoble, France; ludwig@esrf.fr
3. INSA Lyon, MATEIS, University of Lyon, UMR 5510 CNRS, F-69621 Lyon, France
* Correspondence: henry.proudhon@mines-paristech.fr; Tel.: +33-160-763-070
† Current address: Xnovo Technology ApS, Theilgaards Alle 9, 1th, 4600 Køge, Denmark.

Received: 10 September 2018; Accepted: 10 October 2018; Published: 18 October 2018

Abstract: In this paper, we present a comprehensive 4D study of the early stage of plastic deformation in a polycrystalline binary AlLi alloy. The entire microstructure is mapped with X-ray diffraction contrast tomography, and a set of bulk grains is further studied via X-ray topotomography during mechanical loading. The observed contrast is analyzed with respect to the slip system activation, and the evolution of the orientation spread is measured as a function of applied strain. The experimental observations are augmented by the mechanical response predicted by crystal plasticity finite element simulations to analyze the onset of plasticity in detail. Simulation results show a general agreement of the individual slip system activation during loading and that comparison with experiments at the length scale of the grains may be used to fine tune the constitutive model parameters.

Dataset: 10.5281/zenodo.1412401

Keywords: polycrystal plasticity; X-ray diffraction imaging; topotomography; in situ experiment; finite element simulation; lattice curvature; rocking curve

1. Introduction

Determining microstructure-property relationships is an essential engineering problem and is directly linked to our ability to observe both the microstructure and the deformation/failure mechanisms concurrently. Electron back scattered diffraction (EBSD), which provides crystal orientation maps with sub-micrometer spatial resolution, remains a key tool, but is limited to the specimen surface [1]. To measure and interpret the strain field produced within individual grains, digital image correlation can be used provided a small-enough speckle can be produced at the specimen surface [2,3]. In this regard, subsequent analysis using numerical methods such as finite elements has also proved to be a powerful tool to interpret experimental results [4–6], but ultimately remains limited if the underlying material volume is not known [7]. In this paper, a new method combining in situ mechanical testing, three-dimensional (3D) bulk X-ray inspections and the crystal plasticity finite elements method (CPFEM) is used to study how plasticity proceeds in individual grains of a polycrystalline sample.

In the last 10 years, one particular focus of the 3D imaging community has been on obtaining reliable three-dimensional grain maps. Since most structural materials are polycrystalline and the mechanical properties are determined by their internal microstructure, this is a critical issue. For instance, considering slip transmission in crystal plasticity problems or small tortuous cracks evolving in a 3D grain network, it is a recurring challenge to assess the bulk mechanisms from surface observations only. Knowing how the different grains of the microstructure are arranged below the surface is therefore essential. Therefore, there has been considerable effort to develop characterization techniques at the meso-scale, which can image typically 1 mm^3 of material with a spatial resolution in the order of the micrometer.

Among 3D characterization, an important distinction exists between destructive and non-destructive techniques. Serial sectioning relies on repeated 2D imaging (which may include several modalities) of individual slices where a thin layer of material is removed between each observation [8]. Considerable progress on that side has been made in the last decade, bringing high quality measurements in 3D of grain sizes and orientations, but also detailed grain shapes and grain boundary characteristics. The price to pay with serial sectioning remains, however, the destruction of the sample. In parallel, accessing crystallographic information in the bulk of polycrystalline specimens was subsequently achieved by using the high penetration power of hard X-rays and leveraging diffraction contrast. This led to the development of a variety of 3D X-ray diffraction techniques (3DXRD; see [9,10]) enabling the characterization of millimeter-sized specimens by tracking the diffraction of each individual crystal within the material volume while rotating the specimen. Among them, the near field variant called X-ray diffraction contrast tomography (DCT) uses an extended box beam to illuminate the specimen and allows for simultaneous reconstruction of both the sample microstructure visible in X-ray absorption contrast and the crystallographic grain microstructure as determined from the diffraction signals with a single tomographic scan provided the grains have a limited orientation spread [11–13]. The typical acquisition time is one hour on a high brilliance beam line such as ID11at ESRF, which makes it the fastest non-destructive grain mapping technique.

Non-destructive imaging allows one to observe both the microstructure and the deformation/failure mechanisms in situ (4D studies). However, resolving 3D grain shapes by near-field diffraction imaging requires reducing the sample to detector distance to a few millimeters, which has for a long time drastically limited mechanical 4D studies due to the space constraints. Recent progresses with mechanical stress rigs solved this issue and opened new perspectives to study the deformation and fracture of polycrystalline materials [14,15].

Another key challenge is to link 3D microstructure characterization tools with computational models in order to predict engineering mechanical properties. This can be done using synthetically-generated images, but it requires using sophisticated models to ensure that the microstructure is representative, in particular of the tail distributions (critical when looking at fracture processes) [16]. Another approach is to use as-measured 3D microstructures [17]. The major advantage of this route is directly comparing experiment and mechanical simulation at the grain scale. For mechanical problems, continuum crystal plasticity (either using the finite elements or the spectral method to solve the equilibrium) has proved to be a powerful tool to interpret experimental results obtained in the deformation of metallic polycrystals [4,18–20]. Large-scale 3D polycrystalline simulations can be performed with sufficient local discretization to predict the transgranular plastic strain fields. One of the issues of this type of models is the material parameters' identification. This is mostly due to the fact that identification is classically done by minimizing a cost function versus macroscopic tests' response, which makes the problem ill posed. Some recent attempts directly used local strain measurements in the identification dataset in order to identify material model parameters [21], but have remained limited to surface studies so far.

The mechanical behavior of polycrystalline alloys is especially important for structural materials. A few key examples are Al alloys used for transports in general, Ti alloys used in aerospace where specific strength and high performance is particularly needed or Ni base alloys when looking for very high temperature performances. In this regard, advancing our understanding on the deformation of such materials will lead to lighter, stronger and more reliable parts.

In this paper, we present a new 4D study, using a combination of X-ray diffraction contrast tomography, topotomography and phase contrast tomography to study polycrystal plasticity in an aluminum-lithium alloy. Experimental and simulation methods are detailed in Section 2. The main experimental results are then presented in Section 3.1 for the observed slip system activation and

Section 3.2 for the measured crystal lattice orientation. Simulation results and comparison with the experiments are described in Section 3.3. In the Conclusion Section, the main results are summarized, and some future directions of research are suggested.

2. Materials and Methods

2.1. Experimental Setup for Topotomography

A plate made of Al-Li 2.5% wt purchased from MaTeck was rolled to 45% reduction and recrystallized 20 min at 530 °C to tune the grain size at 100 µm. This step ensured the number of grains in the illuminated section was well below the DCT limit (typically a few thousands) and that the initial state was as defect free as possible to study the onset of plasticity. The material was then aged 4 h at 100 °C to form very small Al_3Li precipitates with an expected size of 8 nm [22]. At this size, the precipitates remain shearable by the dislocations, which will promote a planar and localized slip as the main deformation mechanism. Small dog-bone tomographic tension specimens (0.5 mm × 0.5 mm minimal cross-section) were cut using EDM in the middle of the plate for the experiment. A 8.5-mm radius in the gauge length produces a small stress concentration and ensures that the first plastic events will occur in the observed region.

The experiment was carried out at the material science beam line (ID11) at the ESRF. The X-ray beam was produced by an in-vacuum undulator and collimated to a size of about 0.7 × 0.7 mm by means of an X-ray transfocator [23]. The beam energy was set to 40 keV with a relative bandwidth of 3×10^{-3}. The diffractometer installed on ID11 has been designed for this particular variant of diffraction imaging experiments, which requires the alignment of a scattering vector parallel to the tomographic rotation axis [24,25]. The scattering vector $\underline{G}$ is defined as $\underline{G} = \underline{K} - \underline{X}$ with $\underline{X}$ the incident wave vector and $\underline{K}$ the diffracted wave vector, both with a norm of $1/\lambda$. Let us call (p, q, r) the components of the scattering vector $\underline{G}_s$ expressed in the sample coordinate system for a set of (hkl) planes:

$$\underline{G}_s = g^{-1} B \begin{pmatrix} h \\ k \\ l \end{pmatrix} = \begin{pmatrix} p \\ q \\ r \end{pmatrix} \quad (1)$$

where g is the orientation matrix of the crystal and B accounts for the lattice geometry (for cubic structures like aluminum, it reduces to the lattice parameter times the identity).

Following [26,27] and accounting for the four diffractometer rotation angles $(\Theta, \omega, \phi, \chi)$, via the four associated rotation matrices (T, Ω, Φ, X), the Bragg diffraction condition in 3D can be written as:

$$\frac{2\sin^2(\theta_{Bragg})}{\lambda} = - \left[T\Omega\Phi X \begin{pmatrix} p \\ q \\ r \end{pmatrix} \right]_1 \quad (2)$$

where the rotation matrix X associated with the diffractometer angle χ should not be confused with the incident wave vector $\underline{X}$.

In a topotomographic experiment, the scattering vector is aligned with the rotation axis of the tomographic rotation stage ω (see Figure 1a). Circles χ and ϕ are used to set $\underline{G}$ parallel to the rotation axis ω (matrices X and Φ, respectively), while Θ is used to tilt the whole setup (including the tomographic rotation axis around Y by the Bragg angle θ_{Bragg} (matrix T). For a known crystal structure and orientation (i.e., B and g are known), it is straightforward to rework Equation (2) to derive the two tilt rotation values (χ, ϕ):

$$\chi = \arctan\left(-\frac{p}{r}\right) \quad (3)$$

$$\phi = \arctan\left(\frac{q}{-p\sin(\chi) + r\cos(\chi)}\right) \quad (4)$$

Equation (2) is then fulfilled for all possible values of ω. Note that in the DCT case, T, Φ and X vanish, and the equation is only verified for 2 particular values, solutions of a quadratic equation in ω known as the rotating crystal problem [26].

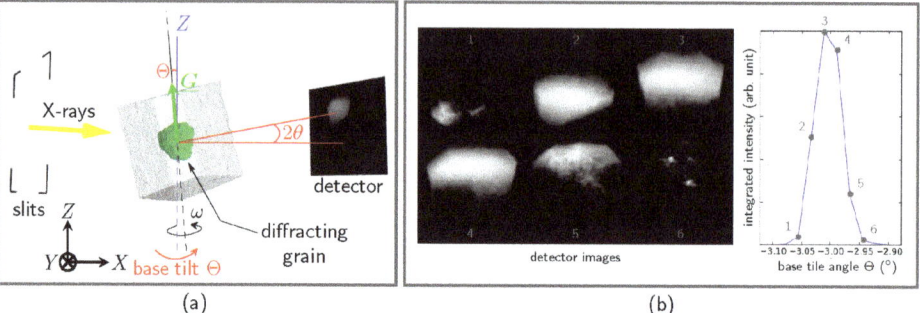

Figure 1. Schematics of the topotomographic alignment: (**a**) a scattering vector is put on the rotation axis (ω), and the whole setup is tilted by the nominal value $\Theta = \theta_{Bragg}$; (**b**) example of integrated detector intensity to build the rocking curve.

It is important to understand clearly the difference between Θ and θ_{Bragg}, as they differ in nature. Θ is the value of the tilt applied by the base tilt motor, whereas θ_{Bragg} is a material parameter associated with a given d_{hkl} spacing and wavelength λ: $\theta_{Bragg} = \arcsin \lambda/2d$.

Now, in the ideal case of a perfect crystal and quasi-monochromatic plane wave illumination, the entire grain would diffract for the position of the base tilt $\Theta = \theta_{Bragg}$, and only a simple rotation around ω would be needed to collect the topotomographic dataset (see Figure 1a). This is in practice never the case, as the inner mosaicity of the grain and dispersion effects require rocking the base tilt (this is again a rotating crystal problem, which is covered in more detail in Section 2.5). A topotomographic (TT) dataset can therefore be seen as a collection of rocking curves and associated stacks of 3D projection data and can be used in two different ways. First, integrated projection topographs corresponding to projections of the entire crystal volume are obtained by summing the intensity over the base tilt Θ. Inspection of these topographs (see Figure 1b) allows for direct identification of the presence of crystalline defects such as slip bands. Second, the width of the rocking curve $I = f(\Theta)$ for a given ω position is a measure of the lattice rotation around the base tilt axis (Y here; see Figure 1c).

2.2. Details of the In Situ Experiment

A small tomographic tension specimen was mounted in the Nanox stress rig, specifically designed to be compatible with both DCT and topotomography acquisition geometries [15]; see Figure 2, left. The machine has a very limited size and weight and, thanks to the load bearing quartz tube, allows 360° visibility in the DCT configuration with the detector as close as 3 mm to the rotation axis. Full visibility is also achieved in the TT configuration with $\theta_{Bragg} \leq 10°$, with the detector as close as 10 mm from the rotation axis, for the complete range of motion of the two inner diffractometer circles ($\pm 20°$ and $\pm 15°$ for Φ and χ, respectively). For given values of θ_{Bragg}, Φ and χ, it is possible to move the TT detector even closer, but this has to be checked manually.

With the specimen inside the stress rig, it is possible to analyze the initial undeformed bulk microstructure (positions, shapes and grain orientations) by a DCT scan. The DCT data are then processed to extract all grain orientations and positions within the gauge length. Later on, the DCT reconstruction is also used as input to perform 3D CPFEM calculations (see Section 2.3). For now, these data are used to select a series of grains for further analysis using topotomography imaging during mechanical loading. Here, the selection was based on the following criteria: (i) a low order reflection must be accessible (note that the two circles Φ, χ used for the topotomographic alignment only have a limited range of motion); (ii) the grain must be located in the bulk of the specimen; (iii) all the selected grains must form a small neighborhood. Grain Numbers 4, 10 and 18 (see Figure 3), located in the central region of the specimen, fit these constraints and were selected for the present study to carry out a series of topotomographic scans during a (interrupted) tensile test; see Figure 2, right.

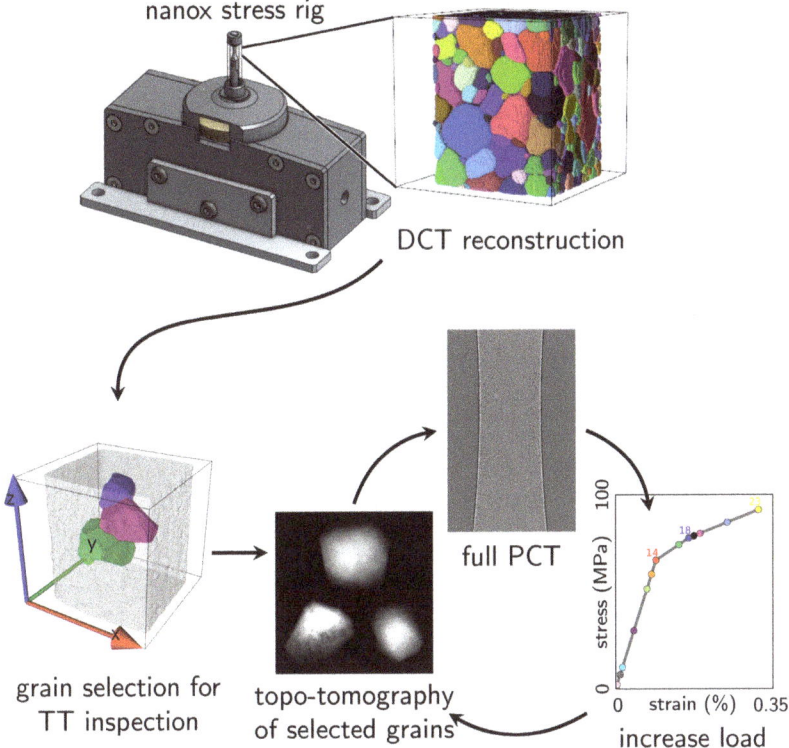

Figure 2. Sequence of the in situ topotomography experiment: the initial microstructure is characterized by DCT and analyzed after mounting the specimen into the Nanox device [15], then 23 complete sequences, each comprising three topotomography and one phase contrast tomography scan, have been recorded at increasing levels of load (the color code used in the tension curve is used later on to plot results at a given load level).

At each of the 23 load steps, a phase contrast tomography of the gauge length (1.5 mm in height) is also recorded and used to extract macroscopic strain information. The essential information (grain orientation, aligned hkl reflection, Bragg angle, diffractometer angles) for each grain is reported in Table 1. Grains 10 and 18 have a similar orientation, as seen in Figure 3c).

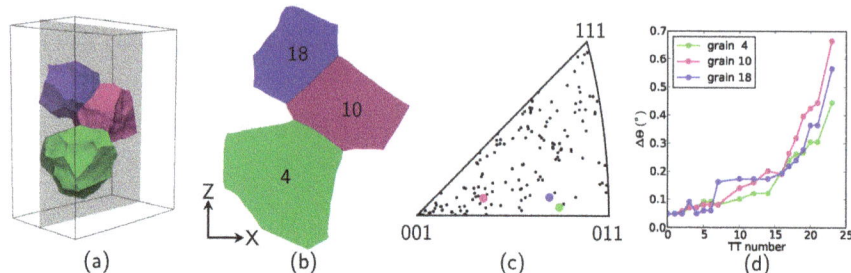

Figure 3. Details on the 3-grain cluster (a) 3D visualization of the grains (b); XZ slice through the 3 grains; (c) inverse pole figure of the gauge length with the 3 grain orientations of interest highlighted; (d) Θ integration range automatically determined at each loading step.

Table 1. Details of the 3 grains selected for TT imaging; the orientation convention is consistent with [28]; and angle values are given in °.

Grain ID	Orientation (Rodrigues)	Aligned Reflection	θ_{Bragg}	(Φ, χ) Values (°)
4	[0.050, −0.305, 0.104]	(202)	6.21	(0.52, −11.04)
10	[−0.028, −0.145, 0.062]	(002)	4.39	(2.14, 16.63)
18	[−0.135, −0.272, −0.333]	(202)	6.21	(−4.81, −12.88)

DCT scans were composed of 3600 equally-spaced projections over 360°, recorded on a high-resolution detector with a transparent luminescent screen optically coupled by a 10× microscope objective to a 2048 × 2048 pixel ESRF Frelon camera, giving an effective pixel size of 1.4 µm. A 0.3-s exposure time has been used, resulting in an acquisition time of about 40 min. The same camera was used to record the PCT scans; the camera traveled back and forth along the X-axis from the DCT position (about 5 mm behind the specimen) to the PCT position, about 105 mm downstream. A second detector system with a 20× objective and an effective pixel size of 0.7 µm was used for topotomographic scan acquisition. The angular range of the rocking curve scans was automatically adjusted after each load increment in order to cover the entire width of the crystal reflection curve for any ω rotation position of the sample. Moreover, the X-ray flux density was further increased by focusing the beam on the area covered by the 3-grain cluster. A continuous motion acquisition procedure with a fixed integration range of 0.1° and 0.5-s exposure time per image was used. Integration gaps caused by the readout time of the CCD detector could be eliminated by operating the system in frame transfer mode. In this mode, only half of the active area is available for image acquisition, whilst the other half is used for temporary storage and readout of the previous frame. This procedure was repeated every 4°, and a complete topotomographic acquisition comprising 90 such rocking scans per grain typically lasted from 10 min up to an hour as the Θ integration range increased during loading. In this experiment, the integration range determination after each load step has been automated by acquiring a coarse TT scan at two values of ω separated by 90°, post-processing the intensity in the image and taking the largest bounds of the two rocking curves, increased by a small amount not to miss any intensity. From the beginning to the end of the experiment, the integration range has been multiplied by a factor of 10 (from 0.05° to 0.5°; see Figure 3d).

2.3. Crystal Plasticity Finite Element Simulations

A finite strain crystal plasticity model, fully described in [29], is used here to compute the mechanical response of the polycrystalline sample under tension. It is based on the multiplicative decomposition $\mathbf{F} = \mathbf{EP}$ of the deformation gradient, $\mathbf{F}$, into an elastic part, $\mathbf{E}$, and a plastic part, $\mathbf{P}$. The multiplicative decomposition is associated with the definition of an intermediate configuration for which the elastic part of the deformation gradient is removed. The intermediate released configuration is uniquely determined up to a rigid body rotation, which is chosen such that the lattice orientation in the intermediate configuration is the same as the initial one. Mandel called it the isoclinic intermediate configuration [30]. As a result, lattice rotation and distortion during elastoplastic deformation are contained in the elastic deformation part $\underline{\underline{E}}$. The transformation $\underline{\underline{E}}$ has a pure rotation part $\underline{\underline{R}}^e$ and a pure distortion part $\underline{\underline{U}}^e$, which can be obtained by the polar decomposition:

$$\underline{\underline{E}} = \underline{\underline{R}}^e \underline{\underline{U}}^e \tag{5}$$

Plastic deformation is the result of slip processes according to a collection of N slip systems, each one characterized by the slip direction $\underline{m}^s$ and the normal to the slip plane $\underline{n}^s$. Note that here, plastic slip is the only deformation mechanism considered. This has been double checked up to 10% strain by performing an in situ tensile test in a scanning electron microscope (not reported here for brevity). In other cases, mechanical twinning or grain boundary sliding may need to be considered. In the intermediate configuration, $\underline{\underline{P}}$ verifies:

$$\underline{\underline{\dot{P}}}\underline{\underline{P}}^{-1} = \sum_{s=1}^{N} \dot{\gamma}^s \underline{m}^s \otimes \underline{n}^s \tag{6}$$

In order to analyze the microplastic behavior of the studied AlLi polycrystal, an elasto-visco-plastic crystal plasticity model was selected. Numerical computations were performed using the Z-set software (http://www.zset-software.com) (see [31]). The slip rate on a given slip system s depends, via a phenomenological power law with two parameters K and n, on how much the resolved shear stress τ^s exceeds the threshold $\tau_0 + r^s$:

$$\dot{\gamma}^s = \left\langle \frac{|\tau^s| - \tau_0 - r^s}{K} \right\rangle^n \text{sign}(\tau^s) \tag{7}$$

Here, τ_0 is the critical resolved shear stress, and r^s, initially zero, increases with increasing plastic strain and hardens the system s through a non-linear isotropic Voce hardening rule, as developed in [32]:

$$r^s = Q \sum_{r=1}^{N} h^{sr}(1 - \exp(-bv^s)) \tag{8}$$

v^s is the cumulative slip and h^{sr} denotes the interaction matrix taking into account the relative influence of slip systems on each other. It includes the self and latent hardening, and only indirect and estimated quantitative information is available about the components of this matrix (see for instance [33,34]). Q and b are 2 material parameters to be determined.

Monotonic tensile tests were performed on five different macroscopic samples, at three strain rates, and a numerical optimization using Z-set implementation was performed to find a suitable parameter set (τ_0, Q, b), as presented in Table 2. The yield stress of the material exhibits an inverse strain rate sensitivity, which prevented identifying the viscosity parameters K (not to be confused with the norm of the diffraction vector $\underline{K}$) and n; instead, sensible values for aluminum alloys have been used. Modeling this effect requires a more complex model, including dynamic strain aging as in [35]. For the interactions between dislocations, coefficients h^{rs} from [36] have been used.

Table 2. Material parameters identified from the macroscopic tensile tests.

K (MPa$^{1/n}$)	n (-)	τ_0 (MPa)	Q (MPa)	b (-)
38	10	10	5.3	763

The experimental grain map is used as input to produce a high fidelity digital clone of the specimen. The entire $L = 1.57$ mm zone, where 3 DCT scans were merged, was used to ensure the boundary condition application is far enough from the grains of interest to avoid any boundary layer effect [37]. Details on how the mesh was produced can be found in [17]. The initial grain boundary surface generated contains a very large number of triangles, and an iterative decimation approach using an edge collapsing algorithm is applied. The surface mesh is filled with tetrahedra controlling the mesh density as a function of the euclidean distance d from the three-grain cluster. This allowed minimizing the computational cost while preserving a rich description of the mechanical fields in the region of interest. The final mesh is composed of 341,687 linear tetrahedra with a gradient in element size (the ratio between the maximum and minimum tetrahedron size is about 4000) visible in Figure 4.

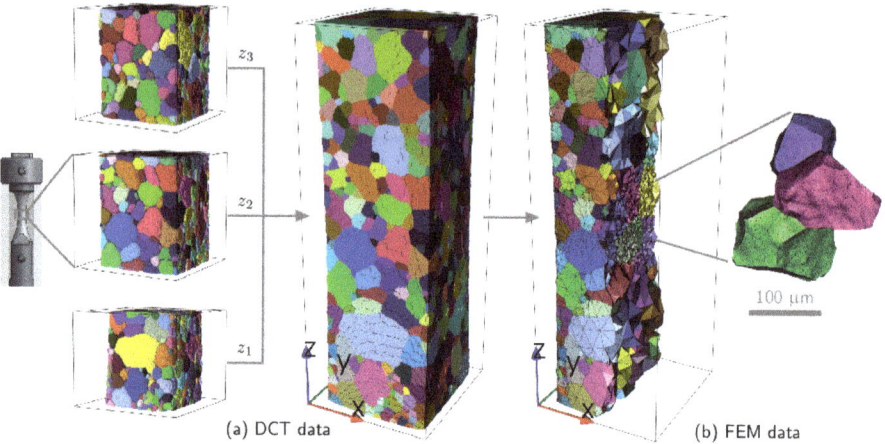

Figure 4. Comparison between the DCT data (**a**) and the mesh generated (**b**); the colors denote the grain numbers, which are consistent from the experiment to the simulation. Note the specimen shape with a radius in the gauge length as mentioned in Section 2.

Dirichlet boundary conditions ($u_z = 0$ on the lower face and $u_z = 15.7$ microns on the upper face for the final deformation step) were imposed to deform the specimen in tension up to 1% total strain in 100 steps. Suitable boundary conditions have been set on the lower surface of the sample to prevent any rigid body motion, and lateral surfaces were free of stress. The steps corresponding to the experimentally-measured strain (for instance 0.32% total strain) can be used for comparison.

2.4. Lattice Rotations

The continuum mechanical approach used here makes it possible to distinguish between the transformation of material and lattice directions. Material lines are made of material points that are subjected to the motion field u. In contrast, lattice directions are not material insofar as they are not necessarily made of the same material points (atoms) in the initial and current configurations due to the passing of dislocations, but keep the same crystallographic meaning. According to the concept of isoclinic configuration, lattice directions are unchanged from the initial to the intermediate configuration. Dislocations passing through a material volume element do not distort nor rotate the lattice, although material lines are sheared. According to the continuum theory of dislocations, statistically-stored dislocations accumulating in the material volume element affect material hardening, but do not change the element shape. Accordingly, an initial lattice direction $\underline{d}^\sharp$ is transformed into $\underline{d}$ by means of the elastic deformation:

$$\underline{d} = \mathbf{E}.\underline{d}^\sharp \tag{9}$$

This important distinction allows one to precisely compute both the local rotation and distortion of the crystal lattice, which will be further used to derive the 3D rocking curve of a grain from its deformed state in the simulation (see Section 2.5).

2.5. Rocking Curves Simulations from CPFEM Data

As explained in Section 2.1, the rocking curve represents the intensity diffracted by the illuminated grain at a given Θ angle. As soon as the crystal deforms, the exact Bragg condition is violated, and the $I(\Theta)$ curve will widen. In crystal plasticity, geometrically necessary dislocations (GND) give rise to gradients of crystal orientation, leading to local modification of the Bragg condition. The problem is therefore to solve the 3D diffraction condition stated in Equation (2) for the angle Θ, with a locally deformed crystal lattice. In Equation (2), the values of (p, q, r), as well as θ_{Bragg} need to be updated for the new lattice geometry. It is therefore possible to use the mechanical fields computed in each element (namely $\underline{E}$ and $\underline{R}^e$) to evaluate locally the Bragg condition (for a given ω value) and to find the corresponding Θ. Building the volume weighted histogram for all elements within the grain will produce a simulated rocking curve (for this value of ω).

3. Results

3.1. Topography Results

X-ray topographs are 2D oblique projections of a crystal. At the onset of plasticity, slip system activity may modify the local Bragg condition within the grain and produce orientation contrasts on the detector. In this section, topographs over a full ω turn for three neighboring grains as a function of the load are analyzed.

As we shall see, the perturbations of the crystal lattice are localized within the slip plane and may only be visible at certain ω angles, called edge-on configuration, when the diffraction direction is contained in the plane (i.e., the tilted slip plane normal $\underline{n}_t$ is perpendicular to $\underline{K}$). Knowing the grain orientation and the tilt geometry, it is straightforward to obtain the two edge-on ω angles for a given slip plane by solving (here, we do not account for the base tilt, as both the left and right side of the equation would be equally affected):

$$(\underline{\Omega}.\underline{n}_t).\underline{K} = 0 \quad \text{with} \quad \underline{K} = \frac{1}{\lambda} \begin{pmatrix} \cos(\theta_{Bragg}) \\ 0 \\ \sin(\theta_{Bragg}) \end{pmatrix} \quad (10)$$

This means solving $[n_t[0]\cos(\omega) - n_t[1]\sin(\omega)]\cos(\theta) + n_t[2]\sin(\theta) = 0$. The two ω values, for a given slip plane, are separated by close to 180°, a value depending on θ_{Bragg}.

Table 3 gathers the ω values for the two slip systems with the highest Schmid factor in the 3three grains calculated with Equation (10). These values will be used to show the topographs in edge-on configuration where the contrast is expected to be maximal for a given slip plane. Note that the values for the two observed slip planes (not to be confused with the two values (ω_1, ω_2) for a given slip plane) are exactly separated by 180 degrees. For Grains 4 and 18, the aligned reflection is (202). Rotating around this axis, the edge-on configurations for the slip planes with the two highest Schmid factors ($1\bar{1}1$) and (111) are 180° apart (they share the $[\bar{1}01]$ zone axis, which is perpendicular to the scattering vector). For Grain 10, the aligned reflection is (002); rotating around this axis, the edge-on configurations for the (111) and ($11\bar{1}$) planes are also 180° apart (they share the $[\bar{1}10]$ zone axis, which is perpendicular to the scattering vector).

The Schmid factor allows estimating the resolved shear stress and is here calculated in the single crystal approximation. It should be noted that it cannot rigorously be applied to the polycrystal case (this value neglects the effect of neighboring grains and any other heterogeneity), although it is very often used to predict slip system activation. It is thus interesting to see in this case how well this indicator performs. In contrast to the estimation using the Schmid factor, the full field finite element simulations performed in this work (see Section 3.3) take into account the interaction of individual grains with their neighbors. This results in strongly non-homogeneous fields of plastic strain and lattice rotation inside the grains. The advantage of the Schmid factor is clearly that it depends only on the crystal orientation (and is thus easily computed on the fly during this type of diffraction experiment). However, the material scientist doing an experiment may rely on a more detailed stress estimation, either using a supplementary far field detector and track the motion of diffraction spots [19] or inferring the stress tensor value from mean field or full field computations using the actual 3D microstructure, as done in the present work.

Table 3. Edge-on ω values (in °) for the two slip systems with the highest Schmid factor in the 3 grains; ω values in bold are used to show the topographs in edge-on configuration in Figure 5.

Grain ID	(hkl)	Schmid Factor	(ω_1, ω_2)
4	(1-11)	0.476	(338.6, **176.3**)
	(111)	0.461	(158.6, 356.3)
10	(111)	0.490	(214.5, **40.8**)
	(1-11)	0.473	(304.5, 130.8)
18	(111)	0.488	(209.1, 46.8)
	(1-11)	0.457	(**29.1**, 226.8)

Integrated topographs at the recorded ω closest to the edge-on values and for selected load levels are shown in Figure 5. Contrast forming bands within the grains are clearly visible and appear first for Grain 10, then Grain 18 and finally for Grain 4.

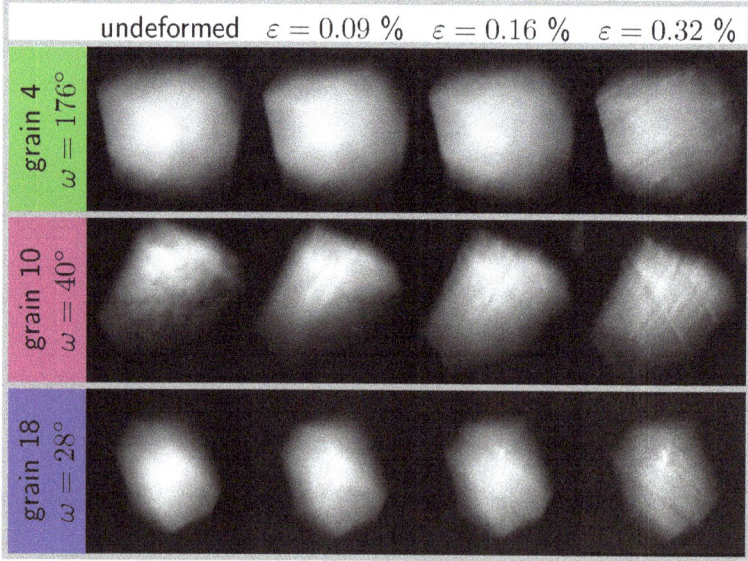

Figure 5. Topographs, integrated over Θ, in edge-on configuration for each grain of the cluster and different load levels; videos of the complete ω set are available as Supplementary Material for Grain 4 in the initial and deformed states.

These sets of bands were parallel; most of the time, they extended through the whole grain, and their number increased as the deformation increased. Going through the whole projection set for each grain shows that two sets of bands were visible at different ω values. For Grain 10, the crystallographic configuration was such that both sets were visible at around 40°. This is, to the best of the author knowledge, the first in situ observation of bulk plasticity in a millimeter sized polycrystalline specimen. From there, the orientation of the bands, their number and location within the grain can be further studied.

Using the initial grain orientation and the tilt geometries for each aligned reflection, it is possible to correlate the angle of those bands to specific crystallographic planes (see Figure 6). For this, a 3D geometrical representation of the grain was built using the DCT reconstruction, and the relevant slip planes had been added inside according to the measured grain orientation (the open source library pymicro [38] was used to this end). The grain can be tilted in the topotomographic condition and rotated to the given edge-on omega angle. Using a parallel projection mode and setting the view in the diffracted beam direction $\boldsymbol{K} = (1, 0, \tan(2\theta_{Bragg}))$ produced the same conditions as when collecting the topographs. All the observed bands had been identified without any ambiguity as projections of the {111} planes; see Figure 6. In each case, the exact slip plane could be identified and further related to the Schmid factor. This demonstrates that the onset of plasticity, from the first slip band to the more advanced state where several slip systems are active, was indeed captured in situ during this experiment.

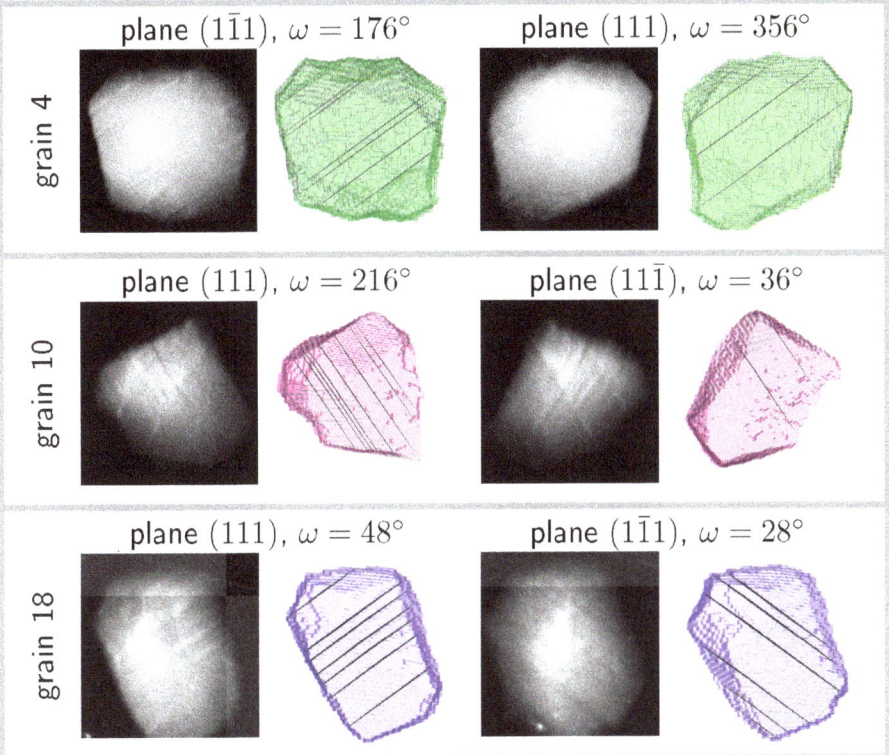

Figure 6. Identification of the bands as slip plane traces visible in the edge-on geometrical configuration in the topographs; here, the two observed active slip planes are shown for one of the two ω angles; the active slip plane locations within the grains have been measured manually and displayed in 3D.

For Grains 4 and 18, the two observed slip planes correspond to the two highest Schmid factors (see Table 3), whereas for Grain 10, they correspond to the first and third highest Schmid factors. The slip system corresponding to the second highest Schmid was not observed to be active in this grain.

Rocking curves are presented in Figure 7, for each grain at the strain levels and for the same ω values as in Figure 5. All three grains exhibited a consistent behavior, with a narrow curve at the beginning, which first shifted to the lower $|\theta_{Bragg}|$ values due to elastic loading (increase of the d_{hkl} interplanar spacing) and then widened considerably when plasticity took place.

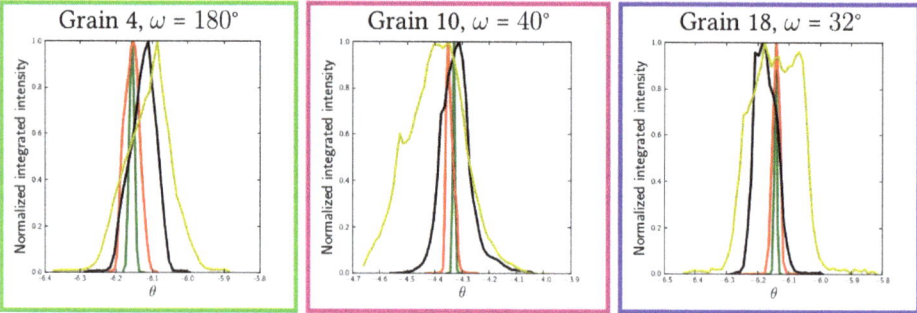

Figure 7. Evolution of the rocking curves for each grain of the cluster during in situ loading; the curve colors (green, red, black, yellow) refer to the load levels in Figure 5, respectively (undeformed, 0.09%, 0.16% and 0.32%).

3.2. 3D Rocking Curves' Results

As grains are aligned in a topotomographic sense, it is possible to measure rocking curves at every ω position. This measure is therefore sensitive not only to the amount of curvature of a crystal, but also to the orientation of this curvature in real space. To quantify the intragranular orientation spread revealed by a rocking curve, we introduced the width of the rocking curve at 10% of the peak of the normalized intensity, denoted as full width of the effective misorientation (FWEM). Although the effective misorientation describes the change in Bragg condition due to both orientation and lattice spacing variations, in practice, the orientation effect is largely predominant. Therefore, this value is a direct (qualitative) measure of the orientation spread, around the axis defined by the base tilt.

The FWEM was measured every 4° (for each ω position) and is plotted at all different load levels in Figure 8 to observe its evolution with increasing plasticity. An interesting dumbbell-shaped curve was consistently obtained for the three grains. The curve widened in a preferential direction linked to the active slip systems within the grain. One can observe that the FWEM was similar for Grains 4 and 18, which have the same combination of active slip systems. The orientation of the curves for Grain 10 was different and exhibited a clear reorientation of the preferential direction towards the end of the loading sequence. This may be linked to changes in the relative activity of the dominant slip system(s) during deformation, but would require further analysis.

The shape of the curve can be understood more clearly considering the idealized case of a strain-free crystal bent by an amount $\Delta\Theta$ by geometrically necessary dislocations. In the kinetic approximation, this configuration would produce a figure made of two tangent circles giving a very limited FWEM in the bending axis direction (almost no lattice rotation) and of exactly $\Delta\Theta$ perpendicular to it. The FWEM can be seen as the limit of the Bragg condition when rocking the base tilt Θ and can be computed using the rotating crystal equation solving for Θ for a known ω (see Figure 8, bottom right).

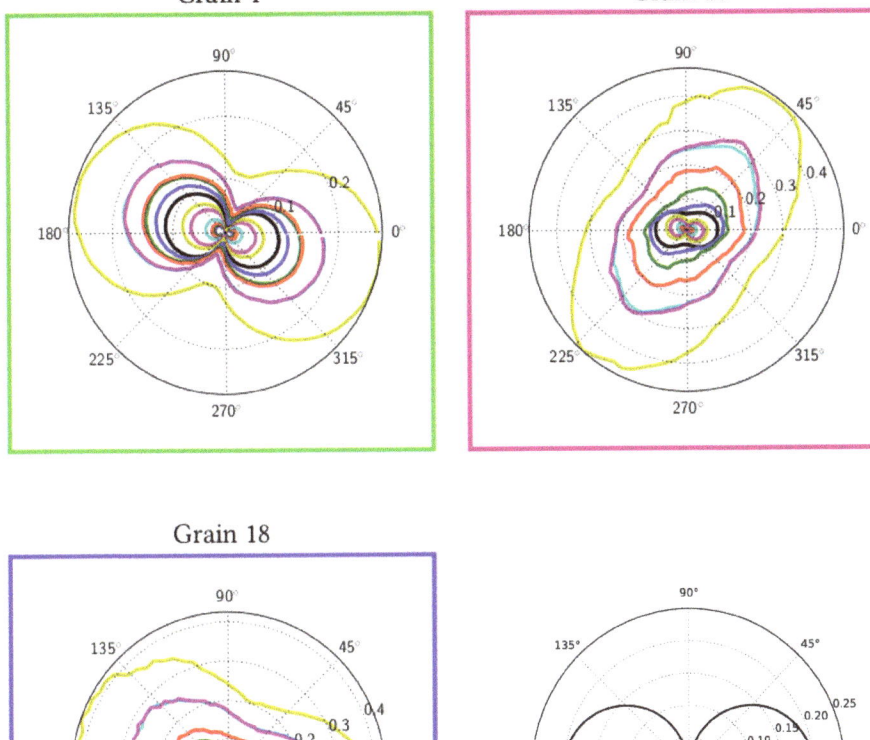

Figure 8. In situ FWEM as a function of ω, for each grain of the cluster (the color code matches the one used for the tensile curve in Figure 2) and example of the FWEM in the case of a crystal ideally bent around a single axis by 0.2° (bottom right).

3.3. Comparison with CPFEM Simulations

The first comparison is the slip activity computed with the numerical model (seen Section 2.3) for each grain with the activity visible on the topographs for a particular ω position (see Section 3.1). The CPFEM simulations give access to the active slip systems as opposed to the slip planes only, which were identified experimentally on the topograph. The slip activity was captured by averaging the amount of slip for a given slip system within the grain and can be compared to the qualitative information obtained on the topographs (intensity of the contrast and number of bands).

Accumulated plasticity (all slip system contributions) within the bulk is shown in Figure 9a, and the slip activity for each of the three grains is detailed in Figure 9b. For Grains 4 and 18, which have close orientations and behave similarly, the model predicted double slip with systems (111)[0-11] and (1-11)[011], in agreement with the experimentally observed slip planes, and no other slip activity. The situation is different in Grain 10, which also showed double slip experimentally, but with planes (111) and (11-1), whereas the model predicts plastic activity on three planes (111), (1-11) and (11-1). It is interesting to see that despite the general agreement, the details of the slip system activation are far from perfect. The discrepancy may be attributed to the parameters of the constitutive laws, especially regarding the interaction matrix coefficients h^{rs}. These parameters were determined from macroscopic tensile curves and from the literature. A detailed parametric study is necessary to analyze the impact of these parameters values on the activation of slip systems in that grain. This can be seen as an opportunity to use such experimental data collected at the length scale of the grains and in the bulk to enrich identification datasets and solve the long-standing issue of crystal plasticity material parameter identification [39].

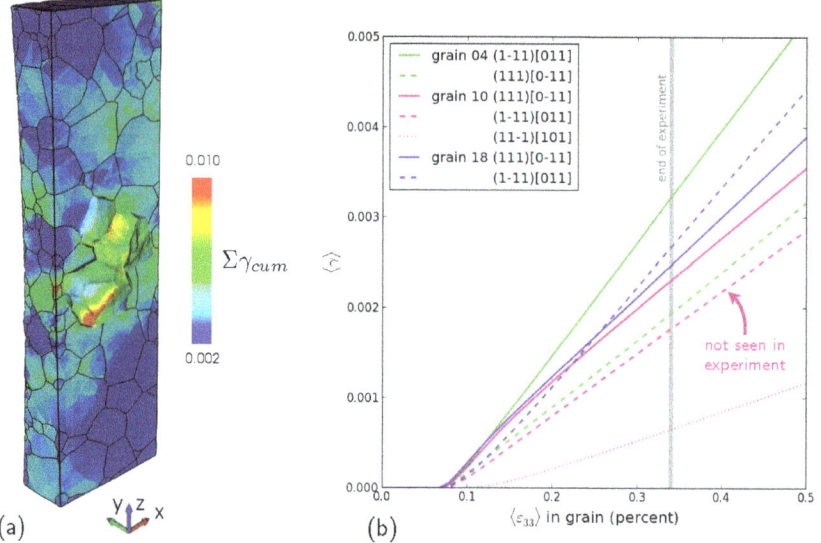

Figure 9. Predicted plastic activity: (**a**) view of the interior of the specimen showing the accumulated plasticity at the load level corresponding to the end of the experiment; (**b**) average plastic activity in the three grains of the cluster; the most active slip system is represented with a solid line and the second most active with a dashed line, while the grain color code is used.

Using the simulated mechanical fields within the grains, rocking curves can be simulated as described in Section 2.5. The generated rocking curves for each grain of the cluster are plotted in Figure 10, at $\omega = 165°$. The obtained behavior was consistent with the experimental observations. The rocking curve first shifted to lower Θ values due to the increase of the interplanar distance during loading. A very small amount of broadening due to the heterogeneous elastic strain field within the grain was also observed. When plasticity sets in and slip systems start to be active, the curve widening was more pronounced, and the shape changed, which can be related to the tendency to form subgrain-like regions. This effect had been studied both theoretically and experimentally, for instance by [40–42].

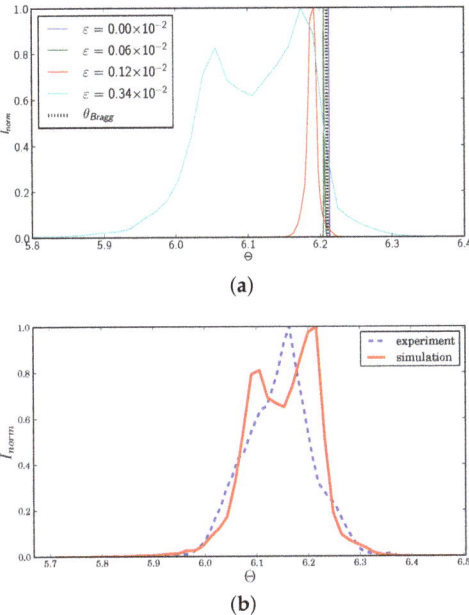

Figure 10. (a) Simulated rocking curves for Grain 4 at $\omega = 165°$ at four different strain levels; (b) comparison between experimental and simulated rocking curves at $\varepsilon = 0.34 \times 10^{-2}$.

Finally, a quantitative comparison between experimental FWEM and simulated FWEM is presented in Figure 11. Simulated curves are plotted for an applied strain of $\varepsilon_{33} = 0.0034$, and experimental surfaces are plotted for the last step of the topotomography experiment. Both curves were in very good agreement for Grain 4 and Grain 18, for both shape and orientation. For Grain 10, the surface had the right amplitude, but not the right direction. As explained previously, the direction of the dumbbell-shaped curve was linked to the precise combination of active slip systems within the grain, and the discrepancy between the predicted and observed slip systems was presumably the cause of the mismatch in orientation for this case.

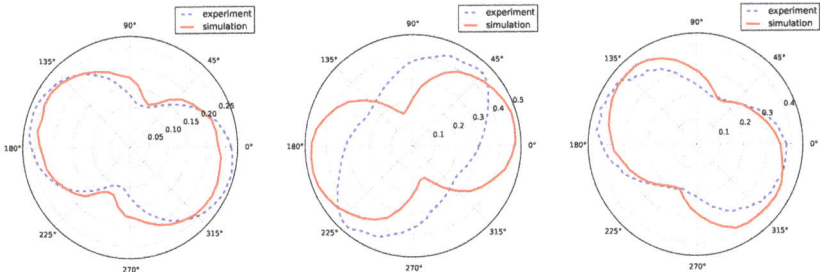

Figure 11. Comparison between experimentally simulated full width of the effective misorientation (FWEM) at $\varepsilon = 0.34 \times 10^{-2}$ for the three investigated grains.

4. Conclusions

An in situ topotomographic experiment has been carried out at a synchrotron facility to collect a very detailed 4D dataset to study the onset of plastic slip under tension in an aluminum lithium alloy. Initial DCT imaging allowed measuring the initial microstructure and selecting a set of three grains in the bulk of the specimen for further investigations. Upon loading, incipient plasticity was observed non-destructively in the form of band contrast in X-ray topographs, which were further related geometrically to active slip planes in the bulk.

Information collected from rocking curves was presented in the form of polar plots illustrating the anisotropy of mean lattice curvature. The polar plots have a characteristic dumbbell shape, which can be attributed to lattice bending with respect to some specific axis, which could be determined in the experiment for the three grains during deformation.

Crystal plasticity simulations were carried out to compare the prediction obtained with a classical continuum model with our experimental observations. Essential features, such as slip system activation and average misorientation per grain, are well captured by the model, although some discrepancies remain for one grain. A major contribution of the work is the comparison of experimental and simulated misorientation polar plots showing characteristic dumbbell shapes quantifying the anisotropy of lattice curvature.

A striking feature of the presented experiment is the direct observation of inelastic deformation mechanisms in the bulk (plastic slip, but also twinning or phase transformation are possible). The detected slip system activity could be used in the material parameter identification procedure instead of macroscopic tests only. This would require a specific treatment (for instance using grain averaged quantities) since the full field calculation is costly and not well suited for an optimization routine.

One of the limitations of the present work is that only one specimen could be tested in the given beam-time. Building on this experiment and using the developed automation algorithms, future work will target more specimens and more grains per specimen to obtain more statistics and study slip transmission in more detail.

The adaptation of advanced reconstruction techniques [13] to this type of combined DCT and topotomography acquisitions might allow one to retrieve finer details of the orientation fields and their evolution at increasing levels of applied strain. This could be used to study more complex inelastic mechanisms in metals or to extract more dependable constitutive parameters minimizing the discrepancies between experimental observations and numerical simulations on the digital twins of the tested specimen.

Supplementary Materials: The following are available online at www.mdpi.com/1996-1944/11/10/2018/s1: Video S1: Full set of integrated topographs over 360 degrees (every 4°) for Grain 4 in the undeformed state; Video S2: Full set of integrated topographs over 360 degrees (every 4°) for Grain 4 in the deformed state.

Author Contributions: H.P. and W.L. conceived of and designed the experiments. N.G., W.L. and H.P. performed the experiments. N.G. and H.P. analyzed the data. S.F. contributed the crystal plasticity model. N.G. ran the simulations. H.P. and N.G. wrote the paper. W.L. and S.F. provided comments on the manuscript.

Funding: MINES Paristech is acknowledged for funding the PhD of N.G.; this research received an additional funding by Association Instituts Carnot grant number MIN ANR CA 6263.

Acknowledgments: The authors thank the ESRF for providing the beam time for this experiment under proposal MA2285.

Conflicts of Interest: The authors declare no conflict of interest.

Abbreviations

The following abbreviations are used in this manuscript:

EDM	Electro-discharge machining
DCT	Diffraction contrast tomography
TT	Topotomography
PCT	Phase contrast tomography
CPFEM	Crystal plasticity finite element method

References

1. Wilkinson, A.J.; Britton, T.B. Strains, planes, and EBSD in materials science. *Mater. Today* **2012**, *15*, 366–376. [CrossRef]
2. Chen, Z.; Daly, S.H. Active Slip System Identification in Polycrystalline Metals by Digital Image Correlation (DIC). *Exp. Mech.* **2016**, *57*, 115–127. [CrossRef]
3. Stinville, J.; Echlin, M.; Texier, D.; Bridier, F.; Bocher, P.; Pollock, T. Sub-Grain Scale Digital Image Correlation by Electron Microscopy for Polycrystalline Materials during Elastic and Plastic Deformation. *Exp. Mech.* **2016**, *56*, 197–216. [CrossRef]
4. Roters, F.; Eisenlohr, P.; Hantcherli, L.; Tjahjanto, D.D.; Bieler, T.R.; Raabe, D. Overview of constitutive laws, kinematics, homogenization and multiscale methods in crystal plasticity finite-element modeling: Theory, experiments, applications. *Acta Mater.* **2010**, *58*, 1152–1211. [CrossRef]
5. Signor, L.; Villechaise, P.; Ghidossi, T.; Lacoste, E.; Gueguen, M.; Courtin, S. Influence of local crystallographic configuration on microcrack initiation in fatigued 316LN stainless steel: Experiments and crystal plasticity finite elements simulations. *Mater. Sci. Eng. A* **2016**, *649*, 239–249. [CrossRef]
6. Guan, Y.; Chen, B.; Zou, J.; Britton, T.B.; Jiang, J.; Dunne, F.P.E. Crystal plasticity modelling and HR-DIC measurement of slip activation and strain localization in single and oligo-crystal Ni alloys under fatigue. *Int. J. Plast.* **2017**, *88*, 70–88. [CrossRef]
7. Zeghadi, A.; N'Guyen, F.; Forest, S.; Gourgues, A.F.; Bouaziz, O. Ensemble averaging stress–strain fields in polycrystalline aggregates with a constrained surface microstructure—Part 1: Anisotropic elastic behavior. *Philos. Mag.* **2007**, *87*, 1401–1424. [CrossRef]
8. Echlin, M.P.; Straw, M.; Randolph, S.; Filevich, J.; Pollock, T.M. The TriBeam system: Femtosecond laser ablation in situ SEM. *Mater. Charact.* **2015**, *100*, 1–12. [CrossRef]
9. Poulsen, H.F. An introduction to three-dimensional X-ray diffraction microscopy. *J. Appl. Crystallogr.* **2012**, *45*, 1084–1097. [CrossRef]
10. Borbély, A.; Kaysser-Pyzalla, A.R. X-ray diffraction microscopy: Emerging imaging techniques for nondestructive analysis of crystalline materials from the millimetre down to the nanometre scale. *J. Appl. Crystallogr.* **2013**, *46*, 295–296. [CrossRef]
11. Ludwig, W.; King, A.; Reischig, P.; Herbig, M.; Lauridsen, E.; Schmidt, S.; Proudhon, H.; Forest, S.; Cloetens, P.; du Roscoat, S.R.; et al. New opportunities for 3D materials science of polycrystalline materials at the micrometre lengthscale by combined use of X-ray diffraction and X-ray imaging. *Mater. Sci. Eng. A* **2009**, *524*, 69–76. [CrossRef]
12. Ludwig, W.; King, A.; Herbig, M.; Reischig, P.; Marrow, J.; Babout, L.; Lauridsen, E.M.; Proudhon, H.; Buffière, J.Y. Characterization of Polycrystalline Materials Using Synchrotron X-ray Imaging and Diffraction Techniques. *JOM* **2010**, *62*, 22–28. [CrossRef]
13. Viganò, N.; Tanguy, A.; Hallais, S.; Dimanov, A.; Bornert, M.; Batenburg, K.J.; Ludwig, W. Three-dimensional full-field X-ray orientation microscopy. *Sci. Rep.* **2016**, *6*, 20618. [CrossRef] [PubMed]
14. Schuren, J.C.; Shade, P.A.; Bernier, J.V.; Li, S.F.; Blank, B.; Lind, J.; Kenesei, P.; Lienert, U.; Suter, R.M.; Turner, T.J.; et al. New opportunities for quantitative tracking of polycrystal responses in three dimensions. *Curr. Opin. Solid State Mater. Sci.* **2015**, *19*, 235–244. [CrossRef]
15. Guéninchault, N.; Proudhon, H.; Ludwig, W. Nanox, a miniature mechanical stress rig designed for near-field X-ray diffraction imaging techniques. *J. Synchrotron Radiat.* **2016**, *23*, 1474–1483. [CrossRef] [PubMed]
16. Turner, D.M.; Kalidindi, S.R. Statistical construction of 3-D microstructures from 2-D exemplars collected on oblique sections. *Acta Mater.* **2016**, *102*, 136–148. [CrossRef]
17. Proudhon, H.; Li, J.; Reischig, P.; Guéninchault, N.; Forest, S.; Ludwig, W. Coupling Diffraction Contrast Tomography with the Finite Element Method. *Adv. Eng. Mater.* **2016**, *18*, 903–912. [CrossRef]
18. McDowell, D.L.; Dunne, F.P.E. Microstructure-sensitive computational modeling of fatigue crack formation. *Int. J. Fatigue* **2010**, *32*, 1521–1542. [CrossRef]
19. Miller, M.P.; Dawson, P.R. Understanding local deformation in metallic polycrystals using high energy X-rays and finite elements. *Curr. Opin. Solid State Mater. Sci.* **2014**, *18*, 286–299. [CrossRef]
20. Erinosho, T.; Collins, D.; Wilkinson, A.; Todd, R.; Dunne, F. Assessment of X-ray diffraction and crystal plasticity lattice strain evolutions under biaxial loading. *Int. J. Plast.* **2016**, *83*, 1–18. [CrossRef]

21. Guery, A.; Hild, F.; Latourte, F.; Roux, S. Identification of crystal plasticity parameters using DIC measurements and weighted FEMU. *Mech. Mater.* **2016**, *100*, 55–71. [CrossRef]
22. Brechet, Y.; Livet, F. Low cycle fatigue of binary Al-Li alloys: III-coalescenceof δ' precipitates in fatigue: X-ray low angle scattering investigation. *J. Phys. Colloq.* **1987**, *48*, C3-717–C3-719. [CrossRef]
23. Vaughan, G.B.M.; Wright, J.P.; Bytchkov, A.; Rossat, M.; Gleyzolle, H.; Snigireva, I.; Snigirev, A. X-ray transfocators: Focusing devices based on compound refractive lenses. *J. Synchrotron Radiat.* **2011**, *18*, 125–133. [CrossRef] [PubMed]
24. Ludwig, W.; Cloetens, P.; Härtwig, J.; Baruchel, J.; Hamelin, B.; Bastie, P. Three-dimensional imaging of crystal defects by 'topo-tomography'. *J. Appl. Crystallogr.* **2001**, *34*, 602–607. [CrossRef]
25. Ludwig, W.; Lauridsen, E.M.; Schmidt, S.; Poulsen, H.F.; Baruchel, J. High-resolution three-dimensional mapping of individual grains in polycrystals by topotomography. *J. Appl. Crystallogr.* **2007**, *40*, 905–911. [CrossRef]
26. Busing, W.R.; Levy, H.A. Angle calculations for 3- and 4-circle X-ray and neutron diffractometers. *Acta Crystallogr.* **1967**, *22*, 457–464. [CrossRef]
27. Poulsen, H.F. *Three-Dimensional X-Ray Diffraction Microscopy—Mapping Polycrystals and Their Dynamics*; Springer Tracts in Modern Physics; Springer: Berlin, Germany, 2004; Volume 205.
28. Rowenhorst, D.; Rollett, A.D.; Rohrer, G.S.; Groeber, M.; Jackson, M.; Konijnenberg, P.J.; De Graef, M. Consistent representations of and conversions between 3D rotations. *Model. Simul. Mater. Sci. Eng.* **2015**, *23*, 083501. [CrossRef]
29. Besson, J.; Cailletaud, G.; Chaboche, J.L.; Forest, S.; Blétry, M. *Non-Linear Mechanics of Materials*; Springer Science & Business Media: Berlin, Germany; 2010; Volume 167.
30. Mandel, J. Equations constitutives et directeurs dans les milieux plastiques et viscoplastiques. *Int. J. Solids Struct.* **1973**, *9*, 725–740. [CrossRef]
31. Besson, J.; Foerch, R. Large scale object-oriented finite element code design. *Comput. Methods Appl. Mech. Eng.* **1997**, *142*, 165–187. [CrossRef]
32. Meric, L.; Poubanne, P.; Cailletaud, G. Single Crystal Modeling for Structural Calculations: Part 1—Model Presentation. *J. Eng. Mater. Technol.* **1991**, *113*, 162–170. [CrossRef]
33. Franciosi, P.; Berveiller, M.; Zaoui, A. Latent hardening in copper and aluminum single crystals. *Acta Metall.* **1980**, *28*, 273–283. [CrossRef]
34. Wu, T.Y.; Bassani, J.L.; Laird, C. Latent hardening in single crystals–I. Theory and experiments. *Proc. R. Soc. Lond. A* **1991**, *435*, 1–19. [CrossRef]
35. Marchenko, A.; Mazière, M.; Forest, S.; Strudel, J.L. Crystal plasticity simulation of strain aging phenomena in α-titanium at room temperature. *Int. J. Plast.* **2016**, *85*, 1–33. [CrossRef]
36. Tabourot, L.; Fivel, M.; Rauch, E. Generalised constitutive laws for f.c.c. single crystals. *Mater. Sci. Eng. A* **1997**, *234–236*, 639–642. [CrossRef]
37. Héripré, E.; Dexet, M.; Crépin, J.; Gélébart, L.; Roos, A.; Bornert, M.; Caldemaison, D. Coupling between experimental measurements and polycrystal finite element calculations for micromechanical study of metallic materials. *Int. J. Plast.* **2007**, *23*, 1512–1539. [CrossRef]
38. Pymicro: A Python Package to Work with Material Microstructures and 3D Data Sets. Available online: https://github.com/heprom/pymicro (accessed on 15 October 2018).
39. Shi, Q.; Latourte, F.; Hild, F.; Roux, S. Backtracking Depth-Resolved Microstructures for Crystal Plasticity Identification—Part 2: Identification. *JOM* **2017**, *69*, 2803–2809. [CrossRef]
40. Barabash, R.I.; Klimanek, P. X-ray scattering by crystals with local lattice rotation fields. *J. Appl. Crystallogr.* **1999**, *32*, 1050–1059. [CrossRef]
41. Pantleon, W.; Wejdemann, C.; Jakobsen, B.; Lienert, U.; Poulsen, H. Evolution of deformation structures under varying loading conditions followed in situ by high angular resolution 3DXRD. *Mater. Sci. Eng. A* **2009**, *524*, 55–63. [CrossRef]
42. Borbély, A.; Ungár, T. X-ray line profiles analysis of plastically deformed metals. *C. R. Phys.* **2012**, *13*, 293–306. [CrossRef]

© 2018 by the authors. Licensee MDPI, Basel, Switzerland. This article is an open access article distributed under the terms and conditions of the Creative Commons Attribution (CC BY) license (http://creativecommons.org/licenses/by/4.0/).

Article

Observation of Morphology Changes of Fine Eutectic Si Phase in Al-10%Si Cast Alloy during Heat Treatment by Synchrotron Radiation Nanotomography

Shougo Furuta [1], Masakazu Kobayashi [1,*], Kentaro Uesugi [2], Akihisa Takeuchi [2], Tomoya Aoba [1] and Hiromi Miura [1]

[1] Department of Mechanical Engineering, Toyohashi University of Technology, Toyohashi 441-8580, Japan; furuta@str.me.tut.ac.jp (S.F.); aoba@me.tut.ac.jp (T.A.); miura@me.tut.ac.jp (H.M.)
[2] Research & Utilization Division, Japan Synchrotron Radiation Research Institute, Hyogo 679-5198, Japan; ueken@spring8.or.jp (K.U.); take@spring8.or.jp (A.T.)
* Correspondence: m-kobayashi@me.tut.ac.jp; Tel.: +81-532-44-6706

Received: 3 July 2018; Accepted: 25 July 2018; Published: 28 July 2018

Abstract: A series of three-dimensional morphology changes of fine eutectic Si-particles during heat treatment have been investigated in Self-modified and Sr-modified Al-10%Si cast alloys by means of synchrotron radiation nanotomography utilizing a Fresnel zone plate and a Zernike phase plate in this study. The coral-like shape particles observed in Sr-modified cast alloy fragmented at branch and neck during heat treatment at 773 K. The fragmentation occurred up to 900 s. After that, the fragmented particles grew and spheroidized by Ostwald ripening. On the other hand, rod-like shaped eutectic Si-particles observed in self-modified cast alloy were larger in size compared with the particle size in Sr-modified cast alloy. Separation of eutectic Si-particles in Self-modified cast alloy occurred up to approximately 900 s, which was similar tendency to that in Sr-modified cast alloy. However, it was found that the morphology change behavior was very complex in rod-like shape Si-particles. The three-dimensional morphology changes of fine eutectic Si-particles in both cast alloys, specifically fragmentation and spheroidizing, can be connected to changes in mechanical properties.

Keywords: particle morphology; heat treatment; aluminum cast alloy; mechanical properties; Ostwald ripening; nanotomography; phase-contrast imaging

---

1. Introduction

Since the building of large synchrotron radiation facilities throughout the world in the latter half of the twentieth century, the performance of synchrotron radiation tomography has gradually improved up to the present day [1–3]. Currently, in the Japanese synchrotron radiation facilities, SPring-8, three-dimensional non-distractive observation with a spatial resolution of 50–160 nm is available constantly in the imaging beamline by using an X-ray focusing device of a Fresnel zone plate [4–6]. Phase-contrast imaging techniques have been also developed for those samples for which visualizations are difficult by X-ray absorption contrast (i.e., these densities are very close) [7]. Furthermore, improvement of the performance of X-ray 2D detector system has rapidly shortened scanning time. Therefore, synchrotron radiation tomography can be used for various studies in various fields.

The advantages of X-ray tomography are that three-dimensional morphologies are obtained, and that the observation is non-destructive. In studies of structural materials, material behaviors changing over time can be visualized, for instance, damage and fracture mechanisms [7–14], fatigue and crack propagation phenomena [15–18] and so on. We can deeply understand various phenomena affected by microstructures from a series of observed images. In Al-Si cast alloys, it is well known that eutectic Si-particle strongly affects mechanical properties. Many studies with regard to the morphology and distribution of eutectic Si-particles have been conducted to date [19–21], because the spheroidizing of Si-particles brings, in particular, ductility improvement by heat treatment. However, most of the research had been performed on the basis of 2D observation by polishing of a heat-treated sample after cross section cutting. Although three-dimensional evaluation also exists using Focus Ion Beam tomography [22], unfortunately this method is destructive.

In the application of hypoeutectic Al–Si alloys for automobile parts which require sufficient toughness, Si-particle refinement is applied by adding trace Sr to improve the mechanical properties [23–25]. The addition of trace Sr prevents aluminum phosphide, AlP which become the nuclei of coarse Si particles [26], and then changes the solidification process of hypoeutectic Al-Si alloys [27]. Note that the origin of phosphorus is the impurity of Si. Si-particles modified by trace Sr addition become very fine at less than 1 µm. A Sr-modified hypoeutectic Al-Si alloy demonstrates excellent mechanical properties. Furthermore, with applying heat treatment to the alloy, its strength and ductility can be controlled. The changes of Si-particles morphology during a heat treatment are considered as follows; firstly, Si-particles with necking divide into parts by Plateau–Rayleigh instability [28,29]. This separation is a change which decreases system energy quickly. Next, the fragmented Si-particles grow into spherical shapes by diffusion-controlled Ostwald ripening to reduce their surface energy.

By contrast, self-modification (self-refinement) of eutectic Si-particles is also possible by killing an impurity element of P, which is contained in Si and forms AlP as the solidification nuclei of Si. This P-free solidification process has been reproduced by phase-field model simulation by Eiken [27]. The morphologies of eutectic Si-particles which are formed in the different solidification processes—self-modification and Sr-modification—are different. Synchrotron radiation nanotomography has revealed that the morphologies of Si-particles in self-modification and Sr-modification are of a rod-like shape and coral-like shape, respectively, by casting self-modified and Sr-modified samples and investigating practically [30]. Therefore, in this study, to clarify the behavior of morphological changes during heat treatment and the effect of them on mechanical properties, hypoeutectic Al-10%Si alloys were cast using two different solidification processes (Self-modification and Sr-modification) that produce different morphology of eutectic Si-particles (rod-like and coral-like). The changes of mechanical properties were investigated in the prepared samples. The three-dimensional morphology changes of eutectic Si-particles during the heat treatment process were observed in both alloys by using nanotomography with a Fresnel zone plate and a Zernike phase plate.

2. Materials and Methods

Al-10%Si alloy was selected as the sample of this study. Two kinds of Al-10%Si alloy, self-modified and a Sr-modified sample, were prepared by gravity casting. It is known that three-dimensional morphology of eutectic Si-particles is different between the two alloys [30]. High purity Al (99.99%) and high purity Si (99.9999%) were melted in a graphite crucible at 993 K in air atmosphere using an electrical resistance furnace (Hamamatsu heat-tech, Hamatsu, Shizuoka, Japan). The molten metal was degassed by hexachloroethane. After the degassing treatment, molten metal was cast into a boat-shaped iron-mold heated at 473 K with a cavity size of 150 mm × 25 mm × 25 mm as a self-modified sample. For the Sr-modified sample, the degassed molten metal was cast into the mold soon after the 100 ppm Sr addition. The chemical compositions of cast alloy samples detected by spark emission spectrometer (OBLF QSN750-II, Witten, Germany) are listed in Table 1. Hereafter, two prepared cast alloys are named as Al-9.8%Si-3ppmP and Al-10.1%Si-4ppmP-108ppmSr on the basis of the result of composition analysis. Photos of microstructures in Al-9.8%Si-3ppmP cast alloy and Al-10.1%Si-4ppmP-108ppmSr cast alloy are shown in Figure 1. The microstructures of both alloys are almost the same in the two-dimensional image. It is difficult to distinguish them.

Table 1. Chemical composition of prepared cast alloys (wt.%).

Sample	Si	P	Sr	Cu	Al
Self-modified alloy	9.8	0.0003	<0.00001	0.08	Bal.
Sr-modified alloy	10.1	0.0004	0.0108	0.07	Bal.

Specimens for nanotomography were cut from the cast ingot. Very small stick-shaped specimens with a section size of 50 μm × 50 μm and length of about 8 mm were manufactured by hand polishing. Five tensile specimens with 19.75 mm^2 section × 30 mm length in a gauge part, which is a half size of JIS No.13 B (JIS Z 2241), were prepared from a position of 2 mm above from the bottom of the cast ingot, then the mechanical properties of the cast alloys were examined by a tensile testing machine (SHIMADZU AG-100 kNX, Kyoto, Japan).

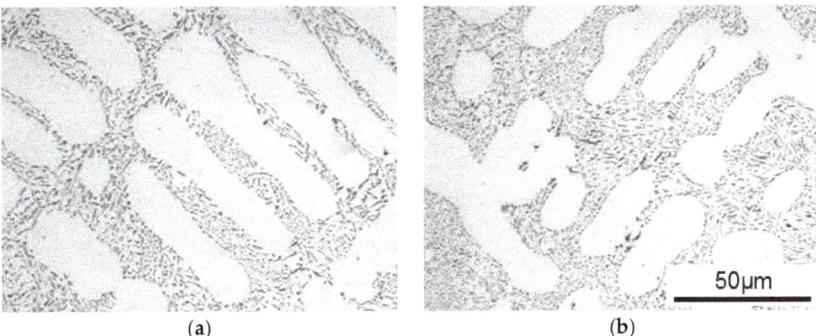

Figure 1. Optical micrographs; (a) Al-9.8%Si-3ppmP cast alloy and (b) Al-10.1%Si-4ppmP-108ppmSr cast alloy.

Synchrotron radiation nanotomography was used for observation of three-dimensional morphology change in eutectic Si-particles during heat treatment. The synchrotron radiation experiment was performed at the undulator beam line of BL47XU in the Japanese synchrotron radiation facility, SPring-8 (Hyōgo, Japan). A schematic illustration of the nanotomography set-up in the experimental hutch is shown in Figure 2. X-ray energy of 8 keV, which was adjusted by a silicon (111) double-crystal monochromator (SPring-8 Standard Monochromator, sKohzu Precision Co.,Ltd, Kawasaki, Kanagawa, Japan), was selected for this observation. A Fresnel zone plate with an outermost zone width of 50 nm was installed as an X-ray objective of an imaging X-ray microscope (NTT-AT, Kawasaki, Kanagawa, Japan). A Zernike phase plate made from tantalum with a thickness of 0.96 μm was also installed at the back focal plane of the Fresnel zone plate. A two-dimensional image detector system consisting of a Gd_2O_2S:Tb scintillator, an optical relay lens and a complementary metal oxide semiconductor camera (Hamamatsu Photonics K.K., C11440-22C, Hamamatsu, Shizuoka, Japan) was used. Since the difference in atomic number between them is only one, there is little X-ray absorption contrast in the Al phase and Si phase as shown in Figure 3a. Therefore, Si-particles in the inside of the aluminum matrix were visualized by phase contrast using a Zernike phase plate as shown in Figure 3b. Exposure time of 250 ms was used and 1800 projections were captured during a 180° rotation, for tomography. Voxel size of (37.8 nm)^3 was achieved in the reconstructed volume image in the set-up of this study.

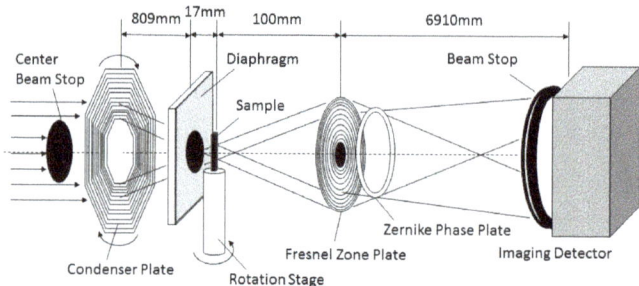

Figure 2. The schematic illustration of set-up of nanotomography in the experimental hutch.

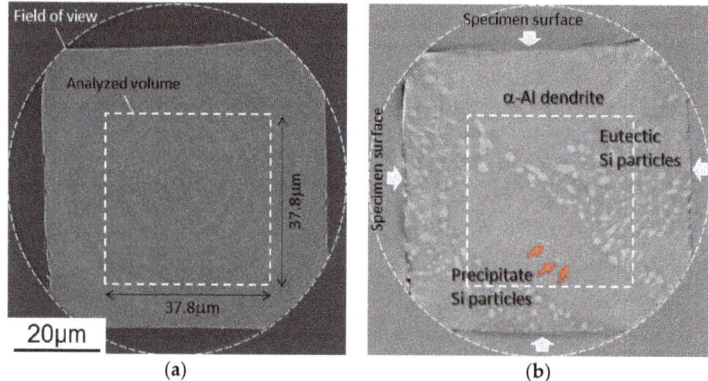

Figure 3. Slice images of nanotomography in Al-10.1%Si-4ppmP-108ppmSr cast alloy heat-treated at 773 K for 7.2 ks. The same slices are shown by (**a**) absorption contrast and (**b**) phase contrast. Field of view and analyze-volume position are indicated by white dashed-line circle and box. Eutectic Si-particles and precipitate Si-particles can be recognized as white objects in (**b**), though specimen surface only can be seen in (**a**).

Experimental procedure for the synchrotron radiation nanotomography observation was simple. The initial state of the sample, i.e., as-cast sample, was scanned. After the first tomography scan, the sample was heat-treated by taking it in and out of a compact air atmosphere furnace maintained at 773 K, and then was tomography scanned repeatedly at the same position at 450 s, 900 s, 1.8 ks, 3.6 ks, 7.2 ks and 14.4 ks. X-ray scanned data were reconstructed into a three-dimensional volume image by a conventional filtered convolution back-projection algorithm. A three-dimensional median filter (3 × 3 × 3) was applied to three-dimensional volume images reconstructed in order to reduce artifacts and image noise. Si-particles observed in the volume images were binarized and segmented with the thresholds value that were decided by comparing the obtained volume images to one another. The result of Si-particles segmentation was checked by visual inspection. Then if wrong connections existed among particles, such connections were carefully corrected one by one. Volume rendering software (VG studio Max 2.0, Volume Graphics, Heidelberg, Germany; and Amira 4.0, Thermo Fisher Scientific, Waltham, MA, USA) was used to visualize the three-dimensional morphology of Si-particles. The analyzed region for the morphology changes of eutectic Si-particles was extracted from a lower effect region of artifacts inside the sample. The analyzed regions were 56.7 μm × 28.4 μm × 27.2 μm and 37.8 μm × 37.8 μm × 41.8 μm in Al-9.8%Si-3ppmP sample and Al-10.1%Si-4ppmP-108ppmSr, respectively. Si-particles within the analyzed regions were segmented and labeled after binarization. Then, volume and surface area were measured for each of the labeled Si-particles.

3. Results

3.1. Mechanical Properties

The stress-strain curves of a tensile test in Al-9.8%Si-3ppmP cast alloy and Al-10.1%Si-4ppmP-108ppmSr cast alloy are shown in Figure 4. The ultimate tensile strength decreases and elongation to failure increases in both samples with increasing heat treatment time. Both samples show almost a similar stress-strain relationship before and after heat treatment, though tensile strength in Al-10.1%Si-4ppmP-108ppmSr cast alloy is slightly higher than that in Al-9.8%Si-3ppmP cast alloy. In both alloys, heat treatment reduces yield stress and work hardening rate mildly. Elongations to both alloys are almost the same with the same heat treatment time. Note that Vickers hardness in Al-9.8%Si-3ppmP cast alloy and Al-10.1%Si-4ppmP-108ppmSr cast alloy were 58.5 HV and 59.6 HV, respectively. There was no difference in the dendrite secondary arm spacing (DASII)—which was approximately 37 μm—in both of the alloys. Changes in ultimate tensile strength and elongation (average of 5 specimens) are shown in Figure 5. By heat treatment, ultimate tensile strength decreases gradually and elongation increases in both alloys. The changes become particularly remarkable after 1.8 ks of heat treatment. With a short period of heat treatment, no differences are seen in either alloys. However, when applying heat treatment for a longer time, the mechanical properties in Al-10.1%Si-4ppmP-108ppmSr cast alloy become superior to that of Al-9.8%Si-3ppmP cast alloy.

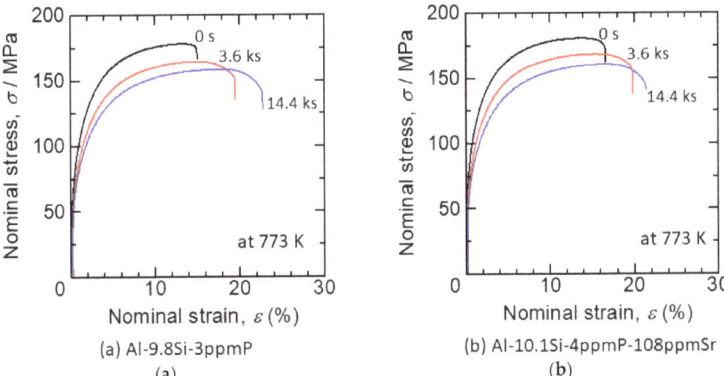

Figure 4. Stress-strain curves of tensile test in (**a**) Al-9.8%Si-3ppmP cast alloy and (**b**) Al-10.1%Si-4ppmP-108ppmSr cast alloy.

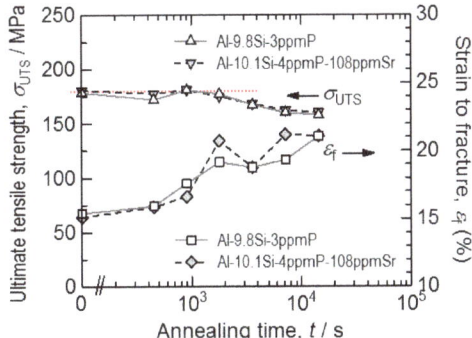

Figure 5. Changes in ultimate tensile strength and elongation during heat treatment at 773 K.

3.2. Morphology Changes of Eutectic Si-Particles

Three-dimensional volume images of eutectic Si-particles in Al-9.8%Si-3ppmP cast alloy, which are obtained by synchrotron radiation nanotomography, are shown in Figure 6. Interior Si-particles are displayed removing the aluminum matrix in the top part of each figure. It can be confirmed that synchrotron radiation nanotomography is high resolution because the figure indicates the changes in a very small region with a size of 28.4 μm × 56.7 μm × 27.2 μm. In the as-cast state (heat treatment time, t = 0 s) as shown in Figure 6a, most of the eutectic Si-particles are of a straight rod-like shape, and a small plate-like shape is also seen partially. It is observed that the rod and plate-like Si-particles are connecting. Particle growth is confirmed during heat treatment up to 14.4 ks as shown in Figure 6b–g. The number of particles decrease gradually during particle growth. Although particle separation that makes particles segment into a small size is also observed, most of the particles maintain a high aspect ratio after 14.4 ks annealing.

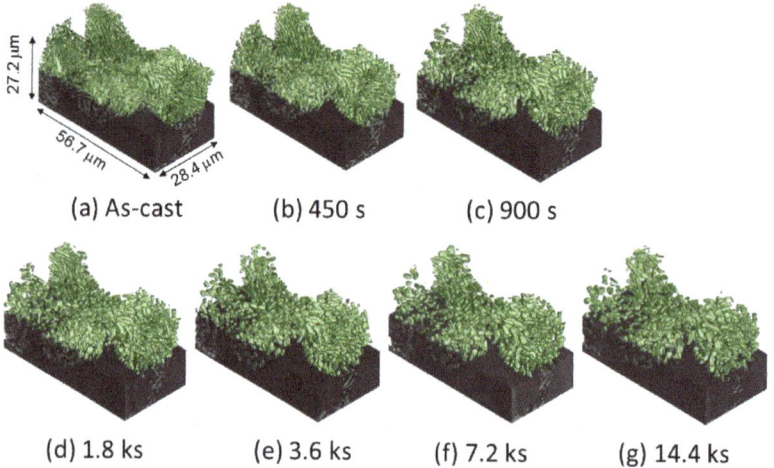

Figure 6. Three-dimensional volume images of eutectic Si-particles in Al-9.8%Si-3ppmP cast alloy. (**a**) As-cast, (**b**) heat-treated at 773 K for 450 s, (**c**) heat-treated at 773 K for 900 s, (**d**) heat-treated at 773 K for 1.8 ks, (**e**) heat-treated at 773 K for 3.6 ks, (**f**) heat-treated at 773 K for 7.2 ks, (**g**) heat-treated at 773 K for 14.4 ks.

Figure 7 shows a three-dimensional volume image of Al-10.1%Si-4ppmP-108ppmSr cast alloy. In the as-cast state (Figure 7a), fine rod-like Si-particles are observed similar to Al-9.8%Si-3ppmP cast alloy. However, the entire morphology of Si-particles in Sr-modified alloy are that of a coral-like shape with multiple branches. Particle size is slightly finer than that in Al-9.8%Si-3ppmP cast alloy. The Si-particles grow gradually with fragmentation during heat treatment, and spheroidize after 14.4 ks. In Al-10.1%Si-4ppmP-108ppmSr cast alloy, formation of Sr precipitations was confirmed in primary α-Al dendrite during heat treatment. Three-dimensional volume images in (a) as-cast and (b) after 7.2 ks heat-treated are shown in Figure 8. New Si-particles, which are not found in the as-cast state, are observed in the outside region of eutectic phase in which Si-particles are gathering. The presence of precipitate Si-particles is also confirmed in the slice image shown in Figure 3b. It can be concluded that the particles are not Sr compounds but Si because the particles have disappeared in the absorption images shown in Figure 3a.

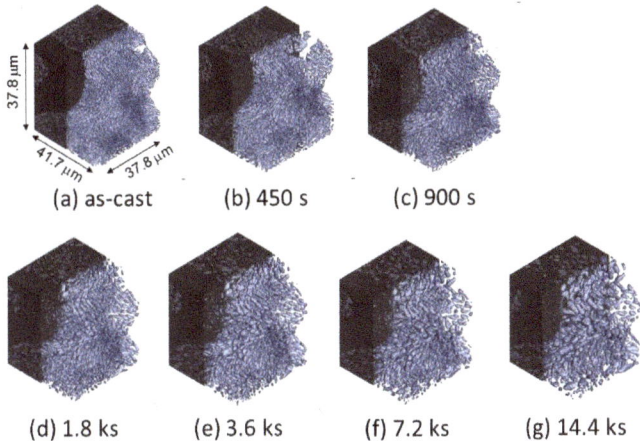

Figure 7. Three-dimensional volume images of eutectic Si-particles in Al-10.1%Si-4ppmP-108ppmSr cast alloy. (**a**) As-cast, (**b**) heat-treated at 773 K for 450 s, (**c**) heat-treated at 773 K for 900 s, (**d**) heat-treated at 773 K for 1.8 ks, (**e**) heat-treated at 773 K for 3.6 ks, (**f**) heat-treated at 773 K for 7.2 ks, (**g**) heat-treated at 773 K for 14.4 ks.

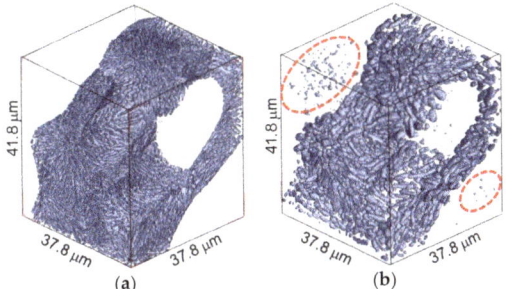

Figure 8. Three-dimensional volume images in Al-10.1%Si-4ppmP-108ppmSr cast alloy; (**a**) as-cast and (**b**) after 7.2 ks heat-treated. New Si-particle precipitate during heat treatment.

Figure 9 indicates changes in total Si-particle volume, number of Si-particles and average Si-particle size (sphere-equivalent diameter) during heat treatment. These statistics were obtained from the microstructures shown in Figures 6 and 7 by three-dimensional image processing analysis. The total Si-particle volume in Al-10.1%Si-4ppmP-108ppmSr cast alloy is larger than that in Al-9.8%Si-3ppmP cast alloy. Volume fraction of Si phase in Al-10%Si alloy should be approximately 11.4%. However, the volume fractions were slightly small in the volumes analyzed and were 7.8% and 8.7% in Al-9.8%Si-3ppmP cast alloy and Al-10.1%Si-4ppmP-108ppmSr cast alloy, respectively. This is due to inhomogeneities of microstructure and the small field of view size in nanotomography. The amount of Si content does not differ in both alloys. The total Si-particle volume in Al-9.8%Si-3ppmP cast alloy looks to slightly decrease during heat treatment. This change is due to particles on the edge of view. Total Si-particle volume is almost constant during heat treatment in both alloys. In the as-cast, the number of Si-particles in Al-10.1%Si-4ppmP-108ppmSr cast alloy is larger than that in Al-9.8%Si-3ppmP cast alloy. This is because total Si-particle volume is large in Al-10.1%Si-4ppmP-108ppmSr cast alloy, and the particle size is small as observed in Figure 7. The number of particles decreases in both alloys during heat treatment. In Al-9.8%Si-3ppmP cast alloy, fragmentation of Si-particles, which is observed in the early stage of heat treatment, causes an increase of number of particles temporarily. As shown in Figure 8, Si-particles precipitate into α-Al dendrites during heat treatment. However, the precipitation has no effect on the number of Si-particles, because the number increase is small compared with the number decrease by particle growth. In Al-9.8%Si-3ppmP cast alloy, average Si-particle size decreases, and then increases. This decrease at the early stage is also due to fragmentation of Si-particles. In case of Al-10.1%Si-4ppmP-108ppmSr cast alloy, a little increase of Si-particle size is found in the early stage of heat treatment, and then the size increases rapidly in the later period of heat treatment.

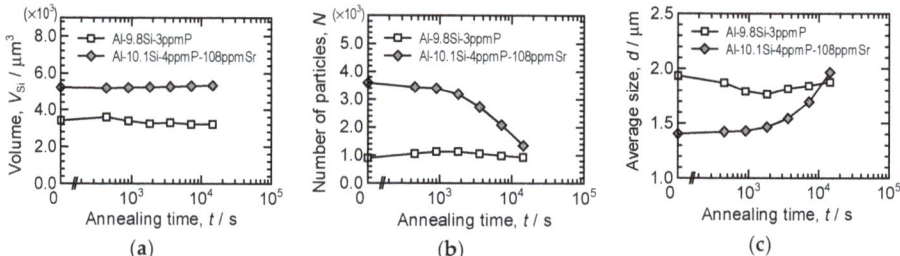

Figure 9. Changes in total Si-particle volume, number of Si-particle and average Si-particle size during heat treatment. Initial total Si-particle volume was 3416.645 μm³ and 5217.278 μm³ in Al-9.8%Si-3ppmP cast alloy and Al-10.1%Si-4ppmP-108ppmSr cast alloy, respectively. (**a**) total Si-particle volume, (**b**) number of Si-particles and (**c**) average Si-particle size.

4. Discussion

Looking at Figure 9, the situations in which particles grow while keeping volume constant are Ostwald ripening. According to classical particle growth theory, a growth of Ostwald ripening [31] is formulated as

$$d^3 - d_0^3 = kt \quad (1)$$

in a diffusion control situation. Here, d and d_0 are the average particle diameter and initial average particle diameter, k is constant and t is annealing time. The relationship between the cube of average particle diameter and annealing time in this study is shown in Figure 10. The early stage of particle growth in Al-9.8%Si-3ppmP cast alloy does not correspond to the growth manner expressed in Equation (1). The Si-particles grow proportionally after 3.6 ks of heat treatment. The particle growth in Al-10.1%Si-4ppmP-108ppmSr cast alloy almost obeys Equation (1) though a little difference is seen in the early stage of growth. It is found that Si-particle growth in Al-10.1%Si-4ppmP-108ppmSr cast alloy is faster than that in Al-9.8%Si-3ppmP cast alloy. A very small difference in particle growth rate had been expected because the alloys were simple binary Al-Si system alloys though there was a difference in 108ppm Sr content. However, a six-times difference is recognized in the comparison with the slopes of fitting lines between both of the alloys in the later stage of growth.

By magnifying a three-dimensional volume image, morphology changes of Si-particles in Al-9.8%Si-3ppmP cast alloy are shown in Figure 11. Actually, Si-particles distribute densely as shown in Figure 11a. To understand the morphological changes of particles easily, an image removing surrounding particles is Figure 11b. In the as-cast, the morphology of Si-particles is mildly complex, possessing a fine rod-like shape, which is elongated along the solidification direction, and a partial small plate-like shape, which is broader than the rod part. While heat treatment is progressing, the Si-particles are divided into plural segments (Figure 11c,d). After that, fragmented particles become gradually round and approach into a sphere-like shape that has the smallest surface area (Particle A in Figure 11e). An elongated Particle B shown in Figure 11e gradually shortens in length, and then becomes close to a sphere-like shape. An elongated Particle C seen in the center of the figures thickens in diameter during heat treatment. However, the tip position does not change so much after 1.8 ks of heat treatment and the small change is observed in particle shortening along longitudinal direction. It is found that the morphology change is slightly different depending on the length of the rod-like shaped particles.

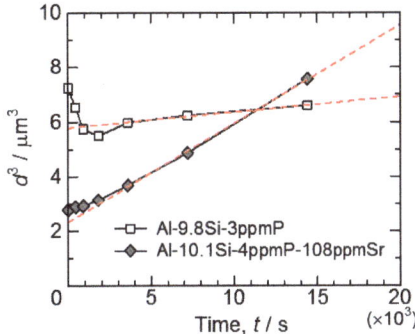

Figure 10. The relationship between cube of average particle size and annealing time.

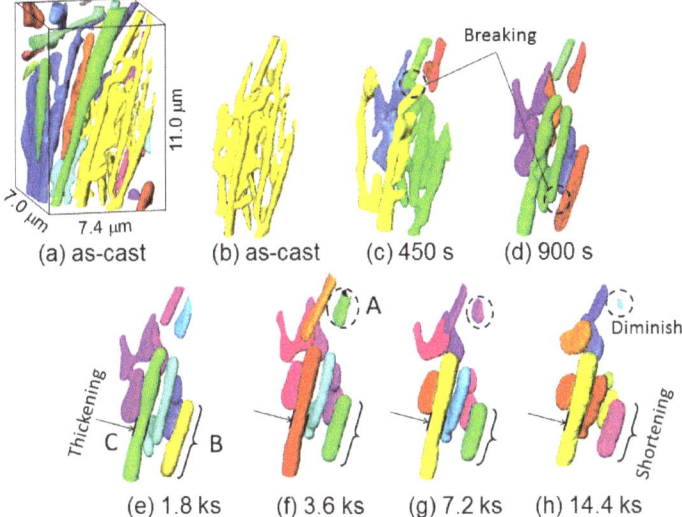

Figure 11. Magnification of three-dimensional volume image, illustrating morphology changes of Si-particles in Al-9.8%Si-3ppmP cast alloy. (**a**) as-cast, (**b**) as-cast (one particle), (**c**) heat-treated at 773 K for 450 s, (**d**) heat-treated at 773 K for 900 s. (**e**) heat-treated at 773 K for 1.8 ks. (**f**) heat-treated at 773 K for 3.6 ks, (**g**) heat-treated at 773 K for 7.2 ks, (**h**) heat-treated at 773 K for 14.4 ks.

Figure 12 shows three-dimensional images viewing Figure 11 from the rear side. Particle D fragmented at 900 s heat treatment seen in Figure 12c becomes a sphere-like shape in (d) 1.8 ks and (f) 3.6 ks of heat treatment. The particle is merged with a U-shape particle growing behind its back. The U-shape particle that absorbed Particle D probably will close a gap and be a sphere-shaped particle if heat treatment continues furthermore. A protuberance indicated by the letter E in Figure 12d is not cut off at the neck, shortens in length gradually and finally is absorbed by a plate-shape particle.

The morphology change of Si-particles is very complex in Al-9.8%Si-3ppmP cast alloy. It was found that the changes during heat treatment were not only just fragmentation and spheroidizing.

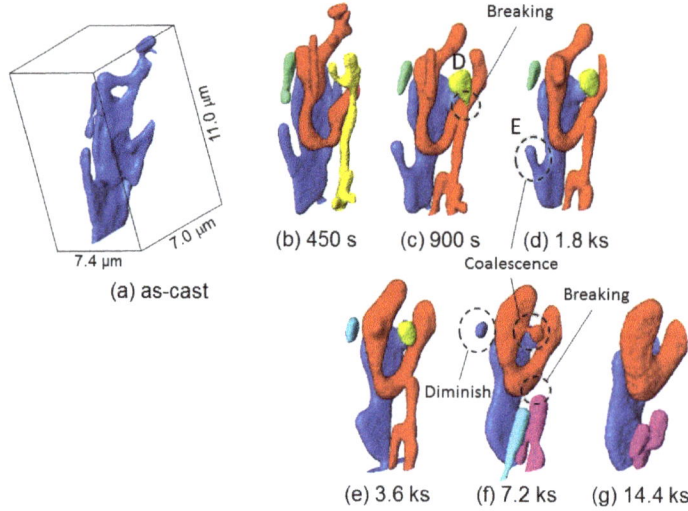

Figure 12. Three-dimensional images viewing Figure 11 from the back side. (**a**) as-cast (one particle), (**b**) heat-treated at 773 K for 450 s, (**c**) heat-treated at 773 K for 900 ks, (**d**) heat-treated at 773 K for 1.8 ks, (**e**) heat-treated at 773 K for 3.6 ks, (**f**) heat-treated at 773 K for 7.2 ks, (**g**) heat-treated at 773 K for 14.4 ks.

Figure 13 shows magnified three-dimensional images of Si-particles in Al-10.1%Si-4ppmP-108ppmSr cast alloy. Figure 13a shows the as-cast state displaying peripheral particles. One particle is shown in Figure 13b by removing the peripheral particles. It can be confirmed that Si-particles in Al-10.1%Si-4ppmP-108ppmSr cast alloy are coral-like complex shapes having numerous branches. As well as Al-9.8%Si-3ppmP cast alloy, separation of Si-particles is found at branches and necks surrounded by dashed lines shown in Figure 13c–e. Frequency of separation in Al-10.1%Si-4ppmP-108ppmSr cast alloy is higher than that in Al-9.8%Si-3ppmP cast alloy because the Si-particles in Al-10.1%Si-4ppmP-108ppmSr cast alloy possess many branches in as-cast. Si-particle segmentations formed by separation thicken gradually. Then, the shape approaches a sphere-like form, with a shortening the length of longitudinal direction. Separation also causes a long trunk of Si-particles, as seen in the center of the figure. The trunk becomes segmented and grows to a sphere-like shape. Such behaviors correspond to those that we had expected. However, the morphology change was clearly different with the observed elongated particles in Al-9.8%Si-3ppmP cast alloy. Therefore, the difference of growth rates in the two cast alloys would be brought about by the difference of growth behavior at long elongated Si-particles. That is, the slow growth rate in Al-9.8%Si-3ppmP cast alloy in heat treatment is due to less separation of long elongated rod-shape particles, which are the main morphological features of eutectic Si-particles by the solidification of Al-9.8%Si-3ppmP cast alloy. In addition, the solidification reaction is different in the two cast alloys as seen in the different eutectic Si-particle morphology. Not only is the eutectic Si-particle size in Al-10.1%Si-4ppmP-108ppmSr cast alloy smaller than that in Al-9.8%Si-3ppmP cast alloy, but the eutectic grain formed in the eutectic reaction is also of a fine size in Al-10.1%Si-4ppmP-108ppmSr cast alloy. Therefore, the contribution of grain boundary diffusion in addition to lattice diffusion could also be a factor.

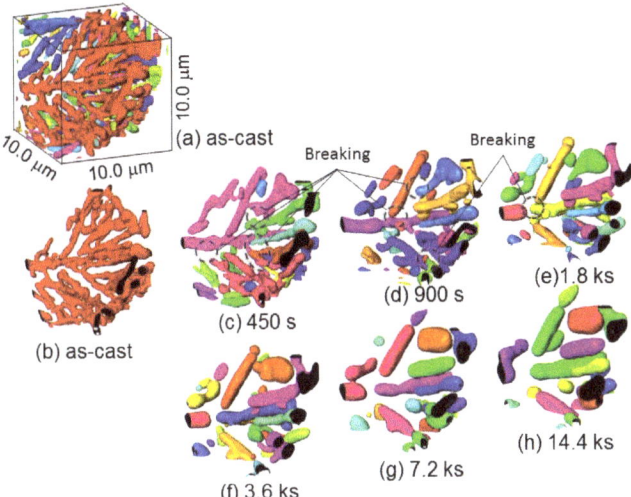

Figure 13. Magnified three-dimensional images of Si-particles in Al-10.1%Si-4ppmP-108ppmSr cast alloy. (**a**) as-cast, (**b**) as-cast (one particle), (**c**) heat-treated at 773 K for 450 ks, (**d**) heat-treated at 773 K for 900 ks, (**e**) heat-treated at 773 K for 1.8 ks, (**f**) heat-treated at 773 K for 3.6 ks, (**g**) heat-treated at 773 K for 7.2 ks, (**h**) heat-treated at 773 K for 14.4 ks.

As results of the detail three-dimensional observation of microstructural change through Figures 11 and 13, it was observed that many Si-particles separate and isolate as small segments in the early stage of heat treatment up to 900 s. Considering the relationship between microstructure and mechanical property, it is found that the mechanical properties shown in Figure 5 also start to change after 900 s heat treatment. The non-destructive observation in this study by means of synchrotron radiation nanotomography supports the idea that the connectivity of the strengthening phase affects strength and elongation as reported recently [32,33]. Further investigation and consideration for this will be possible in image-based simulations that are constructed from three-dimensional volume image tomography.

5. Conclusions

In this study, a series of three-dimensional morphology changes of fine eutectic Si-particles during heat treatment have been investigated in Self-modified and Sr-modified Al-10%Si cast alloys by means of synchrotron radiation nanotomography using a Fresnel zone plate and a Zernike phase plate. The morphology of eutectic Si-particles was rod-like in shape in the self-modified sample of Al-9.8%Si-3ppmP cast alloy. In the Sr-modified sample of Al-10.1%Si-4ppmP-108ppmSr cast alloy, Si-particle was a coral-like shape in the as-cast. The coral-like shape particles observed in Al-10.1%Si-4ppmP-108ppmSr cast alloy fragmented at branch and neck during heat treatment at 773 K. The fragmentation occurred up to 900 s. After that, the fragmented particles grew and spheroidized by Ostwald ripening. The rate of Ostwald ripening in Al-10.1%Si-4ppmP-108ppmSr cast alloy was faster than that in Al-9.8%Si-3ppmP cast alloy. On the other hand, rod-like shaped eutectic Si-particles observed in Al-9.8%Si-3ppmP cast alloy were larger in size compared to the particle size in Al-10.1%Si-4ppmP-108ppmSr cast alloy. In Al-9.8%Si-3ppmP cast alloy, separation of eutectic Si-particles occurred up to approximately 900 s, which was similar tendency to that in Al-10.1%Si-4ppmP-108ppmSr cast alloy. The frequency of separation was low due not to the coral-like shape but the rod-like shape. Three-dimensional morphology changes of fine eutectic Si-particles in both cast alloys, specifically fragmentation and spheroidizing, can be connected to changes in mechanical properties. In the rod-like shape of Si-particles obtained in a self-modified sample of Al-9.8%Si-3ppmP cast alloy, however, it was found that the morphology change behavior was very complex. By non-destructive continuous observation using nanotomography, it was revealed that relatively long rod-shape particles grew slowly without separation. It is also observed that a protuberance was absorbed into a small plate-shape part. Moreover, very complex behavior was observed where a rod-shape particle separated at the neck, spheroidized and was then absorbed by a neighboring larger particle.

Author Contributions: Methodology, K.U and A.T.; Validation, T.A and H.M; Investigation, S.F and M.K.; Resources, S.F, T.A and H.M; Data Curation, K.U and A.T.; Writing—Original Draft Preparation, M.K.; Writing—Review & Editing, T.A and H.M; Visualization, S.F.; Project Administration, M.K.

Funding: The authors appreciate the financial assistance of the Light Metal Educational Foundation.

Acknowledgments: The synchrotron experiment in this study was performed in subject No. 2016B1106 in SPring-8. This research was also partially supported by JST under Industry-Academia Collaborative R&D Program "Heterogeneous Structure control: Toward Innovative Development of Metallic Structural Materials".

Conflicts of Interest: The authors declare no conflict of interest.

References

1. Withers, P.J. X-ray nanotomography. *Mater. Today* **2007**, *10*, 26–34. [CrossRef]
2. Hashimoto, T.; Zhou, X.; Luo, C.; Kawano, K.; Thompson, G.E.; Hughes, A.E.; Skeldon, P.; Withers, P.J.; Marrow, T.J.; Sherry, A.H. Nanotomography for understanding materials degradation. *Scr. Mater.* **2010**, *63*, 835–838. [CrossRef]
3. Maire, E.; Withers, P.J. Quantitative X-ray tomography. *Int. Mater. Rev.* **2014**, *59*, 1–43. [CrossRef]
4. Takeuchi, A.; Uesugi, K.; Takano, H.; Suzuki, Y. Submicrometer-resolution three-dimensional imaging with hard X-ray imaging microtomography. *Rev. Sci. Instr.* **2002**, *73*, 4246–4249. [CrossRef]
5. Takeuchi, A.; Suzuki, Y.; Uesugi, K. Present status of the nanotomography system at BL47XU at SPring-8 and its efficiency improvement using double-condenser optics. *AIP Conf. Proc.* **2011**, *1365*, 301–304.
6. Suzuki, Y.; Takeuchi, A.; Terada, Y.; Uesugi, K.; Mizutani, R. Recent progress of hard x-ray imaging microscopy and microtomography at BL37XU of SPring-8. *AIP Conf. Proc.* **2016**, *1696*, 020013.
7. Takeuchi, A.; Uesugi, K.; Suzuki, Y. Zernike phase-contrast x-ray microscope with pseudo-Kohler illumination generated by sectored (polygon) condenser plate. *J. Phys. Conf. Ser.* **2009**, *186*, 012020. [CrossRef]
8. Weck, A.; Wilkinson, D.S.; Maire, E.; Toda, H. Visualization by X-ray tomography of void growth and coalescence leading to fracture in model materials. *Acta Mater.* **2008**, *56*, 2919–2928. [CrossRef]
9. Toda, H.; Maire, E.; Yamauchi, S.; Tsuruta, H.; Hiramatsu, T.; Kobayashi, M. In situ observation of ductile fracture using X-ray tomography technique. *Acta Mater.* **2011**, *59*, 1995–2008. [CrossRef]
10. Thuillier, S.; Maire, E.; Brunet, M. Ductile damage in aluminium alloy thin sheets: Correlation between micro-tomography observations and mechanical modeling. *Mater. Sci. Eng. A* **2012**, *558*, 217–225. [CrossRef]
11. Landron, C.; Bouaziz, O.; Maire, E.; Adrienz, J. Experimental investigation of void coalescence in a dual phase steel using X-ray tomography. *Acta Mater.* **2013**, *61*, 6821–6829. [CrossRef]
12. Toda, H.; Oogo, H.; Horikawa, K.; Uesugi, K.; Takeuchi, A.; Suzuki, Y.; Nakazawa, M.; Aoki, Y.; Kobayashi, M. The true origin of ductile fracture in aluminium alloy. *Metall. Mater. Trans. A* **2014**, *45*, 765–776. [CrossRef]
13. Hosokawa, A.; Toda, H.; Batres, R.; Li, H.; Kuwazuru, O.; Kobayashi, M.; Yakita, H. Ductile fracture via hydrogen pore mechanism in an aluminum alloy; quantitative microstructural analysis and image-based finite element analysis. *Mater. Sci. Eng. A* **2016**, *671*, 96–106. [CrossRef]
14. Gupta, C.; Toda, H.; Fujioka, T.; Kobayashi, M.; Hoshino, H.; Uesugi, K.; Takeuchi, A.; Suzuki, Y. Quantitative tomography of hydrogen precharged and uncharged Al-Zn-Mg-Cu alloy after tensile fracture. *Mater. Sci. Eng. A* **2016**, *670*, 300–313. [CrossRef]
15. Marrow, T.J.; Buffière, J.-Y.; Withers, P.J.; Johnson, G.; Engelberg, D. High resolution X-ray tomography of short fatigue crack nucleation in austempered ductile cast iron. *Int. J. Fatigue* **2004**, *26*, 717–725. [CrossRef]

16. Herbig, M.; King, A.; Reischig, P.; Proudhon, H.; Lauridsen, E.M.; Marrow, J.; Buffière, J.-Y.; Ludwig, W. 3-D growth of a short fatigue crack within a polycrystalline microstructure studied using combined diffraction and phase-contrast X-ray tomography. *Acta Mater.* **2011**, *59*, 590–601. [CrossRef]
17. Dezecot, S.; Buffière, J.-Y.; Koster, A.; Maurel, V.; Szmytka, F.; Charkaluk, E.; Dahdah, N.; Bartali, A.; El Limodin, N.; Witz, J.-F. In situ 3D characterization of high temperature fatigue damage mechanisms in a cast aluminum alloy using synchrotron X-ray tomography. *Scr. Mater.* **2016**, *113*, 254–258. [CrossRef]
18. Teranishi, M.; Kuwazuru, O.; Gennai, S.; Kobayashi, M.; Toda, H. Three-dimensional stress and strain around real shape Si particles in cast aluminum alloy under cyclic loading. *Mater. Sci. Eng. A* **2016**, *678*, 273–285. [CrossRef]
19. Li, H.J.; Shivkumar, S.; Luo, X.J.; Apelian, D. Influence of modification on the solution-treatment response of cast Al-Si-Mg alloy. *Cast Met.* **1989**, *1*, 227–234. [CrossRef]
20. Apelian, D.; Shivkumar, S.; Sigworth, G. Fundamental aspects of heat treatment of cast Al-Si-Mg alloys. *AFS Trans.* **1989**, *97*, 727–742.
21. Lados, D.A.; Apelian, D.; Wang, L. Solution treatment effects on microstructure and mechanical properties of Al-(1 to 13 pct)Si-Mg cast alloys. *Metall. Mater. Trans. B* **2011**, *42*, 171–180. [CrossRef]
22. Lasagni, F.; Lasagni, A.; Marks, E.; Holzapfel, C.; Mücklich, F.; Degischer, H.P. Three-dimensional characterization of 'as-cast' and solution-treated AlSi12(Sr) alloys by high-resolution FIB tomography. *Acta Mater.* **2007**, *55*, 3875–3882. [CrossRef]
23. Dahle, A.K.; Nogita, K.; Zindel, J.W.; McDonald, S.D.; Hogan, L.M. Eutectic nucleation and growth in hypoeutectic Al-Si alloys at different strontium levels. *Metall. Mater. Trans. A* **2001**, *32*, 949–960. [CrossRef]
24. McdDonald, S.D.; Dahle, A.K.; Taylor, J.A.; StJhon, D.H. Eutectic grains in unmodified and strontium-modified hypoeutectic aluminum-silicon alloys. *Metall. Mater. Trans. A* **2004**, *35*, 1829–1837. [CrossRef]
25. McDonald, S.D.; Nogita, K.; Dahle, A.K. Eutectic nucleation in Al-Si alloys. *Acta Mater.* **2004**, *52*, 4273–4280. [CrossRef]
26. Liang, S.-M.; Schmid-Fetzer, R. Phosphorus in Al-Si cast alloys: Thermodynamic prediction of the AlP and eutectic (Si) solidification sequence validated by microstructure and nucleation undercooling data. *Acta Mater.* **2014**, *72*, 41–56. [CrossRef]
27. Eiken, J.; Apel, M.; Liang, S.-M.; Schmid-Fetzer, R. Impact of P and Sr on solidification sequence and morphology of hypoeutectic Al-Si alloys: Combined thermodynamic computation and phase-field simulation. *Acta Mater.* **2015**, *98*, 152–163. [CrossRef]
28. Aageson, L.K.; Johnson, A.E.; Fife, J.L.; Voorhees, P.W.; Miksis, M.J.; Poulsen, S.O.; Lauridsen, E.M.; Marone, F.; Stampanoni, M. Universality and self-similarity in pinch-off of rods by bulk diffusion. *Nat. Phys.* **2010**, *6*, 796–800. [CrossRef]
29. Aageson, L.K.; Johnson, A.E.; Fife, J.L.; Voorhees, P.W.; Miksis, M.J.; Poulsen, S.O.; Lauridsen, E.M.; Marone, F.; Stampanoni, M. Pinch-off of rods by bulk diffusion. *Acta Mater.* **2011**, *59*, 4922–4932. [CrossRef]
30. Furuta, S.; Kobayashi, M.; Uesugi, K.; Takeuchi, A.; Aoba, T.; Miura, H. Investigation of three-dimensional morphology changes of the eutectic Si particles affected by trace P and Sr in Al-7%Si cast alloys by means of synchrotron nano-tomography. *Mater. Charact.* **2017**, *130*, 237–242. [CrossRef]
31. Fan, D.; Chen, S.P.; Chen, L.-Q.; Voorhees, P.W. Phase-field simulation of 2-D Ostwald ripening in the high volume fraction regime. *Acta Mater.* **2002**, *50*, 1895–1907. [CrossRef]
32. Requena, G.; Garcés, G.; Asghar, Z.; Marks, E.; Staron, P.; Clotens, P. The effect of the connectivity of rigid phase on strength of Al-Si Alloy. *Adv. Eng. Mater.* **2011**, *13*, 674–684. [CrossRef]
33. Kruuglova, A.; Engstler, M.; Gaiselmann, G.; Stenzel, O.; Shimidt, V.; Roland, M.; Diebels, S.; Mücklich, F. 3D connectivity of eutectic Si as a key property defining strength of Al-Si alloys. *Comput. Mater. Sci.* **2016**, *120*, 99–107. [CrossRef]

© 2018 by the authors. Licensee MDPI, Basel, Switzerland. This article is an open access article distributed under the terms and conditions of the Creative Commons Attribution (CC BY) license (http://creativecommons.org/licenses/by/4.0/).

Article

Investigation of the Foam Development Stages by Non-Destructive Testing Technology Using the Freeze Foaming Process

Johanna Maier [1,*], Thomas Behnisch [1], Vinzenz Geske [1], Matthias Ahlhelm [2], David Werner [2], Tassilo Moritz [2], Alexander Michaelis [2] and Maik Gude [1]

1. Institute of Lightweight Engineering and Polymer Technology (ILK), Technische Universität Dresden, Holbeinstr. 3, 01307 Dresden, Germany; thomas.behnisch@tu-dresden.de (T.B.); vinzenz.geske@tu-dresden.de (V.G.); maik.gude@tu-dresden.de (M.G.)
2. Fraunhofer Institute for Ceramic Technologies and Systems (IKTS), Winterbergstraße 28, 01277 Dresden, Germany; matthias.ahlhelm@ikts.fraunhofer.de (M.A.); david.werner@ikts.fraunhofer.de (D.W.); tassilo.moritz@ikts.fraunhofer.de (T.M.); alexander.michaelis@ikts.fraunhofer.de (A.M.)
* Correspondence: johanna.maier@tu-dresden.de; Tel.: +49-351-42508

Received: 12 November 2018; Accepted: 4 December 2018; Published: 6 December 2018

Abstract: With a novel Freeze Foaming method, it is possible to manufacture porous cellular components whose structure and composition also enables them for application as artificial bones, among others. To tune the foam properties to our needs, we have to understand the principles of the foaming process and how the relevant process parameters and the foam's structure are linked. Using in situ analysis methods, like X-ray microcomputed tomography (μCT), the foam structure and its development can be observed and correlated to its properties. For this purpose, a device was designed at the Institute of Lightweight Engineering and Polymer Technology (ILK). Due to varying suspension temperature and the rate of pressure decrease it was possible to analyze the foam's developmental stages for the first time. After successfully identifying the mechanism of foam creation and cell structure formation, process routes for tailored foams can be developed in future.

Keywords: Freeze Foaming; in situ computed tomography; non-destructive testing; bioceramics

1. Introduction

The two conventional processes for manufacturing ceramic cellular foam structures are the replica, as well as the space holder method [1,2]. These methods use organic scaffolds, which have to be burnt out. A novel manufacturing route for ceramic foam structures, called Freeze Foaming, that avoids the use of organic additives, has been developed by the Fraunhofer Institute for Ceramic Technologies and Systems (IKTS) [3,4]. The cell structure of a sample manufactured using Freeze Foaming is defined by a pressure-induced and pressure-controlled foaming process, followed by subsequent freeze drying, of a ceramic suspension in a vacuum. There are two different foaming agents—as ambient pressure drops, a reduced boiling point leads to the evaporation of water out of the aqueous suspension. The other one is air that is introduced during the manufacturing of the suspension. While the pressure is reduced (and the foam expands), the suspension's temperature follows the line of equilibrium in the phase diagram of water to the triple point. Since the pressure is reduced further, the temperature falls beneath the equilibrium temperature in the triple point of our suspension, which causes our created structure to be instantly frozen, and dries via sublimation. This freezing step can result in cryogenic structures similar to typical freeze cast structures [5,6] and accounts for the microporosity of foamed structures. Possible applications of foams and porous parts made using Freeze Foaming encompass a wide spectrum, including biomedical uses, like artificial bones [7–9] as well as carrier materials for catalytic converters [10], biosensors and drugs, or thermal [11] and acoustic insulators [12]. Freeze Foaming enables the processing of biocompatible materials while offering the unique possibility of creating foams exhibiting a multimodal pore-size distribution and interconnectivity. These factors offer good conditions for the cultivation of organic cells. Previous work [3,4,7,13] has shown a particular suitability for an application in artificial bones due to their special structural properties.

As the foaming is influenced by a complex interaction of several process and material parameters, further research into the foam formation during Freeze Foaming is needed. A reproducible manufacturing of tailored foams with a specified structure is not possible as of now. To adjust the properties according to applications, and develop process and quality guidelines, it is important to further examine the influence of relevant process parameters on the foam's structure. With the presented work, the authors aim at finding a solution to a tunable shaping method, which enables the manufacturing of highly controllable pore structures to be used, e.g., as bone-mimicking scaffolds for an ever-older population [14].

To that end, an in situ µCT extension for an existing scanner was developed, which allows the analysis of different steps of the very foaming process during the manufacturing. Conventionally, examining changes in a sample using X-ray computed tomography is done by scanning before and after a change in state or structure [15,16]. The conventional method does not enable observation of the foam development of Freeze Foams. In material research focusing on damage and degradation analysis, significant improvements were made in the last years, due to progressive in situ techniques [15,17,18].

In the investigations from [19], the sintering process of ceramics could be analyzed with the use of CT. It was sufficient to analyze the process every 30 min to get a statement about the sintering theories. Further investigations in the field of in situ analyses are described in [20]. In so-called in situ X-ray nanotomography systems, pixel sizes of 100 nm are achieved in 20 s, with the use of a focus size of 50 nm. The sintering stages of metals and ceramics were also analyzed. In the investigations presented here for the analysis of the formation process of Freeze Foams, such a resolution is not necessary. In the investigations from [21], the hydration of gypsum plaster setting was investigated with in situ X-ray tomography. The scan duration was 200 s. However, the entire structure formation of the Freeze Foaming process (i.e., foaming and freezing) takes only about 60 s. Therefore, a novel CT setup had to be developed, firstly, in order to visualize the foaming process per se and, secondly, to introduce measures making tomographic image acquisition possible. The process had to be designed in such a way that it could be stopped at certain process steps and fixed for the CT imaging. This entire experimental setup—a new controllable laboratory freeze dryer in a computer tomography scanner—is one of the fundamental novelties of this work. In the first phase, a process-optimized testing device was developed [22]. It is suitable for 2-dimensional examinations using X-ray radiography (for real-time observation of the foaming progress), as well as three-dimensional scans to evaluate structural phenomena. Using the now-reproducible manufacturing of a model suspension [23], detailed results of in situ foam structure analysis are presented.

2. Materials and Methods

The ceramic suspensions used in this work are composed of water and dispersion agent (Dolapix CE 64, Co. Zschimmer & Schwarz Mohsdorf GmbH & Co. KG, Burgstädt, Germany), added hydroxyapatite powder (Sigma-Aldrich now Merck KGaA, Darmstadt, Germany; BET = 70.01 m^2/g, d_{50} = 2.64 µm), binder (polyvinyl alcohol), and rheological modifier (Tafigel PUR40, Co. Münzing Chemie GmbH, Heilbronn, Germany) [23]. The choice of suspension and composition was derived from preliminary tests on the basis of different suspensions which, after Freeze Foaming, resulted in reproducible foam structures [23]. The detailed manufacturing process of the suspension is described in [24]. For the investigations of this contribution and with regard to its possible influence on the foaming process and structure formation, three ceramic suspensions with different temperatures were used (5, 23, and 40 °C).

An in situ device, to be used inside a v | tome | x L450 (General Electric, Cincinnati, OH, USA), was developed during the first phase of research [22]. It allows the material to be subjected to phenomenological analysis, and for detection and characterization of pores during the foaming process. To examine developmental steps during the formation of the foam's structure, the device has to be leak-proof under vacuum (Figure 1). Using different foaming molds, the device can be used for either X-ray radiography (2D) or gaining spatial information (µCT, 3D) about the foam's structure. The resolution was set to 22 µm/vx using an acceleration voltage of 100 kV, and a beam current of 300 µA. To fix the foaming suspension for the time of the CT scan (720 projections with 250 ms exposure time each, 3 min total measurement time), the foaming is stopped using a pressure control system with dedicated software, developed in-house, and an adjustable bypass. The pressure is kept at a constant level for the duration of the CT, in order to stabilize the structure. The vacuum chamber itself is rotationally symmetrical, and made of low-absorbing polymer to ensure optimal image quality. The choice of polymethylmethacrylate (PMMA) for the chamber prevented stabilizing the foam by means of externally freezing, as the material's thermal conductivity is very low.

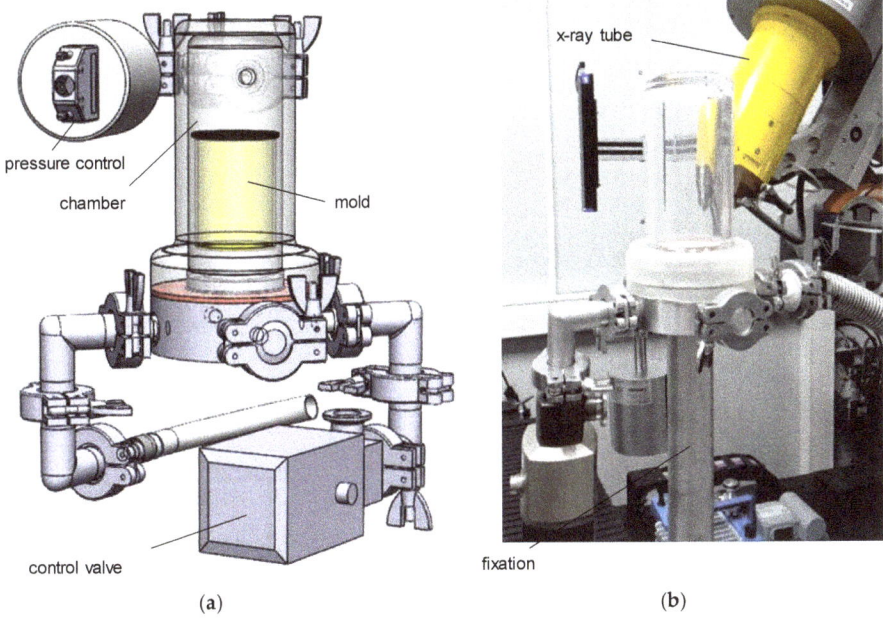

Figure 1. In situ µCT device: CAD model (**a**) and mounting situation (**b**).

In previous examinations [22], the pressure reduction rate's influence on the foam structure—and especially the orientation of pores—has been shown. Due to the concave nature of the bottom of the mold, foaming led to a high number of samples exhibiting large pores near the bottom, which distorted the results of the foam analysis. Therefore, a new mold with a flat bottom was designed (Figure 2). Furthermore, water vapor emitted from the suspension decreased the pressure reduction rate in the lower pressure range of 25 mbar and below (near equilibrium of water vapor at 20 °C). The reduced pressure drop led to a higher amount of coalescence effects in the finished foam structure. To accelerate the pressure reduction, a cold trap has been used. For this purpose, a cylinder made from aluminum, with channels, was manufactured. It was cooled down with liquid nitrogen and placed on top of the mold (Figure 2, right picture). The water vapor condenses on its surface, which significantly reduces the time to fall below the triple point.

Figure 2. Improvements to the experiment.

Besides the process analysis compression tests, in situ μCT scans under compressive load were also conducted on the conducted foams. In situ compressive scans were performed using a Finetec FCTS 160 IS (Garbsen, Germany) with an acceleration voltage of 50 kV and a beam current of 250 μA. The resolution was around 11 μm/vx, and the exposure time 625 ms. Mechanical testing of prepared cylindrical samples took place using an universal testing machine Zwick 1475 (Ulm, Germany). A preload of 2 N and a traverse speed of 2 mm/min were chosen.

3. Results

3.1. Radiographical Evaluation of the Freeze Foaming Process

By acquiring two 2D pictures per second (500 ms exposure time) a real-time observation and analysis of the foaming process is possible. Due to the superposition of structural phenomena, the thickness of the sample was reduced to 5 mm for radiographically evaluations. To qualify the changes between the steps, manual tracking is conducted by overlaying each picture with a grid, and following the movement of distinct points in the sample. The resulting coordinates can be converted to changes in actual values for the size or height of the foam, and can be correlated with the pressure at the moment of picture acquisition. This method of evaluation was applied for three different sample temperatures and a pressure reduction rate of 50 and 10 mbar/s, respectively. As an example, the plotted results for 10 mbar/s are shown in Figure 3, as experiments at different pressure reduction rates behave similarly.

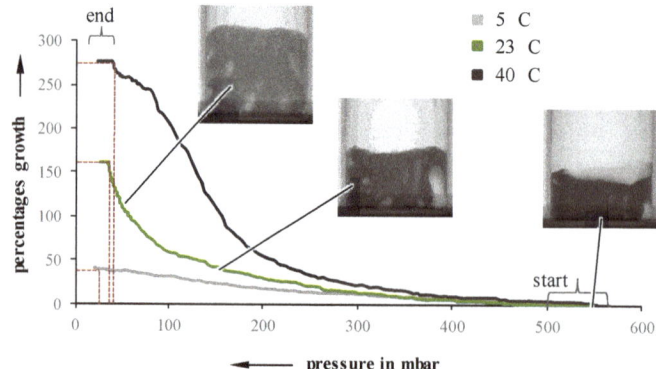

Figure 3. Percentage growth as a function of pressure for 5, 23, and 40 °C, at 10 mbar/s.

Independent of the initial suspension temperature, foaming starts between 450 and 550 mbar. However, the growth, as well as the pressure at which the foaming stops, are highly influenceable by the initial temperature. A lower temperature leads to an inhibited foaming, which results in a lower overall growth. Even though the foaming process itself continues up to lower pressures, the lower foaming rate cannot be compensated. For example, a suspension with a pressure reduction rate of 10 mbar/s and an initial temperature of 5 °C stops foaming at 15 mbar, while a 40 °C suspension already stops at 35 mbar. Suspensions undergoing 50 mbar/s exhibit a very similar behavior. Looking at foaming rates over pressure, the highest suspension temperature also results in the highest growth rates values at higher pressures (Figure 4).

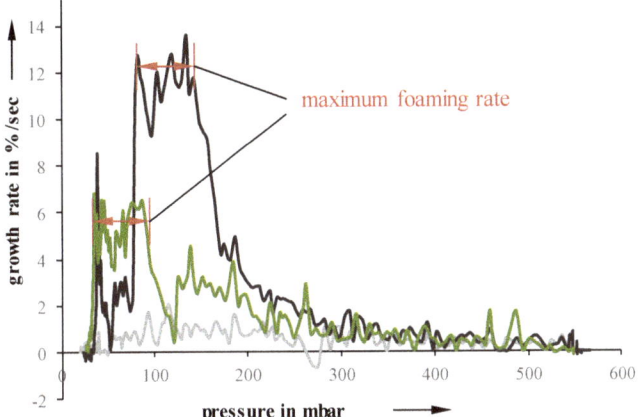

Figure 4. Foaming rate as a function of pressure for different temperatures at 10 mbar/s.

Samples with an initial temperature of 23 °C exhibit a lower maximum at lower pressures while 5 °C samples showed an almost constant foaming rate and, therefore, no identifiable maximum.

To evaluate and verify the findings from the radiographical evaluation, the results were compared to data obtained from the freeze dryer at IKTS (Table 1) [24]. Apart from 5 °C suspensions, their findings support the trend identified at the ILK. For suspensions with an initial temperature of 5 °C, the IKTS identified a foaming rate maximum at pressures between 40 and 60 mbar. However, the freeze dryer is equipped with an additional condenser, which is not available in the in situ device. It is not possible, so far, to achieve the foaming rate maximum and finish the pressure-induced foaming process.

Table 1. Comparison of the pressures of beginning, end, and maximum of foaming at temperatures of 5, 23, and 40 °C by in situ μCTs (ILK) and by freeze dryer (IKTS) [24].

Suspension's Temperature (°C)	Pressure at Beginning of Foaming (mbar)		Pressure at Maximum Foaming Rate (mbar)		Pressure at End of Foaming (mbar)	
	Freeze Dryer	In Situ μCT	Freeze Dryer	In Situ μCT	Freeze Dryer	In Situ μCT
5	500	450–550	40–60	n.d.	10	15
20	400	450–550	80	50–90	20	25
40	400	450–550	80–100	90–150	60	35

Through analysis of radiographic images, edge effects on the foaming process can be verified. Figure 5 illustrates their impact on a 10 mbar/s sample at 23 °C. Both edges, as well as the center, were manually tracked using the method described earlier, and the local growth rates were determined. As expected, the edges exhibit a much slower growth when compared to the center. Possible reasons include wall friction, as well as a drying of the suspension. This behavior is especially observable in suspensions with an initial temperature of 40 °C, which also develops a compact layer on top of the suspension.

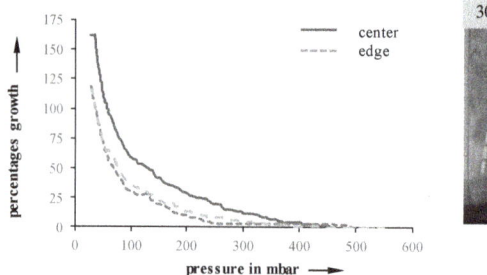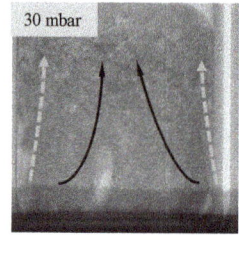

Figure 5. Edge effects on foaming at 23 °C (10 mbar/s).

3.2. CT Evaluation of the Freeze Foaming Process

To examine the developmental stages of foaming, µCT and the improved testing device is used to create a virtual and reconstructed volume (VGStudio 2.0, Volume Graphics GmbH, Heidelberg, Germany) of the foam's structure. During the CT measurements, the sample is rotated 360°, and after an angle of 0.25°, one image is taken. Those pictures can be reconstructed with the program "Phoenix datos", which is generating a 3D model. This model can be imported into "VGStudioMax 3.0" (Volume Graphics GmbH). In this program, it is possible to perform various analyses, such as defect analysis or foam structure analysis. A region of interest (ROI) is selected, and a surface determination is performed automatically. A threshold value is determined to be able to separate material from background. The porosity can be determined from this data.

In order to acquire sufficient CT data, given that Freeze Foaming is a fast process, the exposure time is reduced from 500 to 250 ms. In addition, the number of images for the holding steps is reduced from 1440 to 720. The following parameters were selected for the evaluation of the foam structure: Threshold—80%, Accuracy—Fast; Direction of analysis—Right; Analysis mode—Background; Features—Advanced cell properties.

The first step was a complete foaming at two pressure reduction rates (10 mbar/s and 50 mbar/s). The results and their porosities are shown in Figure 6.

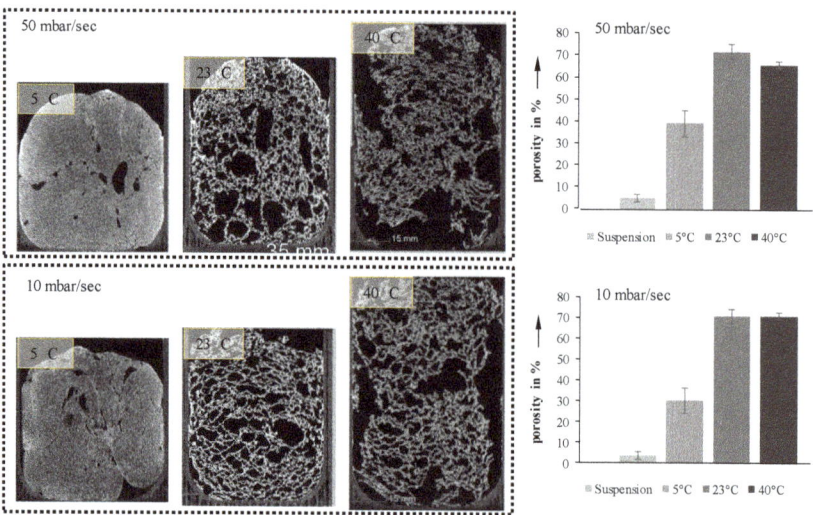

Figure 6. Cross-sections and porosities of completely foamed suspensions at two pressure reduction rates and three temperatures.

Increasing the pressure reduction rate from 10 to 50 mbar/s results in only a slight increase in porosity (Figure 6, right). However, the orientation of the pores seems to be significantly influenced by the pressure reduction rate. Suspensions foamed at 50 mbar/s exhibit vertically elongated pores, while those foamed at 10 mbar/s are oriented more horizontally. Due to the higher reduction rate inflicted on the process-induced air, the velocity of the inflating bubbles increases, thus forming vertical pores. In both cases, this is especially visible for 23 °C samples. Samples with an initial temperature of 40 °C show a decrease in porosity during foaming, due to their low viscosity, and the foam collapses before freezing by reaching the triple point.

3.3. Stages of the Foaming Process

Given the results of radiographic imaging, five holding stages were identified for analyzing the stages of the foaming progress (at 30, 40, 50, 70, and 100 mbar). The pressure reduction rate was adjusted to 10 mbar/s. Cross-sections of those scans are shown in Figure 7 (5 °C) and Figure 8 (23 °C). For each evaluation, three CT scans were executed to observe pores, and their development—the suspension (1000 mbar), the holding stage at its target pressure, and the final foam structure (5 mbar). Air bubbles that have been introduced into the mold during suspension filling have a large influence on the foam structure. They grow even larger during foaming, and develop significantly larger pores. Due to their high viscosity, this growth is inhibited in 5 °C tempered suspensions. Furthermore, the maximum foaming rate takes place at lower pressures (40–60 mbar) [24]. As a large amount of water evaporates, the pressure reduction rate drops, and the time to reach the target pressure of <5 mbar is too long. This process-induced growth inhibition results in a significantly lower porosity. In general, suspensions with an initial temperature of 5 °C exhibit a lower porosity after foaming, due to a higher viscosity and a lower amount of escaping water vapor. On the other hand, suspensions with an initial temperature of 40 °C exhibit a viscosity too low to be stable during the CT scan and, therefore, were not monitored.

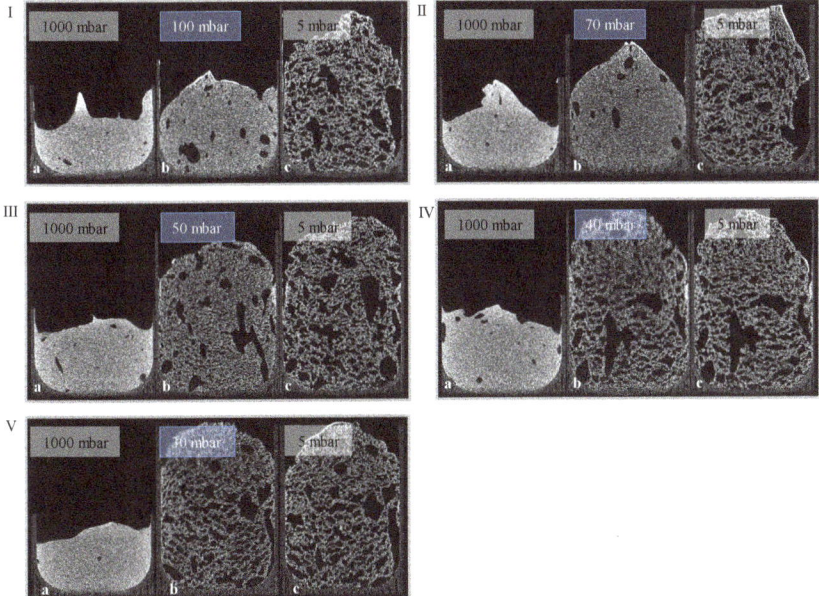

Figure 7. Holding stages (I: 100 mbar, II: 70 mbar, III: 50 mbar, IV: 40 mbar, V: 30 mbar) of a 23 °C tempered suspension with a pressure reduction rate of 10 mbar/s ((**a**) suspension; (**b**) holding stage; (**c**) foam).

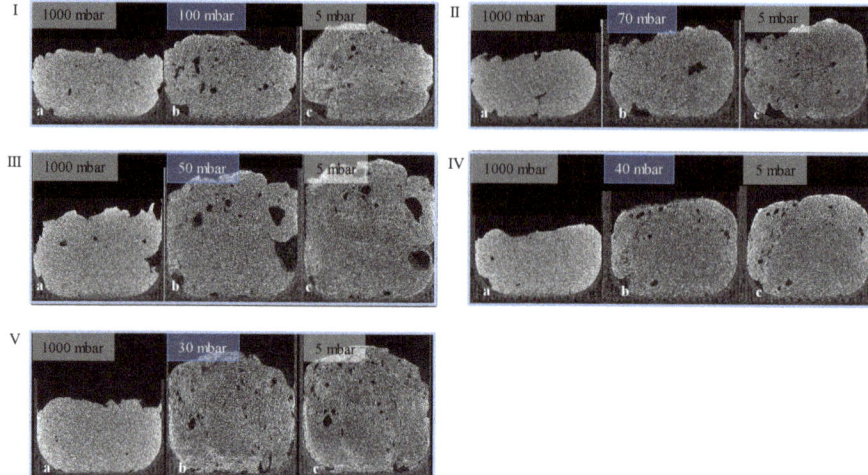

Figure 8. Holding stages (I: 100 mbar, II: 70 mbar, III: 50 mbar, IV: 40 mbar, V: 30 mbar) of a 5 °C tempered suspension with a pressure reduction rate of 10 mbar/s ((**a**) suspension; (**b**) holding stage; (**c**) foam).

Using the software VGStudioMax, which allows access to volume-based data, the pore size distributions were determined for each holding stage. An exemplary distribution for a 23 °C sample is depictured in Figure 9a. The growth starts slowly, accelerates to a maximum, and then slows down again. The lowest variance in pore size distribution can be found at 50 mbar on a 23 °C sample, with a relative curve width (b = d_{90}/d_{10}) of 36.2 ($b_{100mbar}$ = 53.1, b_{70mbar} = 36.2, b_{40mbar} = 153.5, b_{30mbar} = 160.9). When the pressure drops below 50 mbar, the pore size distribution becomes flatter and wider, indicating a ripening process. Figure 9b shows the increasing porosity as a function of the decreasing pressure for the three investigated temperatures. Due to their high viscosity, 40 °C foamed samples could not be investigated with regard to holding stages and foam formation, because the foam structures collapsed during the investigation.

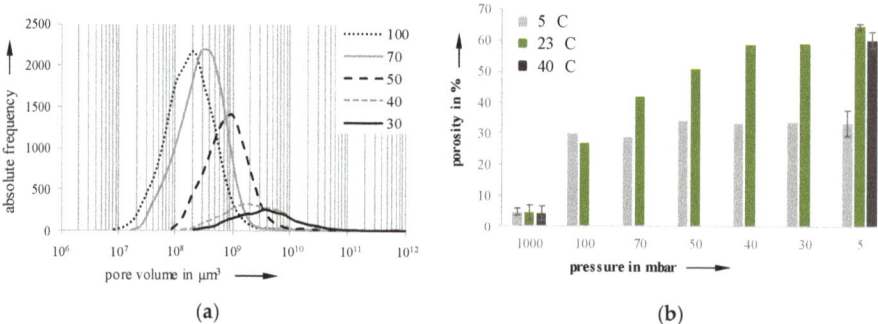

Figure 9. Pore size distribution of a 23 °C sample (**a**) and porosity (**b**) at 5, 23, and 40 °C samples of different holding stages; pressure reduction rate: 10 mbar/s.

3.4. Mechanical Properties

Due to the dependence of porosity and pore size distribution of Freeze Foamed samples on process parameters, mechanical properties should vary as well. To evaluate their behavior under load, cylindrical sintered samples were manufactured and subjected to standardized compression tests. Recorded tension–compression curves are depicted in Figure 10.

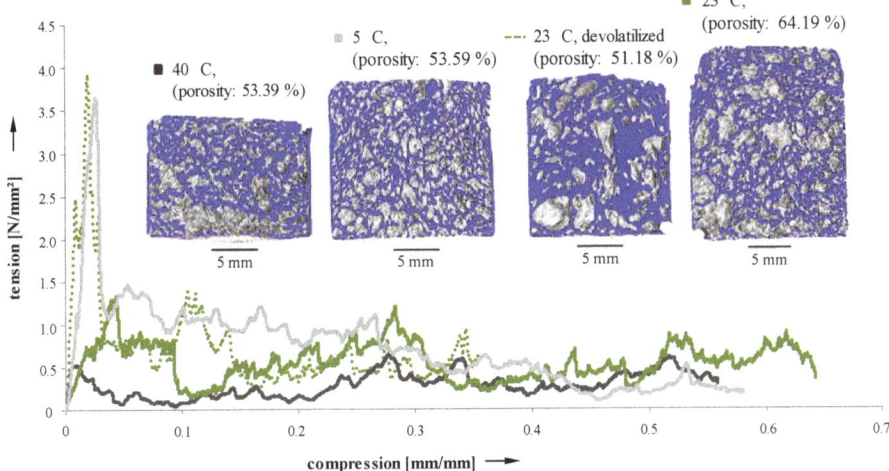

Figure 10. Compression tests on sintered samples (40 °C suspension; 5 °C suspension; devolatilized 23 °C suspension; 23 °C suspension).

The fracture behavior clearly exhibits a dependency on suspension temperature (5, 23, and 40 °C) and pretreatment (devolatilized and not devolatilized suspensions during the manufacturing process). Samples with a narrow pore size distribution [24] (5 °C and 23 °C devolatilized [23]) possess a pronounced maximum of force. On the other hand, specimens with a less uniform distribution of pores [24] (40 °C and 23 °C not devolatilized) show a more constant force level, and only reach about half the maximum force when compared to more homogeneous samples. Mechanical properties of the samples manufactured using Freeze Foaming are strongly influenced by microporosities inside the struts [23]. However, as the resolution of the CT scans were insufficient to examine their structure, they could not be taken into account here.

Furthermore, in situ compression tests, at selected load levels, were conducted to examine failure phenomenology (Figures 11–13).

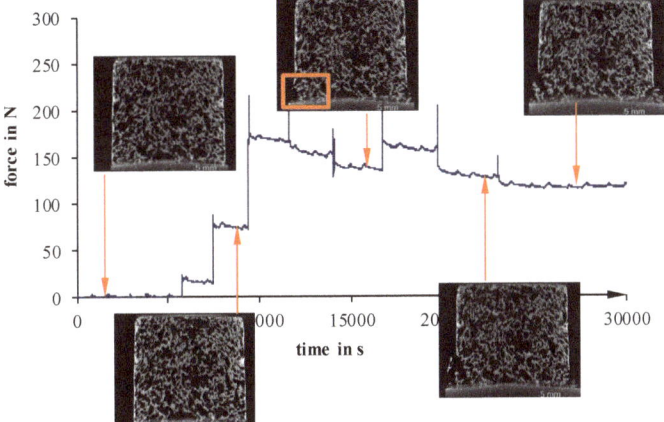

Figure 11. Recorded force–travel curves of compression tests conducted on an in situ μCT; 5 °C.

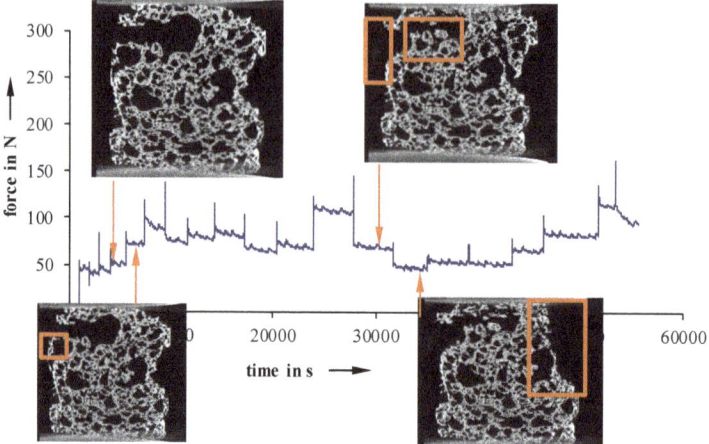

Figure 12. Recorded force-travel curves of compression tests conducted on an in situ µCT; 23 °C not devolatilized.

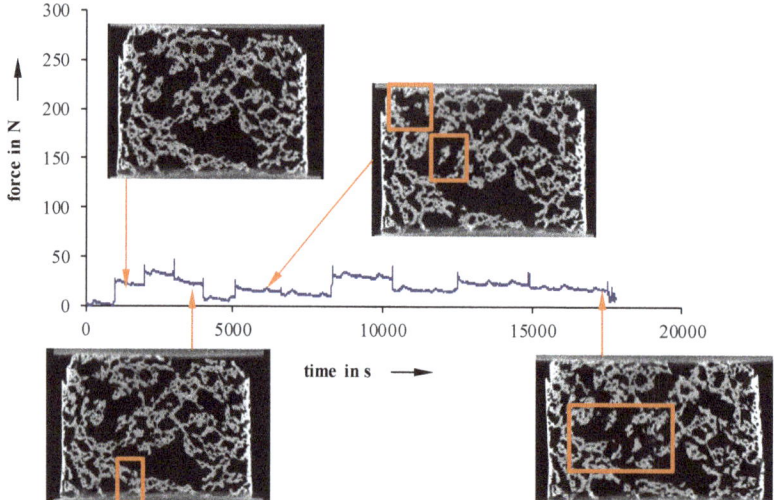

Figure 13. Recorded force-travel curves of compression tests conducted on an in situ µCT; 40 °C.

For more inhomogeneous samples, material failure starts to occur between 25 N (40 °C) and 50 N (23 °C) (Figures 12 and 13). The density of fractures constantly rises, with increasing deformation, until a partial structure failure develops at relatively low load levels of 50 N (40 °C) and 110 N (23 °C not devolatilized). On the other hand, more homogeneous foam structures show a maximum load up to 200 N, even after the first signs of material failure (Figure 11). Detailed examinations show that especially cracks on the surface lead to material failure. This is a sign of an uneven sample surface, and results in a non-uniform load.

4. Conclusions

Biocompatible new materials will become increasingly important in the future. Ceramic structures based on Freeze Foaming allow ecological manufacturing without a need for organic scaffolds. Tailoring these ceramic foams to specific applications, a defined and reproducible adjustment of their structure and mechanical properties is necessary. However, the formation of the foam structure during Freeze Foaming is not yet fully understood. Their manufacturing is influenced by a complex interaction of different steps and material's properties within the process.

Using a novel in situ µCT device, it was possible to examine the foaming process and the stages of the foaming process. Due to an integrated pressure control system, the foaming could be stopped at any applied pressure.

Radiographic imaging gathered information about the beginning, maximum, and end of foaming, depending on the temperature of the suspension. Independent of the temperature, foaming starts between 450 and 550 mbar. An earlier end of foaming was detected when increasing the suspension temperature to 40 °C, due to a higher water vapor partial pressure and a lower viscosity. Suspensions with an initial temperature of 5 °C did not exhibit a foaming maximum in our device, due to their high viscosity.

To observe the pore formation during Freeze Foaming, µCT scans were performed using the new µCT device. Virtual volumes of Freeze Foam scaffolds were created and analyzed. Foaming was executed with varying pressure reduction rates. While the porosity changed only slightly with varying pressure reduction rates, the pores were oriented differently. During foaming, 40 °C tempered suspensions collapsed before reaching the triple point, due to their low viscosity. On the other hand, the growth of 5 °C suspensions were inhibited by their high viscosity. As a result of radiographic examinations, five pressure values were identified as holding stages of interest. Those stages revealed a large influence of air bubbles introduced during mold filling on the final foam structure. Independent of the initial temperature of the suspension, there is a continuous rise in porosity during the foaming process, in general, while the variance of pore size increases. Furthermore, the results of compression testing of sintered samples show a distinct force maximum for 5 °C and 23 °C tempered and devolatilized Freeze Foams. On the other hand, samples with a less homogeneous structure (40 °C and 23 °C not devolatilized) exhibit a force plateau and a maximum force about half that of samples.

Approaches for the defined production of Freeze Foams have been achieved. However, the complexity of the Freeze Foaming process requires more experiments and evaluation, in order to truly control the pore structure and, thus, make them more suitable for larger industries and applications.

Author Contributions: T.B. and J.M. are the DFG-funded project managers from TU-DD ILK, writing the manuscript and working on the experiments presented in this paper; V.G. is working on the X-ray and is concerning on the experiments; M.A. and D.W. are the DFG project's manager and editors from the Fraunhofer IKTS, scientifically conceiving, guiding and supervising experiments and evaluations; T.M., A.M., and M.G. helped with higher-ranked issues and questions.

Funding: This research has been funded by DFG, grant number 310892168.

Conflicts of Interest: The authors declare no conflict of interest. The founding sponsors had no role in the design of the study; in the collection, analyses, or interpretation of data; in the writing of the manuscript, and in the decision to publish the results.

References

1. Kormarneni, S.; Pach, L.; Pidugu, R. Porous-alumina ceramics using bohemite and rice flour. *Mater. Res. Soc. Symp. Proc.* **1995**, *371*, 285–290.
2. Schwartzwalder, K.; Arthur, V.S. Method of making porous ceramics articles. U.S. Patent 3090094A, 21 May 1963.
3. Moritz, T. Lightweight Green Compact and Molded Article Made of a Ceramic and/or Powder-Metallurgical Material, and the Method for the Production of Thereof. Ger. Patent E102008000100, 18 January 2008.
4. Ahlhelm, M.; Moritz, T. Synthetic Bone Substitute Material and Method for Producing the Same. Eur. Patent EP2682137A3, 1 February 2013.
5. Deville, S.; Saiz, E.; Tomsia, A.P. Freeze casting of hydroxyapatite scaffolds for bone tissue engineering. *Biomaterials* **2006**, *27*, 5480–5489. [CrossRef] [PubMed]

6. Deville, S. Freeze-Casting of Porous Biomaterials: Structure, Properties and Opportunities. *Materials* **2010**, *3*, 1913–1927. [CrossRef]
7. Ahlhelm, M.; Schwarzer, E.; Scheithauer, U.; Moritz, T.; Michaelis, A. Novel ceramic composites for personalized 3D-structures. *J. Ceram. Sci. Technol.* **2017**, *8*, 91–100.
8. Best, S.M.; Porter, A.E.; Thian, E.S.; Huang, J. Bioceramics: Past, present and for the future. *J. Eur. Ceram. Soc.* **2008**, *28*, 1319–1327. [CrossRef]
9. Hench, L. Bioceramics: From concept to clinic. *J. Am. Ceram. Soc.* **1991**, *74*, 1487–1510. [CrossRef]
10. Adker, J.; Standke, G. Offenzellige Schaumkeramik. *Keram. Z.* **2003**, *55*, 786–792.
11. Ahlhelm, M.; Fruhstorfer, J.; Moritz, T.; Michaelis, A. The Manufacturing of Lightweight Refractories by Direct Freeze Foaming Technique. *Interceram* **2011**, *60*, 394–398.
12. Schneider, H.; Schreuer, J.; Hildmann, B. Structure and properties of mullite—A review. *J. Europ. Ceram. Soc.* **2008**, *28*, 329–344. [CrossRef]
13. Ahlhelm, M.; Zybell, K.; Gorjup, E.; Briesen, H.V.; Moritz, T.; Michaelis, A. Freeze-foaming: A new approach to manufacture ceramic cellular structures allowing the ingrowth and differentiation of mesenchymal stem cells. In Proceedings of the 13th CCT Ceramics, Cells and Tissues, Faenza, Italy, 17–20 May 2011.
14. Baino, F.; Ferraris, M. Learning from Nature: Using bioinspired approaches and natural materials to make porous bioceramics. *Int. J. Appl. Ceram. Technol.* **2017**, *14*, 507–520. [CrossRef]
15. Hufenbach, W.; Gude, M.; Ullrich, H.-J.; Czulak, A.; Danckak, M.; Böhm, R.; Zscheyge, M.; Dohmen, E.; Geske, V. Computer tomography-aided non-destructive and destructive testing in composite engineering. *J. Compos. Theory Pract.* **2012**, *12*, 279–284.
16. Hufenbach, W.; Böhm, R.; Gude, M.; Berthel, M.; Hornig, A.; Rucevskis, S.; Andrich, M. A test device for damage characterisation of composites based on in situ computed tomography. *Compos. Sci. Technol.* **2012**, *72*, 1361–1367. [CrossRef]
17. Scott, A.E.; Mavrogordato, M.; Wright, P.; Sinclair, I.; Spearing, S.M. In situ fibre fracture measurement in carbon–epoxy laminates using high resolution computed tomography. *Compos. Sci. Technol.* **2011**, *71*, 1471–1477. [CrossRef]
18. Hufenbach, W.; Gude, M.; Böhm, R.; Hornig, A.; Berthel, M.; Danczak, M.; Geske, V.; Zscheyge, M. In situ CT based damage characterisation of textile-reinforced CFRP composites. In Proceedings of the 15th European Conference on Composite Materials (ECCM 15), Venedig, Italy, 24–28 June 2012.
19. Feng, X.U.; Hu, X.F.; Yu, N.I.U.; Zhao, J.H.; Yuan, Q.X. In situ observation of grain evolution in ceramic sintering by SR-CT technique. *Trans. Nonferr. Met. Soc. China* **2009**, *19*, s684–s688.
20. Villanova, J.; Daudin, R.; Lhuissier, P.; Jauffres, D.; Lou, S.; Martin, C.L.; Labouré, S.; Tucoulou, R.; Martinez-Criado, G.; Salvo, L. Fast in situ 3D nanoimaging: A new tool for dynamic characterization in materials science. *Mater. Today* **2017**, *20*, 354–359. [CrossRef]
21. Adrien, J.; Meille, S.; Tadier, S.; Maire, E.; Sasaki, L. In-situ X-ray tomographic monitoring of gypsum plaster setting. *Cem. Concr. Res.* **2016**, *82*, 107–116. [CrossRef]
22. Maier, J.; Behnisch, T.; Geske, V.; Ahlhelm, M.; Werner, D.; Moritz, T.; Michaelis, A.; Gude, M. Investigation of foam structure formation in the Freeze Foaming process based on in-situ computed tomography. *Res. Phys.* **2018**, *11*, 584–590. [CrossRef]
23. Ahlhelm, M.; Werner, D.; Maier, J.; Abel, J.; Behnisch, T.; Moritz, T.; Michaelis, A.; Gude, M. Evaluation of the pore morphology formation of the Freeze Foaming process by in situ computed tomography. *J. Eur. Ceram. Soc.* **2018**, *38*, 3369–3378. [CrossRef]
24. Ahlhelm, M.; Werner, D.; Kaube, N.; Maier, J.; Abel, J.; Behnisch, T.; Moritz, T.; Michaelis, A.; Gude, M. Deriving Principles of the Freeze-Foaming Process by Nondestructive CT Macrostructure Analyses on Hydroxyapatite Foams. *Ceramics* **2018**, *1*, 65–68. [CrossRef]

© 2018 by the authors. Licensee MDPI, Basel, Switzerland. This article is an open access article distributed under the terms and conditions of the Creative Commons Attribution (CC BY) license (http://creativecommons.org/licenses/by/4.0/).

Article

Variation of the Pore Morphology during the Early Age in Plain and Fiber-Reinforced High-Performance Concrete under Moisture-Saturated Curing

Miguel A. Vicente [1,2,*], Jesús Mínguez [1] and Dorys C. González [1,2]

1. Department of Civil Engineering, University of Burgos, c/Villadiego, s/n, 09001 Burgos, Spain; jminguez@ubu.es (J.M.); dgonzalez@ubu.es (D.C.G.)
2. Parks College of Engineering, Aviation & Technology, Saint Louis University, 3450 Lindell Blvd, Saint Louis, MO 63103, USA
* Correspondence: mvicente@ubu.es; Tel.: +34-947-259-523

Received: 13 November 2018; Accepted: 19 March 2019; Published: 24 March 2019

Abstract: In this paper, two concrete mixtures of plain concrete (PC) and steel fiber-reinforced high-performance concrete (SFRC) have been scanned in order to analyze the variation of the pore morphology during the first curing week. Six cylinders of 45.2-mm diameter 50-mm height were performed. All of the specimens were kept in a curing room at 20 °C and 100% humidity. A computed tomography (CT) scan was used to observe the internal voids of the mixtures, and the data were analyzed using digital image processing (DIP) software, which identified and isolated each individual void in addition to extracting all of their geometrical parameters. The results revealed that the SFRC specimens showed a greater porosity than the PC ones. Moreover, the porosity increased over time in the case of SFRC, while it remained almost constant in the case of PC. The porosity increased with the depth in all cases, and the lowest porosity was observed in the upper layer of the specimens, which is the one that was in contact with the air. The analysis of the results showed that the fibers provided additional stiffness to the cement paste, which was especially noticeable during this first curing week, resulting in an increasing of the volume of the voids and the pore size, as well as a reduction in the shape factor of the voids, among other effects.

Keywords: pore morphology; voids; fiber-reinforced concrete; CT scan technology; DIP software

1. Introduction

Concrete is a porous material in nature. The content of pores is very wide between different concrete mixtures. In most cases, the concrete elements show an inevitable porosity, which is what remains inside concrete mixtures after being vibrated to eliminate all the entrained air (using, for example, surface or needle vibrators, etc.). In other cases, the concrete mixture is specially designed to include a certain porosity level, depending on particular needs.

Porosity has a strong influence on the behavior of fresh concrete, because it modifies the rheology of the mixture [1–3]. Moreover, porosity modifies the mechanical properties and the durability of hardened concrete. Among other effects, porosity has a relevant impact on permeability [4–7], the behavior under freeze-thaw cycles [8,9], the behavior under fire [10], and the behavior under fatigue loading [11–13].

The porosity of concrete must not necessarily be a problem. In fact, in some cases, porosity is a sought property. For example, an increase in porosity leads to an increase in the flowability of concrete, which is a needed requisite for pumpable concretes [14]. Pervious concrete is a good example of how porosity is a sought-after property. In this case, concrete pavement is designed with an extremely high volume of pores, in order to assure that roads remain dry when it is raining, even during extreme downpour, increasing the safety of the road [4,6]. This same idea can be used for ultralight concretes [15].

Another situation where air voids are useful is in the case of concrete elements subjected to freeze–thaw cycles [8,9]. Under these cycles, porosity improves the resistance of the concrete. In fact, concrete elements placed in regions with extreme freeze–thaw cycles events must be designed with a minimum threshold of porosity.

On the other hand, there has been an increasing use of fiber-reinforced concrete. It is a very suggestive solution because of the reduction of the labor cost, especially if it is combined with self-compacting concrete. In most cases, fibers are used to improve the mechanical behavior of concrete: they reduce cracking [16–18], improve the fatigue life [19–22], increase the tension strength capacity of concrete [23,24], improve the behavior under freeze–thaw cycles [25,26], and extend the fatigue life [27,28].

A different case involved the use of plastic fibers to improve the behavior of concrete under fire [29,30]. In this case, the strategy is that plastic fibers melt under a relatively low temperature, which is significantly below the temperature that fire typically starts to result in spalling in conventional concrete, and thus the internal overpressure caused by water vapor is dissipated.

However, in all the research works mentioned above, it has been implicitly assumed that fibers do not modify the concrete matrix, i.e., the microstructure of concrete matrix and, in particular, the pore morphology (voids content, pore size distribution, shape of the voids, etc.) is not affected by the presence of fibers.

The voids inside concrete can be classified into micropores (size less than 1 µm), mesopores (size between 1–10 mm), and macropores (size greater than 10 mm) [31]. Several methods can be found in the literature to analyze the pore structure. The traditional ones are nitrogen absorption and mercury intrusion porosimetry (MIP) [32,33]. These methods show two main limitations. First, they can only provide the pore size distribution, but not the pore distribution, shape, etc. Second, these techniques can only provide information about open porosity, and not about closed porosity.

Currently, the use of computed tomography (CT) scan technology is being used to analyze, in general, the microstructure of concrete and, in particular, the pore structure. Most of the research conducted has focused on fiber-reinforced concrete, and hence fiber orientation [12,34,35]. However, in the last years, there has been a growing interest in the internal pore structure, and several works have been published in this area [4,36–43].

Using CT scan technology, it is possible to visualize all the pores of the concrete samples, and not only the open porosity, but also the closed porosity. CT scan technology provides a lot of useful information of each individual void, such as the position, volume, length, etc. With this information, it is possible to determine several geometrical parameters, such as the shape factor, among others. Moreover, it is possible to obtain several correlations, such as the spatial distribution, among others.

In addition, when the scanning process is carried out daily during the first curing week of the specimens, it is possible to analyze the evolution of all the geometrical parameters of the voids over time.

This information can be used as a basis to establish the correlation between the porosity of the concrete and its macroscopic response.

The CT scan technology is a powerful tool; to date, no other technology can provide this information about the internal microstructure of concrete.

In this paper, the CT scan is used in order to detect the voids inside two different concrete mixtures: plain and fiber-reinforced concrete, and also to study the evolution of the voids over time, during the first curing week. In both cases, the concrete paste is the same, and the only difference is the presence of steel fibers. Using post-processing routines especially developed by the authors, it is possible to analyze the pore morphology in both cases: porosity, pore size distribution, pore shape, etc., and its variations with time during the early ages of concrete. The results show that plain and fiber-reinforced concrete mixtures have initial differences in pore morphology, and they also exhibit a different variation with time. The final result is that both concretes have very different pore morphology at the end of the studied time. The results also reveal the two mechanisms behind the differences between the final pore morphology of plain and steel fiber-reinforced concrete.

This paper is structured as follows: the experimental procedure is presented in Section 2, the results of the tests are described and discussed in Section 3, and finally the conclusions are found in Section 4.

2. Experimental Program

In this section, the materials, the manufacturing procedure, and the scanning procedure are described.

2.1. Materials

A total of six cylinders that were 45.2 mm in diameter 50 mm in height were manufactured. Three of them were of plain high-performance concrete (PC), and the rest were steel fiber-reinforced high-performance concrete (SFRC). The mixing proportions by weight, in both cases, were: 1:2.00:0.015:0.31:0.035 (cement:fine aggregate:nanosilica:water:superplasticizer). The fiber-reinforced concrete was reinforced with 7.8 kg/m^3 of steel fibers Dramix OL 8/.16 (BEKAERT, Kortrijk, Belgium), which means that concrete had a fiber volume fraction of 0.1%. Fibers were 8 mm in length and 0.16 mm in diameter (aspect ratio 50). According to the technical information provided by the supplier, their tensile strength was 3000 MPa, and their modulus of elasticity was 200 GPa. The nanosilica used was MasterRoc MS 685 (BASF, Ludwigshafen am Rhein, Germany). The superplasticizer used was Glenium 52 (BASF, Ludwigshafen am Rhein, Germany). Siliceous aggregate was used, with a nominal maximum aggregate size of 4 mm. Portland cement with a high initial strength of CEM I 52.5 R was used.

Additionally, two prisms that were 40 × 40 × 160 mm were manufactured; one per mixture. A total of three cubes with 40-mm edges were obtained from each prism, and they were tested under compression in order to characterize the compression strength (according to EN 196-1:2016). The concrete compressive strength f_c' was 68.9 MPa with a standard deviation of 1.9 MPa.

2.2. Specimen Manufacturing Process

The six cylinders above mentioned were cast in the same number of Polyvinyl chloride (PVC) molds with a 45.2-mm inner diameter, 50-mm outer diameter, and 50-mm height. At the bottom of each pipe, a PVC disc with a 60-mm diameter and 3-mm thickness was welded, in order to ensure a watertight joint (Figure 1). The concrete was built using a cement mortar mixer, and the manufacturing process followed the standard EN 196-1:2016 [44]. The molds were filled in two parts using a small aluminum scoop to form the specimens without applying vibration. However, some small punches were applied on the side of the molds, once it was filled with mortar, to help the mortar expel the entrapped air. The upper surface was smoothed with a trowel. Once all of the specimens were cast, they were kept in a moisture curing room, where they remained at 20 °C and 100% humidity.

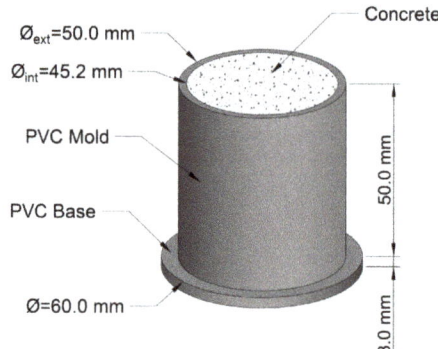

Figure 1. Mold and specimen.

2.3. Scanning Process

During the first week, a total of five scans were performed of each specimen, belonging to the following ages: one day, two days, three days, four days, and seven days. The CT scan used was a GE Phoenix v | tome | x device (General Electric, Boston, MA, USA), belonging to the 'Centro Nacional de Investigación sobre la Evolución Humana (CENIEH)', in Burgos, Spain. It was equipped with a tube of 300 kV/500 W. This facility emits a cone ray, which is received by an array of detectors. Thereby, the scanning process is fast, and highly accurate scans are produced of equal resolution in the X, Y, and Z axes. The rotation step around the Z-axis was 0.5°, so 720 shots were carried out on each specimen. The CT scan has a post-processing software that provides flat pictures of 2048 × 2048 pixels. Thus, for a section of 45.2-mm diameter, the equipment provides a horizontal resolution of 25 × 25 µm². The vertical distance between the cutting planes was fixed at 25 µm, so the CT scan produced 2000 pictures per specimen, such as the ones shown in Figures 2 and 3. The software assigns a grey level to each voxel (volumetric pixel), varying from zero to 255, where zero means black, and 255 means white, depending on the current density of the matter. Light grey voxels correspond to more dense points and dark grey voxels correspond to less dense points, i.e., pores. The final result of the CT scan is a matrix including X, Y, and Z coordinates of the voxel center of gravity and a number, from zero to 255, regarding the density. The total number of voxels in a specimen is approximately 4.3×10^9. The average time of scanning is around 1 h for each specimen, including the post-processing process. A more detailed explanation of the CT scan technology and the scanning process can be found in [40–45].

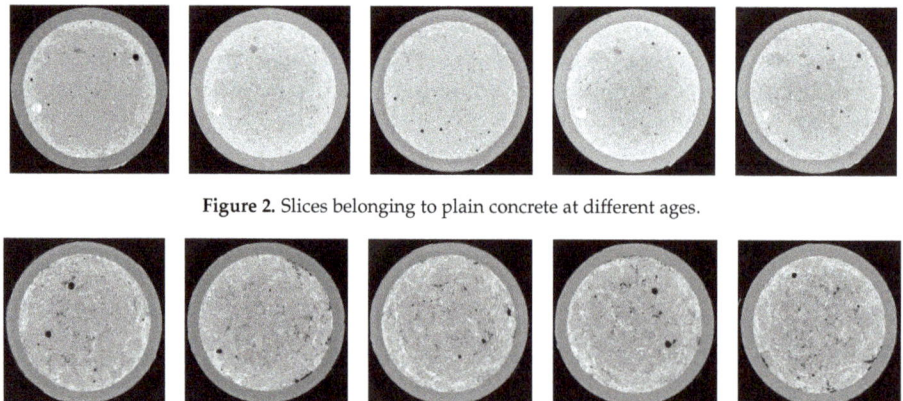

Figure 2. Slices belonging to plain concrete at different ages.

Figure 3. Slices belonging to steel fiber-reinforced concrete at different ages.

Next, the digital image processing (DIP) software AVIZO (FEI Visualization Sciences Group, Hillsboro, OR, USA) was used to identify and isolate each individual void inside the specimen. First, the software identified the voxels belonging to voids, which are the ones showing a grey color below a threshold. In this case, once the histogram of grey distribution was studied, a threshold of 65 was considered (Figure 4).

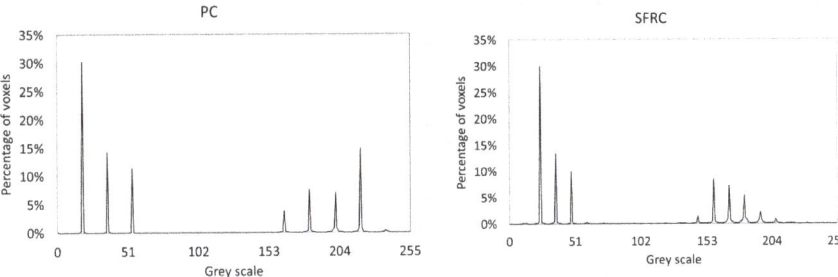

Figure 4. Histograms of grey scale.

Histograms revealed that the images have a great amount of dark-grey voxels. Most of them belonged to the empty space placed outside the mold. So, the next step was to delete all the voxels placed outside the inner face of the mold, because they did not belong to the concrete specimen. Moreover, the histograms of Figure 4 are not valid to obtain the percentage of voids in the specimen.

Then, all the voxels in contact were merged, since they belong to the same void. The software identified and isolated the different voids. The final result of the scanning was a dot matrix containing the Cartesian coordinates X, Y, and Z of the center of gravity of each individual void. Figures 5 and 6 shows the view of one specimen of each mixture along the timeline.

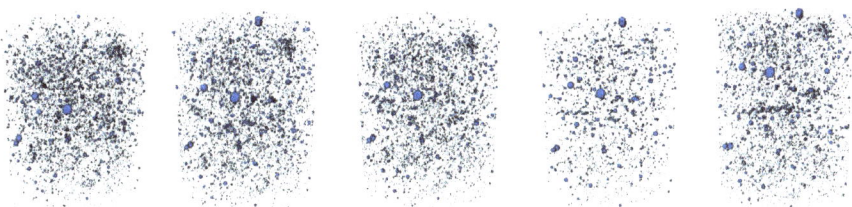

Figure 5. Three-dimensional (3D) views of the voids belonging to plain concrete at different ages. Age increases from **left** to **right**.

Figure 6. 3D views of the voids belonging to steel fiber-reinforced concrete at different ages. Age increases from **left** to **right**.

Moreover, in the case of SFRC, the fiber distribution and orientation were obtained (Figure 7). In this case, the fibers did not move over time.

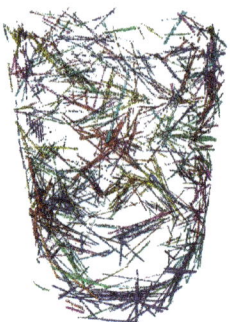

Figure 7. 3D view of the fibers belonging to steel fiber-reinforced concrete.

It can be noticed that a relevant part of the big pores was placed in the region that came in contact with the mold. This is because of the following: first, the concrete showed a wall effect, which led to a greater percentage of the smaller components of the cement paste (filler, cement, and water) in the lateral border. Second, the shrinkage of concrete resulted in a gap between the concrete specimen and molds. This space is initially filled with water and later with air, which can initially be understood as voids, although they cannot be considered 'conventional' voids.

In order to prevent the distortion on the results that can be caused by the voids located in the lateral border of the cylinders, because of the reasons mentioned above, this lateral area was discarded. In particular, the whole cylinder was not studied; instead, only the inner portion of the cylinder was studied, i.e., a cylinder with a diameter that was 90% of the real diameter of the cylinder, i.e., 40.7 mm and 90% of the whole height, i.e., 50 mm (Figure 8).

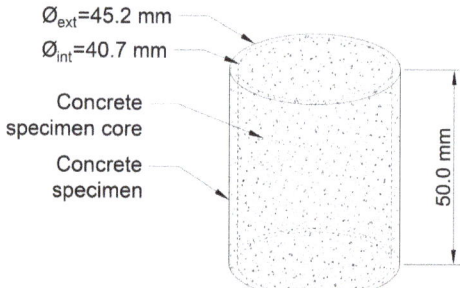

Figure 8. Portion of the concrete to be studied.

In this work, voids with less than three voxels in the largest direction (i.e., 75-µm length, approximately) have been discarded because they are too small, and the CT scan does not provide enough sharpness to clearly identify them. Additionally, pores larger than 10 mm in the largest direction have been discarded, since they are non-representative pores. The results shown in this paper are the average values of the three specimens.

3. Results and Discussion

Using the naked eye, it is not possible to detect significant differences between the specimens and mixtures, nor the evolution of the specimens with time. Instead, digital image processing (DIP) software needs to be used in combination with homemade routines to deeply analyze the data and extract relevant information. Next, this in-depth analysis is shown, and some interesting conclusions are exposed. The values that are presented correspond to the average of the three specimens belonging to the same mixture.

3.1. Total Volume of Voids and Porosity

The first studied parameter is the total volume of the voids and the porosity; this last parameter is defined as the ratio between the volume of voids and the volume of the specimen.

Figure 9 shows the variation of the porosity with the age of concrete in the different mixtures.

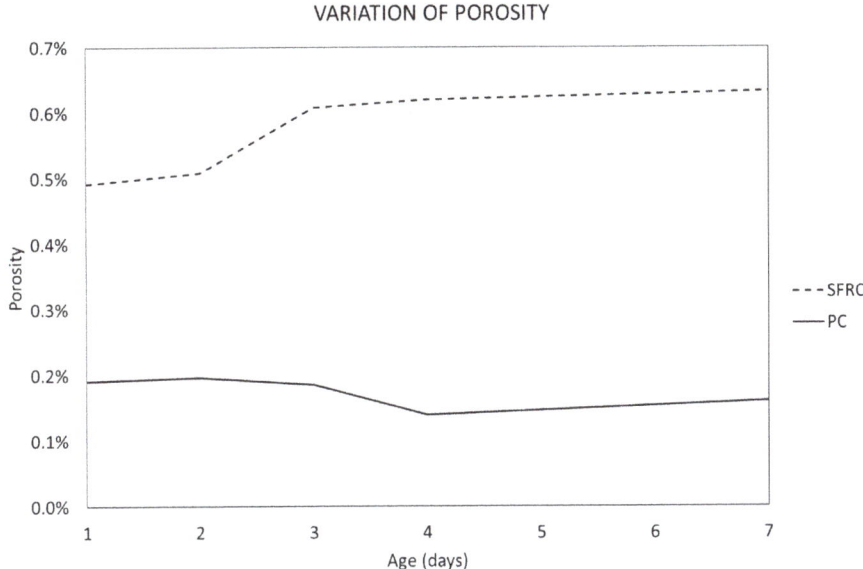

Figure 9. Variation of the porosity with the age of concrete.

Figure 9 shows some interesting conclusions. The first and most striking conclusion is that the porosity is greater in steel fiber-reinforced concrete (SFRC) than in plain concrete (PC). It is concluded that the fibers provoke an increase of the content of entrained air. It is known that the addition of steel fibers to the concrete mixture results in a modification of the properties of fresh concrete, which includes its viscosity, consistency, and porosity, among other properties. To avoid this, usually the concrete mix is corrected a bit in order to mitigate these effects. However, the aim of this work is to analyze how the addition of fibers modifies the evolution of porosity over time, during the first curing week, in the same concrete matrix.

The mixture used in this research is very fluid and, in both cases, a low porosity is observed. This result agrees with the ones obtained in other research, where it is demonstrated that fibers reduce the flowability of the mixture [46,47], hindering the removal of entrained air, and leading to a greater porosity.

Regarding the variation of porosity with time, it can be noticed that SFRC specimens show an increase of the porosity along a seven-day timeline. This is a damped process, since the variation is more intense during the first days, and it decreases with time. In the case of PC, the porosity shows a slight decrease over time.

The porosity of concrete is a dynamic phenomenon and two opposite forces are acting on it. First, the curing process is a water-consuming process. At the beginning of this process, most of the voids are full of water, and the CT scan does not recognize them as voids, since they are not air voids. Over time, because of the hydration process, water is reacting with cement grains to create hydration products. Progressively, water is being consumed and the space is occupied by the air, creating internal voids. The consequence is a progressive increase of the porosity in concrete.

Second, the air entrained inside the pores tends to move up and leave the concrete specimen. In addition, this space tends to be occupied by the cement paste when a collapse of the voids occurs. This second phenomenon implies a progressive decrease of the porosity in concrete. From the macroscopic point of view, this phenomenon is called plastic settlement or initial plastic shrinkage.

The SFRC showed a concrete matrix more consistent than that of the PC because of the presence of fiber. Consequently, in the case of SFRC, the first phenomenon prevails over the second one, so the final result is an overall progressive increase of the concrete's porosity. On the contrary, in the case of PC, both phenomena are balanced, and the final result is that the porosity shows a very slight variation [48].

3.2. Variation of the Total Porosity along the Depth

Using the Z coordinate of the center of gravity of each void, it is possible to know the variation of the porosity along the depth. Figures 10 and 11 show the variation of the porosity along the depth with the age of concrete in both mixtures. In all cases, the depth is shown in relative terms, i.e., varying from zero to one, where zero belongs to the exposed surface of the specimens, and one belongs to the other end of the specimens.

Figures 10 and 11 show some interesting results. First, it can be noticed that a progressive increase of the porosity with the depth was observed in all cases. This is especially clear in the case of SFRC, but this tendency is not so clear in the case of PC. However, in both cases, the lowest porosity was observed in the first two-tenths of the specimens.

Second, SFRC showed a progressive increase of the porosity with time for all the depths, except for the 10% upper depth, where a slight decrease was observed. In the case of PC, a weak tendency of the porosity with time was observed; if anything, a slight decrease was also observed with time.

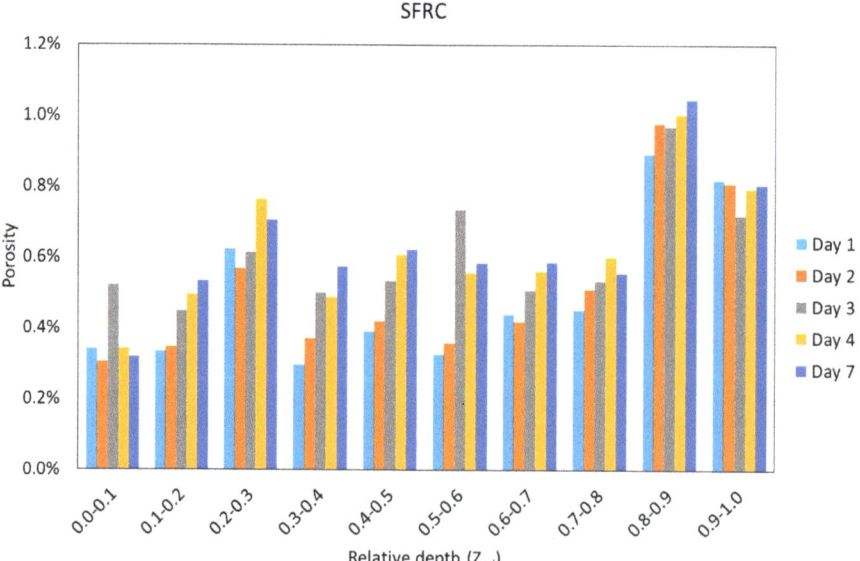

Figure 10. Variation of the porosity in steel fiber-reinforced high-performance concrete (SFRC) according to the depth and age.

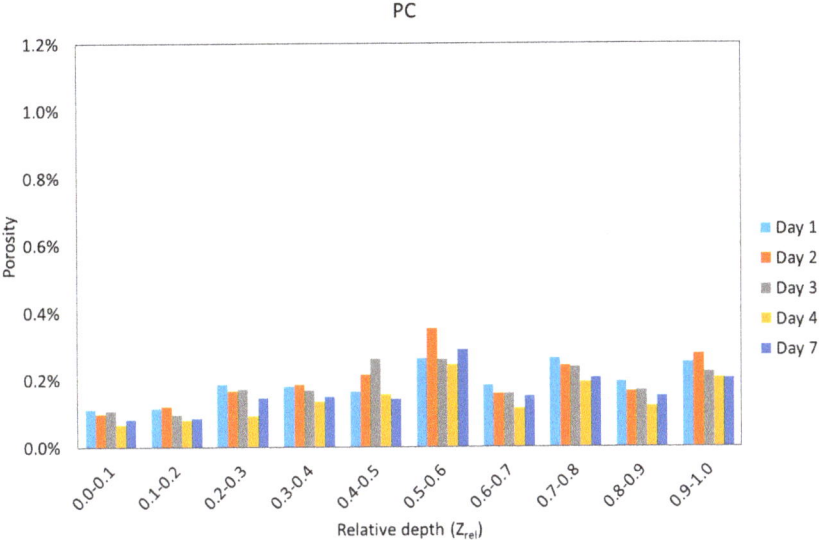

Figure 11. Variation of the porosity in plain concrete (PC) according to the depth and age.

Using the data of the porosity with time, the best fitting line was obtained at each depth, and its slope was extracted. This represents the average value of the porosity variation speed during the first seven days. Figure 12 shows the variation of this parameter along the depth for both mixtures.

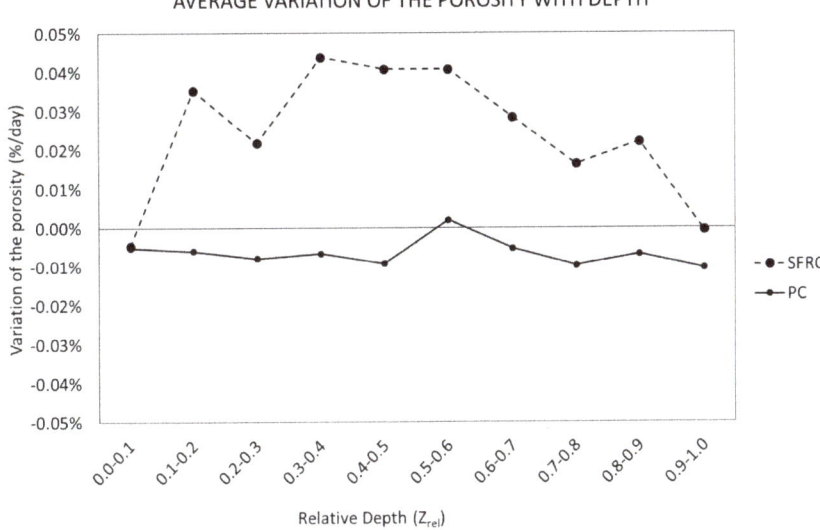

Figure 12. Variation of the porosity variation speed along the depth.

Figure 12 reveals that SFRC specimens showed a progressive increase of the porosity with time for almost all of the depths, except for the upper tenth, where a slight decrease was observed. In the case of PC, a slight decrease was observed for almost all the depths.

Again, the two different mechanisms regarding the void generation can be used to explain the behavior shown in Figure 12. Inside the specimen, when it is not close to the exposed surface, the loss of water is mainly due to hydration. In this case, the progressive reduction of free water, and in consequence, the progressive appearance of voids is strongly related to the creation of the cement matrix. In the case of SFRC, fibers provide extra stiffness to the cement matrix, which prevent the collapse of the voids. On the contrary, the cement matrix of PC specimens is less stiff, and the collapse of the voids happens more easily.

On the other side, close to the exposed surface, the loss of water is mainly due to evaporation, and the risk of a collapsing void increases. In this case, since the specimens are kept in a moisture curing room, the evaporation is almost null, and in consequence, this phenomenon can be observed only really close to the exposed surface.

In the case of SFRC, the average porosity variation speed is approximately 0.02%/day. On the contrary, in the case of PC, the average porosity variation speed is approximately −0.01%/day.

In all cases, the specimens show a slight different behavior around their middle height, i.e., at a relative depth of approximately 0.5. This is because the molds were filled in twice, and a horizontal joint appeared.

3.3. Porosity and Porosimetric Curves

Using the DIP software, it is possible to know the exact geometry of each void. At this point, it is interesting to obtain the volume of each void and its length, which is defined as the maximum distance between two voxels belonging to the same void.

Using this information, it is possible to obtain the pore volume curves and the porosimetric curves. A pore volume curve is defined as the graph correlating the length of the void and the total pore volume of the voids with a length equal to or less than this one. On the other hand, the porosimetric curve is defined as the graph correlating the length of the void and the relative pore volume of the voids with a length equal to or less than this one.

Figures 13 and 14 show the pore volume curves of the mixtures at the different ages. On the other side, Figures 15 and 16 show the porosimetric curves of the mixtures at the different ages. In all cases, the maximum length is limited to 10 mm, since larger voids are residual.

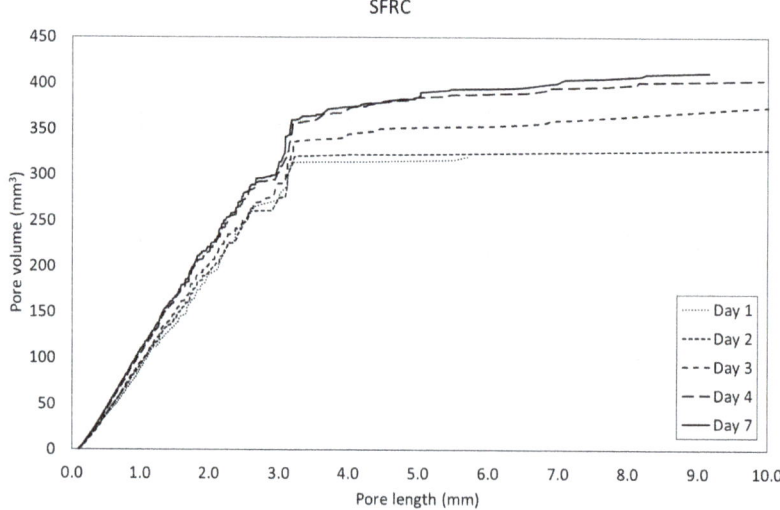

Figure 13. Pore volume curves of SFRC.

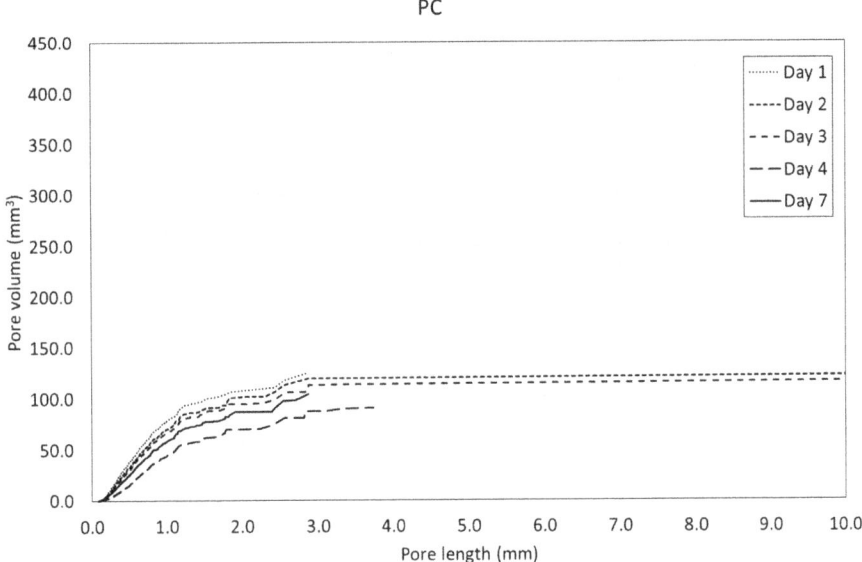

Figure 14. Pore volume curves of PC.

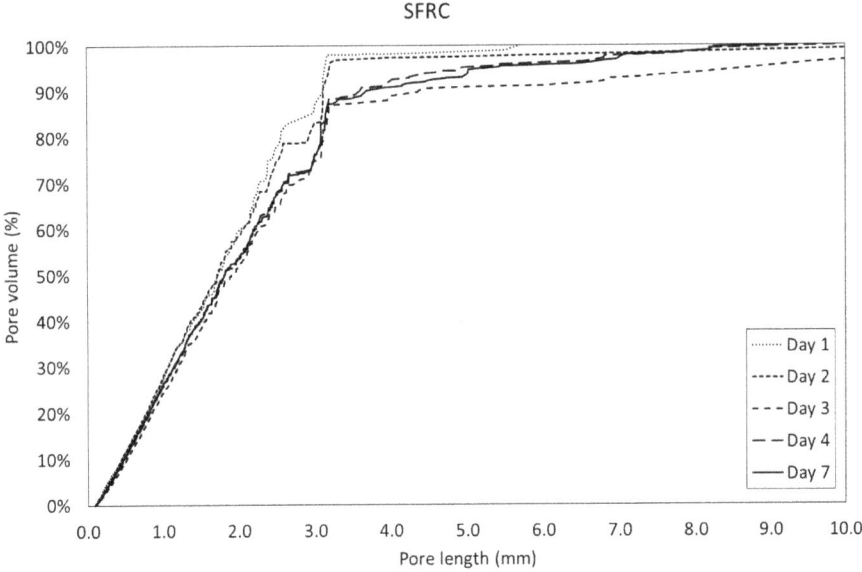

Figure 15. Porosimetric curves of SFRC.

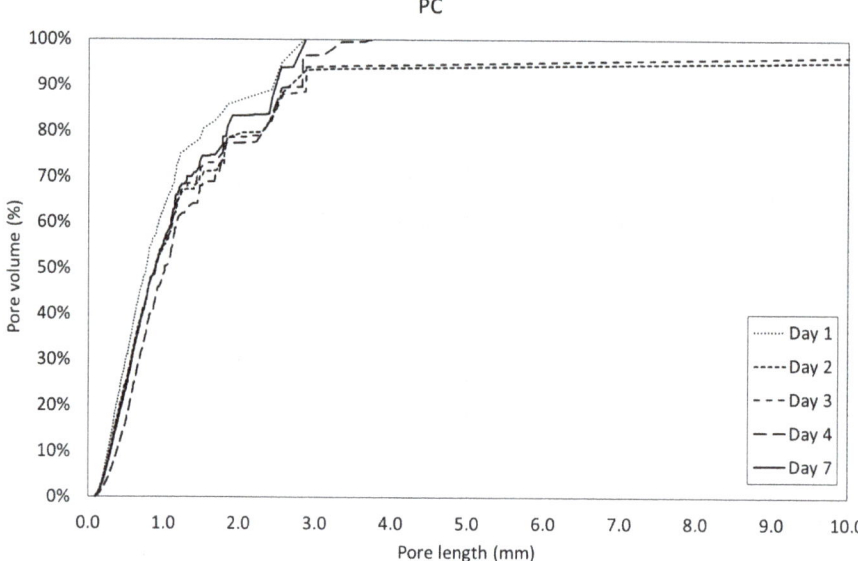

Figure 16. Porosimetric curves of PC.

Figures 13 and 14 reveal that, in general, the pore volume curves show small variations with the age of concrete. In the case of SFRC, the curves belonging to the first days are placed below the ones belonging to the last days, being, in general, homothetic. This means that a progressive increase of the pore volume occurs with time, and it happens for all of the pore sizes.

In the case of PC, the curves belonging to the first days are placed over the ones belonging to the last days, being, in general, homothetic again. This means that a progressive decrease of the pore volume occurs with time, and it also happens for all of the pore sizes.

Figure 15 shows that, in general, the porosimetric curves belonging to the first days are placed over the ones belonging to the last days. This means that SFRC tends to increase the pore size with time.

In the case of PC (Figure 16), this tendency is not so clear. This means that PC tends to keep the pore size constant with time.

An interesting parameter that can be obtained through the porosimetric curves is the nominal maximum pore size (NMPS), which can be defined, similarly to the well-known nominal maximum aggregate size (NMAS), as the pore length corresponding to a cumulative pore volume of 90%. This is a representative value of the pore size distribution, since the remaining 10% of the pore volume belongs to an extremely low number of individual pores (less than 0.05% of the pores, on average). Figure 17 show the variation of the NMPS in both mixtures with time.

Figure 17 reveals that the NMPS is greater in SFRC than in PC at all ages. This means that SFRC not only shows a greater pore volume (as can be observed in Figure 9), but also that the pores are bigger. In the case of SFRC, there is a progressive increase of the NMPS, which agrees with there being a progressive increase of the pore size with the age of concrete (Figure 15). On the contrary, in the case of PC, the curve shows a horizontal tendency, which means that there is not a variation (neither increase nor decrease) of the NMPS.

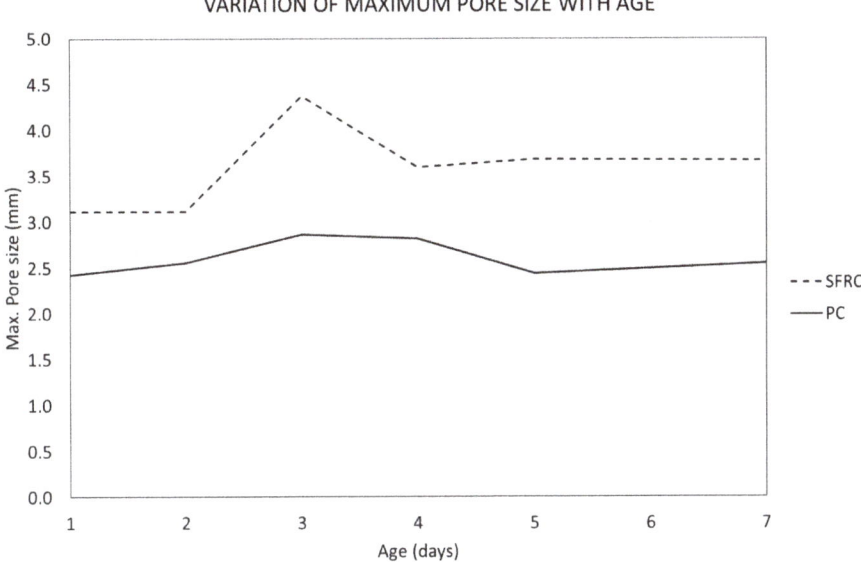

Figure 17. Variation of the nominal maximum pore size (NMPS) with the age.

Both the pore volume curves (Figures 13 and 14) and porosimetric curves (Figures 15 and 16) reveal that, in the case of SFRC, the curves shows an initial part substantially straight up to a pore length of approximately 3 mm. This means that below a 3-mm pore length, there is a substantially uniform volume distribution. Beyond this critical pore length, the slope of the curves decreases drastically, becoming horizontal very quickly. The PC curves show a similar behavior, although the critical pore length can be established at 1 mm.

SFRC shows porosimetric curves flatter than the ones belonging to PC. This can be explained in terms of the stiffness of the cement matrix. In the case of SFRC, fibers provide additional stiffness to the cement paste, and they prevent the voids from collapsing. The greater the void, the greater the instability of the concrete. As a consequence, SFRC is able to withstand a greater percentage of larger voids than PC.

3.4. Variation of the Porosity and Porosimetric Curves along the Depth

Voids are not uniformly distributed along the depth. The distance to the exposed surface, where the loss of water occurs, has a strong influence on the pore size distribution. Next, the porosimetric curves of all the days at the different depths are shown (Figures 18 and 19).

In the case of SFRC (Figure 18), it can be observed that in the first two tenths of the depth, there is an intense variation of the porosity with the age of concrete. This is because of the water interchange with the environment (mainly evaporation). This activity starts to decrease beyond day four, showing almost identical graphs beyond this day.

Something similar can be observed at the middle height of the specimens. This is because the molds were filled in two parts, creating a small horizontal joint in this area.

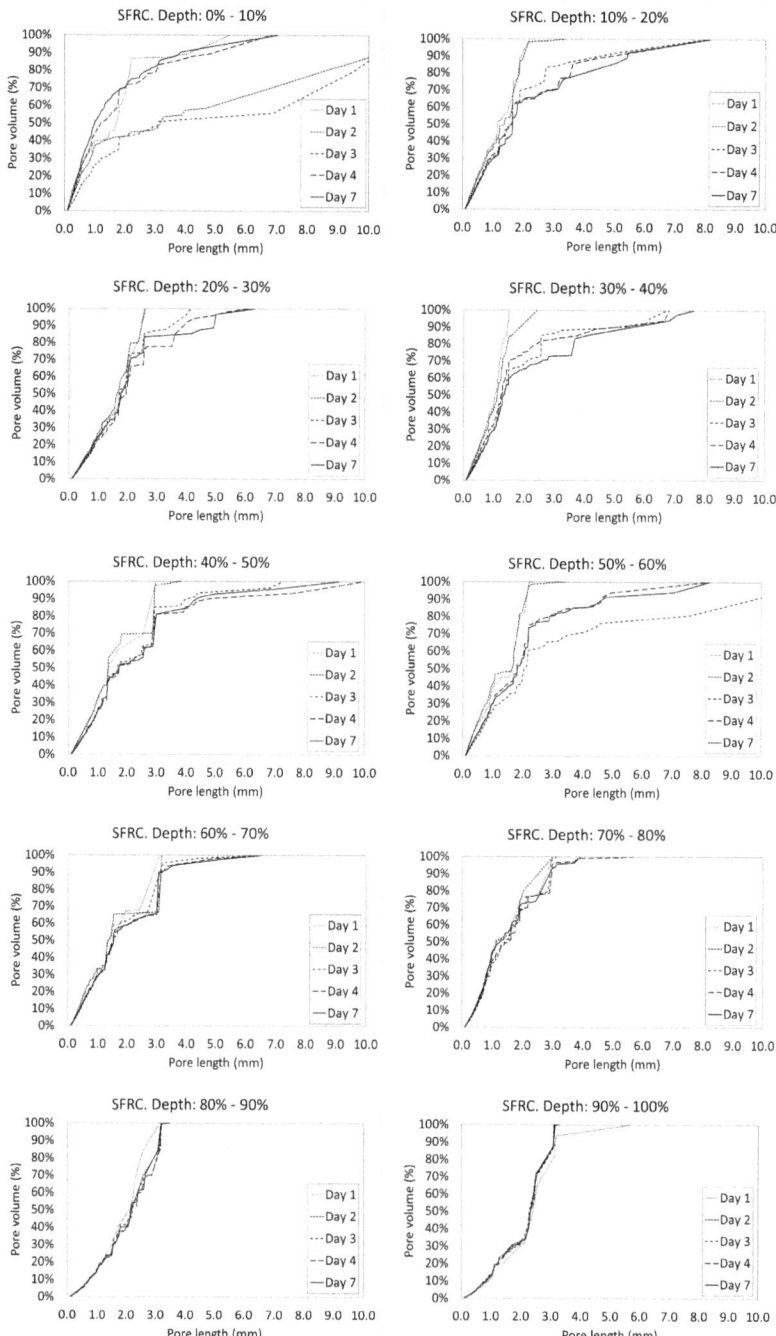

Figure 18. Porosimetric curves of SFRC at different ages and depths.

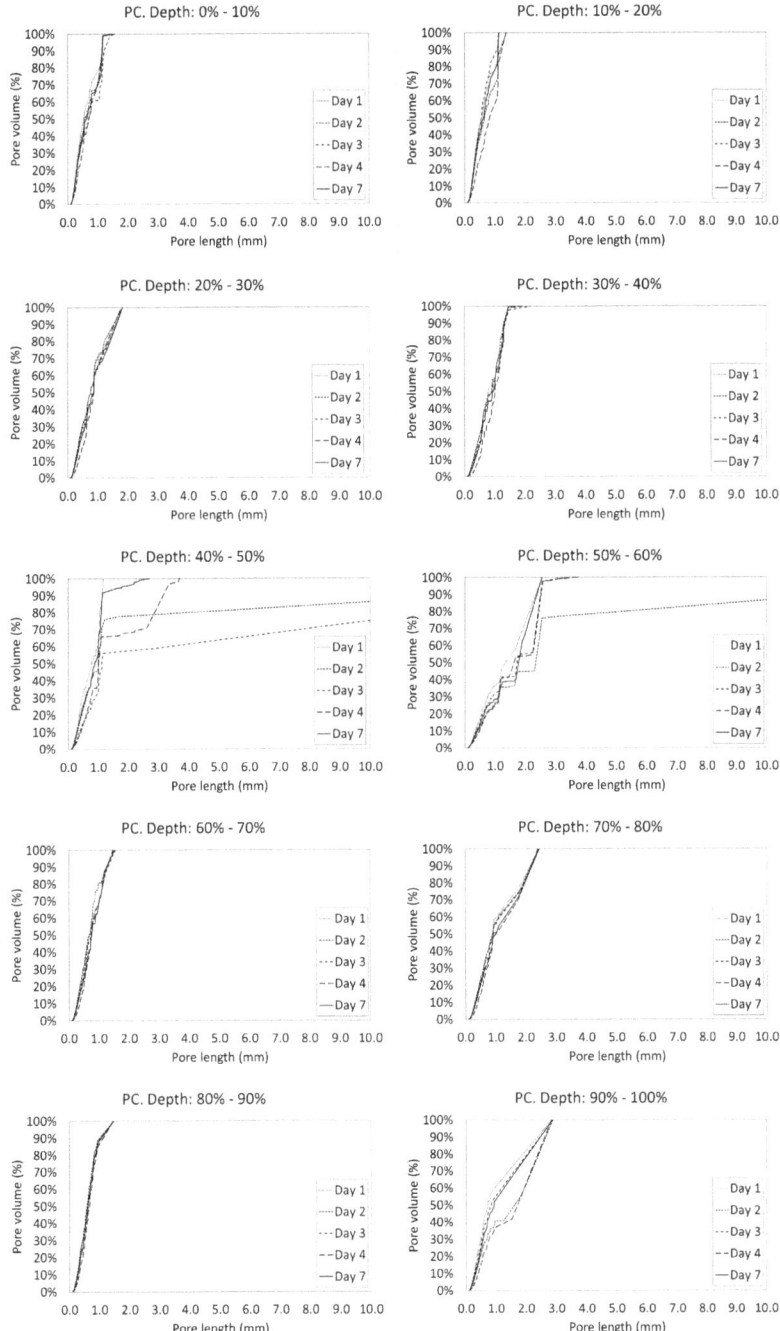

Figure 19. Porosimetric curves of PC at different ages and depths.

Similarly to what happens in Figure 15, in the case of SFRC, all the curves are substantially bilinear. The first part is approximately a straight line with a big slope up to a critical pore length. Beyond this value, the slope of the curve decreases drastically. In all cases, lines belonging to the first part of the curve show a similar slope. However, the critical pore length varies with the depth, from 2 to 3 mm. This means that the percentage of greater pores decreases with the depth. It can be explained in terms of the hydrostatic pressure of fresh concrete, which increases with the depth. There is an inverse relation between the hydrostatic pressure of fresh concrete and the maximum stable void volume. Voids smaller than this critical volume are stable, but voids greater than that tend to collapse, becoming several smaller voids.

In the case of PC, a relevant temporary variation around the middle height of the specimens is observed. However, this phenomenon is not observed in the upper tenths. Except for the first day (where voids are full of free water and they are not detected by a CT scan), the tendency is toward a progressive reduction of the greater pores.

In this case, almost all of the curves are substantially straight. The slope of the curve is significantly greater than that in the case of SFRC, which is directly related to the viscosity and stiffness of the cement paste [46,47]. In this case, the lower stiffness of the cement paste of the PC causes the large pores to be much more unstable than in the case of SFRC, and they tend to collapse, becoming several smaller voids.

3.5. Shape Factor of the Voids

As explained before, the data provided by the CT scan and the DIP software lead us to know the volume and the length of each void. Using these two data, it is possible to obtain the shape factor of each pore, which is defined as the quotient between the volume of the void and the volume of the sphere circumscribed to the void, as shown in Equation (1) [49]:

$$SF = \frac{V_p}{\frac{1}{6} \cdot \pi \cdot L_p^3} \tag{1}$$

where V_p is the pore volume and L_p is the pore length.

Figures 20 and 21 show the histograms of the shape factor of the different mixtures.

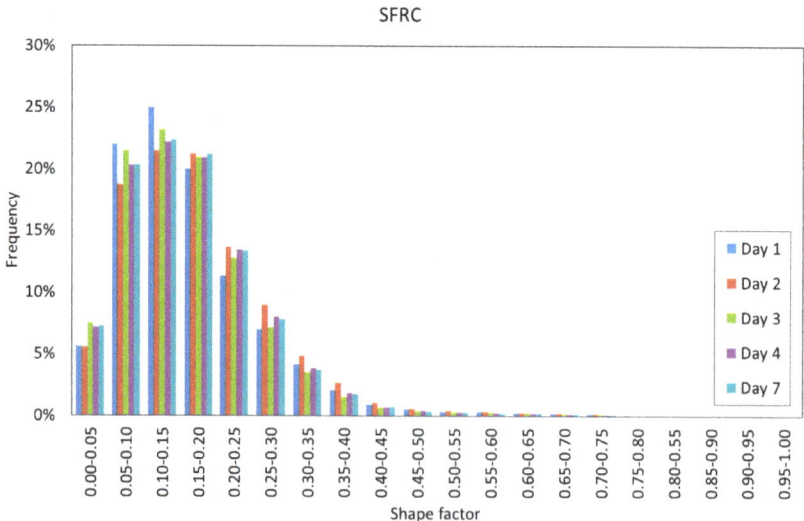

Figure 20. Histogram of the shape factor of SFRC.

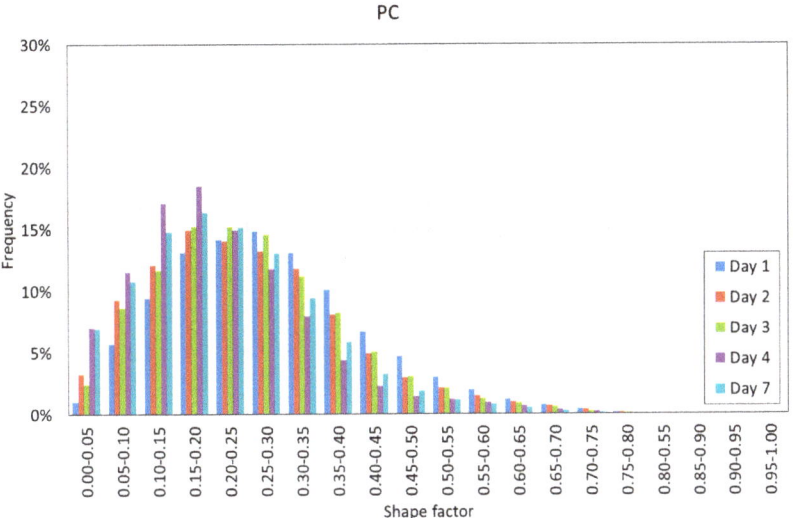

Figure 21. Histogram of the shape factor of PC.

Figures 20 and 21 reveal interesting results. First, it can be observed that in both cases, the voids are far from the spheres since in all of the cases, the shape factor is far from one.

SFRC shows smaller shape factors than PC, showing a mode value between 0.10–0.15, as well as more than 90% of the voids showing a shape factor below 0.30.

PC shows slightly greater shape factors, with a mode value between 0.15–0.25, as well as more than 90% of the voids showing a shape factor below 0.40.

Pores are the less stiff components of the cement paste, and they tend to occupy the space that is not occupied by the rest of the components of the cement paste. These 'free' spaces do not show spherical shapes; instead, they are flaky or elongated. In the case of SFRC, these spaces become even more elongated because of the fibers. Again, fibers modify the pore morphology.

The shape factor shows relevant variations along the depth. Figures 22 and 23 show the histograms of the shape factor at different ages of concrete and at different depths for both SFRC and PC.

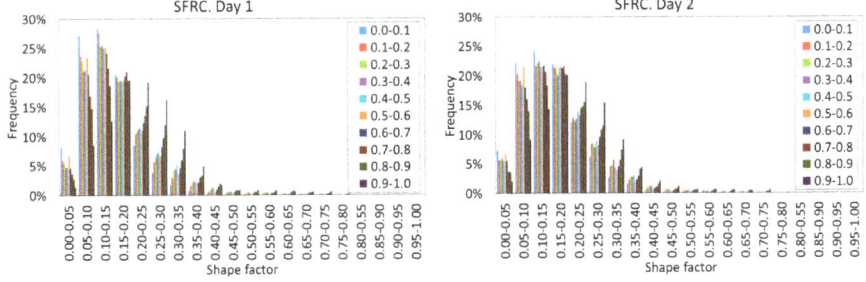

Figure 22. *Cont.*

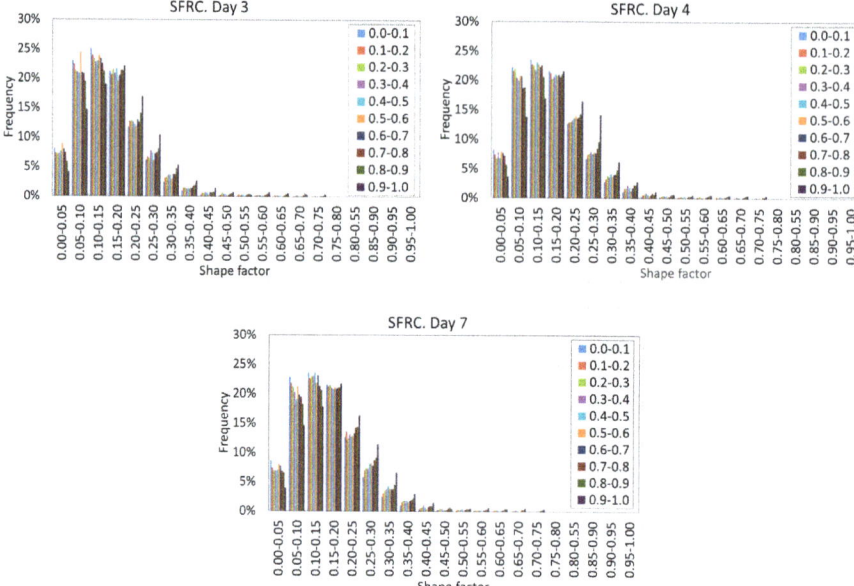

Figure 22. Histograms of the shape factors of SFRC at different depths and ages.

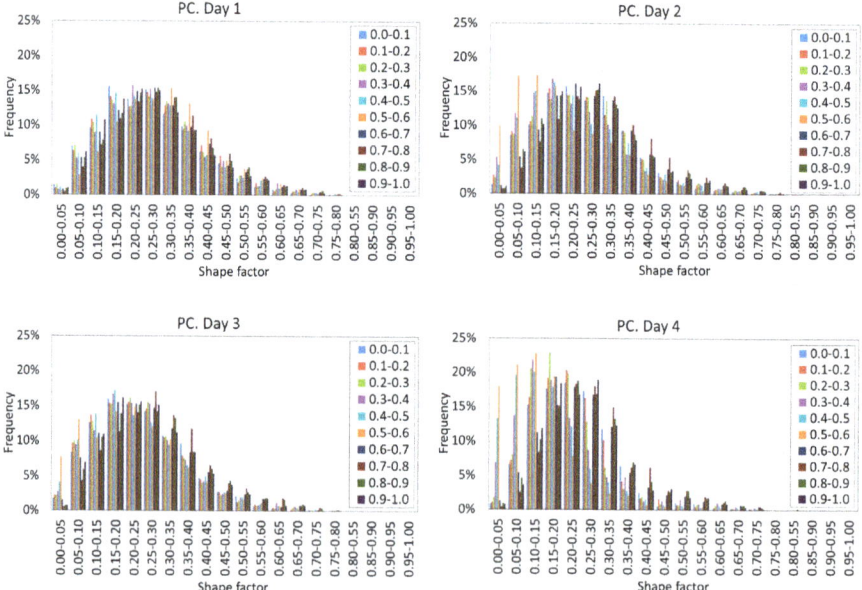

Figure 23. *Cont.*

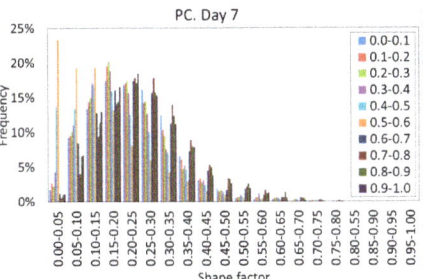

Figure 23. Histograms of the shape factors of PC at different depths and ages.

Figures 22 and 23 reveal that in general, the shape of the histograms remains constant with the age of the concrete, which demonstrates that the shape of the voids does not vary with time.

However, a substantial variation of the shape of the histograms at different depths is observed. When the depth increases, the histograms moves toward the right. This phenomenon can be observed on all the days, although it is more intense at the earliest ages. This is because the increase of the hydrostatic pressure of the fresh concrete promotes more spherical voids, which are more stable under higher hydrostatic pressure values.

This variation cannot be clearly shown in the case of PC because of the collapse of the voids due to the lower stiffness of the cement paste, which redraws the shape of the voids. This can be clearly observed on day four, when a substantial increase of the elongated voids was observed.

4. Conclusions

This paper analyzed the internal pore morphology of two different mixtures through the use of CT scan technology. Both of them had the same matrix (same type and amount of cement, fine aggregate, nanosilica, water, and superplasticizer); the only difference was that one mixture, which was called SFRC, included steel fibers by 0.1% volume, while the other mixture, which was called PC, did not include fibers. The pore morphology was measured during the first week (when most of the curing process occurs) at five different ages, i.e., one day, two days, three days, four days, and seven days.

All of the specimens were kept in a curing room at 20 °C and 100% humidity. The information provided by the CT scan was post-processed using DIP software. Each individual void was identified and isolated, and its main parameters were extracted, which were the X, Y, and Z coordinates of the center of gravity, the volume, and the length, respectively.

Some worthy and interesting results are summarized below:

- The SFRC specimens showed greater porosity than the PC ones. This difference can be observed on all the studied days. Moreover, the porosity increased with the age of the concrete in the case of SFRC, while the porosity remained almost constant in the case of PC.
- Regarding the spatial distribution of the porosity, a progressive increase of the porosity with the depth was observed in all of the cases. This was especially clear in the case of SFRC, but this tendency was not so clear in the case of PC. However, in both cases, the lowest porosity was observed in the first two tenths of the specimens.
- The stiffness of the cement paste was behind most of the observed behaviors. Fibers provided additional stiffness to the cement paste, which allowed it to retain porosity. In the case of PC, where there are no fibers, the stiffness of the fresh concrete paste was small and the voids, especially the big ones, tended to collapse and the air tended to leave the concrete.
- Water loss happens more quickly when it is closer to the exposed surface. Not all of it is used to create cement paste, because some of it evaporates. The consequence is that when concrete is

- closer to the exposed surface, less stiffness and a smaller amount of voids are retained. Even in the case of SFRC, fibers do not provide the extra stiffness in an efficient way.
- The porosimetric curves showed two different stages. The first one belonged to the smaller sizes, from zero to a critical length, where the curves showed a straight line. Beyond this critical length, the slope of the curves decreased drastically. In the case of SFRC, the critical length could be defined as 3 mm, while in the case of PC, the critical length was 1 mm. This behavior could be observed along the depth, although the value of the critical length varied slightly with the depth, especially in the case of SFRC.
- The critical length is, again, strongly related to the stiffness of the fresh cement paste and its capacity to create voids. In the case of SFRC, the extra stiffness provided by the fibers allowed the concrete to retain larger pores up to 3 mm in length. Beyond this value, the pores became more unstable and tended to collapse. In the case of PC, cement paste is less stiff, and the capacity to retain pores is reduced; voids beyond 1 mm in length seem to be difficult to retain, and tend to collapse.
- The fibers changed the concrete porosity through two mechanisms. The first one was to trap air during the concreting process and also increase the viscosity of the mixture, which complicated the expelling of the trapped air during the concreting. The second was to provide an extra stiffness to avoid the collapse of the voids due to water loss.
- The histograms of shape factors showed that the voids were far from spheres, i.e., they were more flaky or elongated. SFRC showed a shape factor smaller than the one shown by PC, which was due to the presence of fibers. In this case, the variation of the shape factor with the age of concrete was insignificant. On the contrary, the shape factor increased with the depth.

The modification of the pore morphology caused by the fiber can potentially affect the macroscopic response of hardened concrete, especially in those properties more directly related to porosity, such as frost resistance, fire resistance, creep, shrinkage, and fatigue, among others.

Author Contributions: J.M. carried out the scanning process and the images analysis, D.C.G. worked on the design of the experiment and on the results analysis and M.A.V. worked on the analysis of the results and wrote the paper.

Funding: This study was funded by the Ministerio de Economía y Competitividad, Gobierno de España (Spain) (grant number BIA2015-68678-C2-R).

Conflicts of Interest: The authors declare that they have no conflict of interest.

References

1. Mehdipour, I.; Khayat, K.H. Understanding the role of particle packing characteristics in rheophysical properties of cementitious suspensions: A literature review. *Constr. Build. Mater.* **2018**, *161*, 340–353. [CrossRef]
2. Lazniewska-Piekarczyk, B. The influence of selected new generation admixtures on the workability, air-voids parameters and frost-resistance of self compacting concrete. *Constr. Build. Mater.* **2012**, *31*, 310–319. [CrossRef]
3. Li, Z. State of workability design technology for fresh concrete in Japan. *Cem. Concr. Res.* **2007**, *37*, 1308–1320. [CrossRef]
4. Chandrappa, A.K.; Biligiri, K.P. Pore Structure Characterization of Pervious Concrete Using X-ray Microcomputed Tomography. *J. Mater. Civ. Eng.* **2018**, *30*, 04018108. [CrossRef]
5. Liu, B.; Luo, G.; Xie, Y. Effect of curing conditions on the permeability of concrete with high volume mineral admixtures. *Constr. Build. Mater.* **2018**, *167*, 359–371. [CrossRef]
6. Akand, L.; Yang, M.; Gao, Z. Characterization of pervious concrete through image based micromechanical modeling. *Constr. Build. Mater.* **2016**, *114*, 547–555. [CrossRef]
7. Torres-Carrasco, M.; Alonso, M.M.; Guarner, P.; Zamora, A.; Puertas, F. Alkali-activated slag concretes. Mechanical and durability behaviour. *Hormigón y Acero* **2018**, *69*, 163–168. (In Spanish) [CrossRef]

8. Ley, M.T.; Welcher, D.; Peery, J.; Khatibmasjedi, S.; LeFlore, J. Determining the air-void distribution in fresh concrete with the Sequential Air Method. *Constr. Build. Mater.* **2017**, *150*, 723–737. [CrossRef]
9. Jin, S.; Zhang, J.; Huang, B. Fractal analysis of effect of air void on freeze-thaw resistance of concrete. *Constr. Build. Mater.* **2013**, *47*, 126–130. [CrossRef]
10. Narayanan, N.; Ramamurthy, K. Structure and properties of aerated concrete: A review. *Cem. Concr. Compos.* **2000**, *22*, 321–329. [CrossRef]
11. Vicente, M.A.; González, D.C.; Mínguez, J.; Tarifa, M.A.; Ruiz, G. Influence of the pore morphology of high strength concrete on its fatigue life. *Int. J. Fatigue* **2018**, *112*, 106–116. [CrossRef]
12. Vicente, M.A.; Ruiz, G.; González, D.C.; Mínguez, J.; Tarifa, M.; Zhang, X. CT-Scan study of crack patterns of fiber-reinforced concrete loaded monotonically and under low-cycle fatigue. *Int. J. Fatigue* **2018**, *114*, 138–147. [CrossRef]
13. Chen, Y.; Wang, K.; Wang, X.; Zhou, W. Strength, fracture and fatigue of previous concrete. *Constr. Build. Mater.* **2013**, *42*, 97–104. [CrossRef]
14. Kim, H.K.; Jeon, J.H.; Lee, H.K. Workability, and mechanical, acoustic and thermal properties of lightweight aggregate concrete with a high volume of entrained air. *Constr. Build. Mater.* **2012**, *29*, 193–200. [CrossRef]
15. Hajimohammadi, A.; Ngo, T.; Mendis, P. Enhancing the strength of pre-made foams for foam concrete applications. *Cem. Concr. Compos.* **2018**, *87*, 164–171. [CrossRef]
16. Zia, A.; Ali, M. Behavior of fiber reinforced concrete for controlling the rate of cracking in canal-lining. *Constr. Build. Mater.* **2017**, *155*, 726–739. [CrossRef]
17. Mazzoli, A.; Monosi, S.; Plescia, E.S. Evaluation of the early-age-shrinkage of Fiber Reinforced Concrete (FRC) using image analysis methods. *Constr. Build. Mater.* **2015**, *101*, 596–601. [CrossRef]
18. Ferrara, L.; Park, Y.-D.; Shah, S.P. A method for mix-design of fiber-reinforced self-compacting concrete. *Cem. Concr. Res.* **2007**, *37*, 957–971. [CrossRef]
19. Gonzalez, D.C.; Moradillo, R.; Mínguez, J.; Martínez, J.A.; Vicente, M.A. Postcracking residual strengths of fiber-reinforced high-performance concrete after cyclic loading. *Struct. Concr.* **2018**, *19*, 340–351. [CrossRef]
20. Parvez, A.; Foster, S.J. Fatigue Behavior of Steel-Fiber-Reinforced Concrete Beams. *J. Struct. Eng.* **2015**, *141*, 04014117. [CrossRef]
21. González, D.C.; Vicente, M.A.; Ahmad, S. Effect of cyclic loading on the residual tensile strength of steel fiber–reinforced high-strength concrete. *J. Mater. Civ. Eng.* **2015**, *27*, 04014241. [CrossRef]
22. Zhang, J.; Stang, H.; Li, V.C. Fatigue life prediction of fiber reinforced concrete under flexural load. *Int. J. Fatigue* **1999**, *21*, 1033–1049. [CrossRef]
23. Minguez, J.; González, D.C.; Vicente, M.A. Fiber geometrical parameters of fiber-reinforced high strength concrete and their influence on the residual post-peak flexural tensile strength. *Constr. Build. Mater.* **2018**, *168*, 906–922. [CrossRef]
24. Bischoff, P.H. Tension stiffening and cracking of steel fiber-reinforced concrete. *J. Mater. Civ. Eng.* **2003**, *15*, 174–182. [CrossRef]
25. Al Rikabi, F.T.; Sargand, S.M.; Khoury, I.; Hussein, H.H. Material properties of synthetic fiber–reinforced concrete under freeze-thaw conditions. *J. Mater. Civ. Eng.* **2018**, *30*, 04018090. [CrossRef]
26. Niu, D.; Jiang, L.; Bai, M.; Miao, Y. Study of the performance of steel fiber reinforced concrete to water and salt freezing condition. *Mater. Des.* **2013**, *44*, 267–273. [CrossRef]
27. Poveda, E.; Ruiz, G.; Cifuentes, H.; Yu, R.C.; Zhang, X. Influence of the fiber content on the compressive low-cycle 585 fatigue behavior of selfcompacting SFRC. *Int. J. Fatigue* **2017**, *101*, 9–17. [CrossRef]
28. Medeiros, A.; Zhang, X.; Ruiz, G.; Yu, R.C.; Velasco, M.S.L. Effect of the loading frequency on the compressive fatigue behavior of plain and fiber reinforced concrete. *Int. J. Fatigue* **2015**, *70*, 342–350. [CrossRef]
29. Yermak, N.; Pliya, P.; Beaucour, A.L.; Simon, A.; Noumowé, A. Influence of steel and/or polypropylene fibres on the behaviour of concrete at high temperature: Spalling, transfer and mechanical properties. *Constr. Build. Mater.* **2017**, *132*, 240–250. [CrossRef]
30. Mazzucco, G.; Majorana, C.E.; Salomoni, V.A. Numerical simulation of polypropylene fibres in concrete materials under fire conditions. *Comput. Struct.* **2015**, *154*, 17–28. [CrossRef]
31. Chen, X.; Xu, L.; Wu, S. Influence of pore structure on mechanical behavior of concrete under high strain rates. *J. Mater. Civ. Eng.* **2016**, *28*, 04015110. [CrossRef]
32. Chen, X.; Wu, S.; Zhou, J. Influence of porosity on compressive and tensile strength of concrete mortar. *Constr. Build. Mater.* **2013**, *40*, 869–874. [CrossRef]

33. Zeng, Q.; Li, K.; Fen-Chong, T.; Dangla, P. Pore structure characterization of cement pastes blended with high-volume fly-ash. *Cem. Concr. Res.* **2012**, *42*, 194–204. [CrossRef]
34. Herrmann, H.; Pastorelli, E.; Kallonen, A.; Suuronen, J.P. Methods for fibre orientation analysis of X-ray tomography images of steel fibre reinforced concrete (SFRC). *J. Mater. Sci.* **2016**, *51*, 3772–3783. [CrossRef]
35. Vicente, M.A.; González, D.C.; Mínguez, J. Determination of dominant fibre orientations in fibre-reinforced high strength concrete elements based on computed tomography scans. *Nondestruct. Test. Eval.* **2014**, *29*, 164–182. [CrossRef]
36. Lu, H.; Peterson, K.; Chernoloz, O. Measurement of entrained air-void parameters in Portland cement concrete using micro X-ray computed tomography. *Int. J. Pavement Eng.* **2018**, *19*, 109–121. [CrossRef]
37. Wang, Y.-S.; Dai, J.-G. X-ray computed tomography for pore-related characterization and simulation of cement mortar matrix. *Ndt & E Int.* **2017**, *86*, 28–35.
38. Moradian, M.; Hu, Q.; Aboustait, M.; Ley, M.T.; Hanan, J.C.; Xiao, X.; Scherer, G.W.; Zhang, Z. Direct observation of void evolution during cement hydration. *Mater. Des.* **2017**, *136*, 137–149. [CrossRef]
39. Lu, H.; Alymov, E.; Shah, S.; Peterson, K. Measurement of air void system in lightweight concrete by X-ray computed tomography. *Constr. Build. Mater.* **2017**, *152*, 467–483. [CrossRef]
40. Vicente, M.A.; Mínguez, J.; González, D.C. The use of computed tomography to explore the microstructure of materials in civil engineering: From rocks to concrete. In *Computed Tomography—Advanced Applications*; Halefoglu, A.M., Ed.; InTech: London, UK, 2017.
41. Yuan, J.; Liu, Y.; Li, H.; Yang, C. Experimental investigation of the variation of concrete pores under the action of freeze-thaw cycles. *Procedia Eng.* **2016**, *161*, 583–588. [CrossRef]
42. Ponikiewski, T.; Katzer, J.; Bugdol, M.; Rudzki, M. Determination of 3D porosity in steel fibre reinforced SCC beams using X-ray computed tomography. *Constr. Build. Mater.* **2014**, *68*, 333–340. [CrossRef]
43. Kim, K.Y.; Yun, T.S.; Choo, J.; Kang, D.H.; Shin, H.S. Determination of air-void parameters of hardened cement-based materials using X-ray computed tomography. *Constr. Build. Mater.* **2012**, *37*, 93–101. [CrossRef]
44. British Standards Institution. *Methods of Testing Cement. Determination of Strength*; EN 196-1:2016; British Standards Institution: London, UK, 2016.
45. Vicente, M.A.; Mínguez, J.; González, D.C. Recent advances in the use of computed tomography in concrete technology and other engineering fields. *Micron* **2019**, *118*, 22–34. [CrossRef] [PubMed]
46. Domingues, A.; Ceccato, M.R. Workability Analysis of Steel Fiber Reinforced Concrete Using Slump and Ve-Be Test. *Mater. Res.* **2015**, *18*, 1284–1290.
47. Mazaheripour, H.; Ghambarpour, S.; Mirmoradi, S.H.; Hosseinpour, I. The effect of polypropylene fibers on the properties of fresh and hardened lightweight self-compacting concrete. *Constr. Build. Mater.* **2011**, *25*, 351–358. [CrossRef]
48. Meng, W.; Khayat, K.H. Effect of hybrid fibers on fresh properties, mechanical properties and autogenous shrinkage of cos-effective UHPC. *J. Mater. Civ. Eng.* **2018**, *30*, 04018030. [CrossRef]
49. Johansson, J.; Vall, J. Jordmaterials Kornform. Inverkan på Geotekniska Egenskaper, Beskrivande Storheter, Bestämningsmetoder. Ph.D. Thesis, Avdelningen för Geoteknologi, Institutionen för Samhällsbyggnad och naturresurser, Luleå Tekniska Universitet, Luleå, Sweden, 2011.

© 2019 by the authors. Licensee MDPI, Basel, Switzerland. This article is an open access article distributed under the terms and conditions of the Creative Commons Attribution (CC BY) license (http://creativecommons.org/licenses/by/4.0/).

Article

Evolution of Electrochemical Cell Designs for In-Situ and Operando 3D Characterization

Chun Tan [1,†], Sohrab R. Daemi [1,†], Oluwadamilola O. Taiwo [1,2], Thomas M. M. Heenan [1], Daniel J. L. Brett [1] and Paul R. Shearing [1,*]

1. Electrochemical Innovation Lab, Department of Chemical Engineering, University College London, Torrington Place, London WC1E 7JE, UK; chun.tan.11@ucl.ac.uk (C.T.); sohrab.daemi.14@ucl.ac.uk (S.R.D.); o.taiwo@imperial.ac.uk (O.O.T.); t.heenan@ucl.ac.uk (T.M.M.H.); d.brett@ucl.ac.uk (D.J.L.B.)
2. Department of Earth Science & Engineering, Faculty of Engineering, Imperial College London, South Kensington Campus, London SW7 2AZ, UK
* Correspondence: p.shearing@ucl.ac.uk
† These authors contributed equally to this work.

Received: 28 September 2018; Accepted: 30 October 2018; Published: 1 November 2018

Abstract: Lithium-based rechargeable batteries such as lithium-ion (Li-ion), lithium-sulfur (Li-S), and lithium-air (Li-air) cells typically consist of heterogenous porous electrodes. In recent years, there has been growing interest in the use of in-situ and operando micro-CT to capture their physical and chemical states in 3D. The development of in-situ electrochemical cells along with recent improvements in radiation sources have expanded the capabilities of micro-CT as a technique for longitudinal studies on operating mechanisms and degradation. In this paper, we present an overview of the capabilities of the current state of technology and demonstrate novel tomography cell designs we have developed to push the envelope of spatial and temporal resolution while maintaining good electrochemical performance. A bespoke PEEK in-situ cell was developed, which enabled imaging at a voxel resolution of ca. 230 nm and permitted the identification of sub-micron features within battery electrodes. To further improve the temporal resolution, future work will explore the use of iterative reconstruction algorithms, which require fewer angular projections for a comparable reconstruction.

Keywords: X-ray tomography; electrochemical cell design; batteries

1. Introduction

Energy storage devices have an increasingly significant role to play in all economic sectors with the de-carbonization of the global economy necessary to meet climate change goals. Rechargeable batteries are one of the key enabling technologies driving the shift to renewable energy with lithium-ion (Li-ion) battery technology becoming the mainstay of applications requiring high energy density such as portable electronics and electric vehicles (EVs). The Li-ion battery has been described as a rocking-chair battery where Li ions are intercalated into a transition metal oxide (TMO) or graphite host, respectively, during discharge and charge processes [1]. Beyond Li-ion technology, other lithium-based rechargeable batteries have been proposed including lithium sulfur (Li-S) and lithium air (Li-air) conversion-type chemistries. These offer theoretical capacities up to an order of magnitude higher and significantly higher energy densities compared to incumbent Li-ion cells [2]. However, their commercialization has been hindered by various design and engineering challenges imposed by mechanistic complexities.

While many competing cell chemistries—involving numerous redox pairs—exist in different stages of development, the basic architecture of the electrodes within most electrochemical devices involve some form of heterogeneous porous media on which electrochemical reactions occur. The microstructure of porous electrode materials often has a profound effect on the performance and lifetime of a cell [3]. Factors such as tortuosity and porosity of the electrodes influence the effective diffusivity of ions within the electrode and with the bulk of the electrolyte, contributing to cell impedance. The electrical conductivity of an electrode is also dependent on its microstructure in terms of the contact area between the different solid phases and is particularly important for cell chemistries involving electronically insulating active material. Heterogeneities within battery electrodes are known to contribute extensively to cell degradation since these induce local variations in current density and the state of charge.

Advances in synchrotron and laboratory-based radiation sources have led to X-ray based techniques such as X-ray absorption spectroscopy (XAS) [4], X-ray diffraction (XRD) [5], transmission X-ray microscopy (TXM) [6], and X-ray computed tomography (XCT) [7,8] becoming widely adopted to investigate the mechanisms behind the operation and degradation of electrochemical devices including lithium-based batteries. These complementary techniques span multiple time-scales and length-scales and provide information about the electronic states (XAS), crystalline states (XRD), and microstructures (TXM) of materials. Many beam and detector configurations exist, and this work will primarily be concerned with micro X-ray computed tomography (micro-CT) under full field illumination from an X-ray beam.

Due to the inherently heterogenous structures and processes occurring within batteries, there is a strong motivation to capture the physical and chemical states of their electrodes in 3D. Based on work on other porous media systems, ex-situ X-ray tomography has become widely adopted for the 3D characterization of battery electrodes to extract microstructural metrics indicative of battery performance including electrode tortuosity, porosity, pore, and particle size distributions [3,9–11].

In addition to these metrics, various authors conducted post mortem micro-CT and nano-CT studies to understand the factors contributing to electrode degradation. Furthermore, post Li-ion technologies typically involve conversion-type chemistries with liquid state or alloying reactions where significant changes typically occur (e.g., large volume change during lithiation of Si anodes) [12]. In these systems, tomographic measurements are of even greater interest in improving the mechanistic understanding of these inherently three-dimensional processes.

The development of in-situ electrochemical cells along with recent improvements in radiation sources have expanded the capabilities of micro-CT as a technique for longitudinal studies: the ability to track the same spatial volume of electrode as a function of some variable effectively eliminates variability inherent in ex-situ studies [13–15]. However, considerable challenges exist in the optimization of in-situ cell designs to achieve spatial and temporal resolutions compatible with the phenomena of interest. In this case, we aim to present an overview of the capabilities of the current state of technology and demonstrate novel tomography cell designs we have developed to push the envelope of spatial and temporal resolution whilst maintaining fidelity to the electrochemical performance of larger format cells.

2. Results and Discussion

The suitability of an in-situ electrochemical cell design for X-ray characterization is largely dependent on size, geometry, and materials used in its construction. Within the scope of lab-based research environments, cells used for pure electrochemical characterization of battery materials take the form of coin, pouch, and cylindrical geometries. X-ray transparent materials (i.e., polyimide) have been introduced to coin and pouch cells by numerous authors for in-situ and operando spectroscopic or imaging applications [16–18] in attempts to reduce the interference of ancillary components in the X-ray beam path. However, in the quest for a finer resolution, highly specialized cell designs are essential to achieve an acceptable signal-to-noise ratio within reasonable acquisition times because of limitations such as flux and detector sensitivity in imaging systems. While operando cell designs often come at the expense of electrochemical performance, the wealth of information obtained from these advanced characterization techniques vastly outweighs the higher cell impedances and lower achievable capacities. Acquisition times are even more important for tomography, which requires the collection of adequate projections at sufficiently fine angular increments to reconstruct. Some parameters that need to be optimized will be discussed in a later section.

The ideal sample for tomographic reconstruction is one that fits fully within the detector field of view (FOV) through all radiographic projections: samples larger than the FOV such as coin and pouch cells will result in the truncation of sinogram data. The horizontal resolution of charged-coupled device (CCD) or flat panel detectors do not typically extend beyond 2048 pixels (equivalent to a FOV of ca. 2 mm at 1 µm pixel size) and it is desirable to fit as much of the cell within the FOV to mitigate out of field artefacts present in interior tomographies. Cylindrical Swagelok-type cell designs [14] ensure attenuation lengths that are uniform on average for all projections as well as a weakly attenuating cell body and are highly desirable for in-situ characterization. A comparison of the properties of common materials used in the construction of Swagelok-type cells is presented in Table 1.

Table 1. Comparison of common materials used in construction of Swagelok-type cells.

Material	Moisture Impermeability	Li-ion Electrolyte Compatibility	Reactivity with Li	X-ray Transparency and Compatibility
Stainless Steel	Excellent	Some grades are compatible	Non-reactive	Poor
Aluminium	Excellent	May be corroded by electrolyte	Alloys with Li	Good
Beryllium	Excellent	May be corroded by electrolyte	Non-reactive	Excellent
Polyimide (Uncoated)	Poor	Compatible	Non-reactive	Excellent
PFA	Good	Compatible	Reacts with metallic Li to form elemental carbon	Transparent but susceptible to radiation damage
PTFE	Good	Compatible	Reacts with metallic Li to form elemental carbon	Transparent but susceptible to radiation damage
PEEK	Good	Compatible	Non-reactive	Good

2.1. Optimization of Tomography Parameters

Imaging parameters for in-situ tomography acquisition are highly dependent on both the sample and instrument configuration and have to be optimized to produce the best possible image and resolution within a reasonable acquisition time. For lab-based instruments, a critical constraint imposed on the design of in-situ cells is sample size since this dictates the exposure time. For high-resolution micro-studies, this is typically approximately ca. 1 min per radiographic projection because of the limited brilliance of lab-based X-ray sources. The variables that have to be considered are discussed below.

2.2. X-ray Beam Energy and Its Effect on Intensity and Transmission

When X-rays interact with a sample, the intensity of the incident beam is attenuated by the sample and the ratio of the transmitted, I, to the incident beam intensities, I_0, is known as transmission, T, and is defined by the equation below.

$$T = \frac{I}{I_0} \tag{1}$$

For a monochromatic beam of known energy, transmission is related to the linear attenuation coefficient, μ, and thickness of a material, t, by the exponential relation known as the Beer-Lambert law.

$$T = \exp(-ut) \tag{2}$$

The linear attenuation coefficient, μ, is a function of the density of the material ρ, total cross-section per atom, σ_{tot}, and the atomic mass of the element of interest, A_r, displayed in Equation (3), which is defined by Hubbell [19].

$$\frac{\mu}{\rho} = \sigma_{tot} \frac{N_A}{A_r} \tag{3}$$

where the total cross-section per atom, σ_{tot}, accounts for each contribution from the principal photon interactions with the material [20] and N_A is the Avogadro constant. Thus, the linear attenuation coefficient, corresponding to material composition and density, can be reconstructed in three-dimensional space with the appropriate inversion algorithm such as the inverse Radon transform. In practice, this is complicated by the use of polychromatic radiation produced by most lab X-ray sources and, in the absence of calibration with a phantom of known composition, micro-CT is most often used to inspect the microstructure within a material with some a priori knowledge of the composition of the sample.

In lab X-ray sources, the intensity and spectrum of the X-ray beam can be controlled by adjusting the X-ray tube voltage and current (i.e., source energy) and by changing the target material. Elements such as Cr, Co, Cu, Mo, Ag, and W are commonly used X-ray targets which each have their own characteristic spectra. In addition to the characteristic emissions of the X-ray target, a broad spectrum of X-rays is emitted via Bremsstrahlung radiation, up to a peak photon energy equivalent to the X-ray tube voltage and a polychromatic beam is produced. While an increase in the X-ray tube voltage and/or current will result in an increase in both incident and transmitted intensities, transmission (i.e., the ratio between incident and transmitted intensities) is not a function of tube current and increases only with tube voltage.

Barring discontinuities in the attenuation coefficients of elements at photon energies close to their specific absorption edges, an increase in mean photon energy will result in an increase in transmission as more photons reach the detector without interacting with the sample. Transmission is a critical variable that influences image quality and Reiter et al. have found that, for an ideal detector, ca. 14% transmission results in the most optimal signal-to-noise ratio [21]. In a multi-component system containing phases of very different attenuation coefficients, a compromise has to be made when selecting the beam energy. This is particularly acute when imaging battery electrodes since transmission varies greatly between the highly attenuating active material particles (consisting of transition metal oxides for Li-ion positive electrodes) and the weakly attenuating carbon and binder phase, which was discussed previously [22].

Another important acquisition parameter is the exposure time per projection since sufficient detector counts are necessary to form a low noise image depending on the dynamic range of the detector. Detector counts are proportional to the transmitted intensity integrated over the exposure time through the image formed on the scintillator. Exposure time is largely independent from transmission and has to be optimized by taking into account two opposing variables: adequate signal-to-noise ratios (long exposure) and minimized blurring induced by sample motion and thermal drift (short exposure).

To determine the optimal acquisition parameters for the PFA and PEEK cells, radiographs were acquired from both cells containing NMC111 electrodes in a half-cell arrangement over a range of X-ray source voltages. Line profiles were drawn across the electrode layer in the radiographs to obtain the graphs presented in Figure 1. Manufacturer's specifications for the ZEISS Xradia Versa 520 laboratory micro-CT instrument used suggested at least 5000 counts and, as seen in Figure 1, this is unachievable even at 120 kV. On the other hand, a factor of ca. 3 improvement in counts is observed with the PEEK cell. Furthermore, transmission across the PEEK electrode at ca. 70 to 80 kV is optimal at around 14%, which indicates that the resulting reconstruction will likely have a good signal-to-noise ratio.

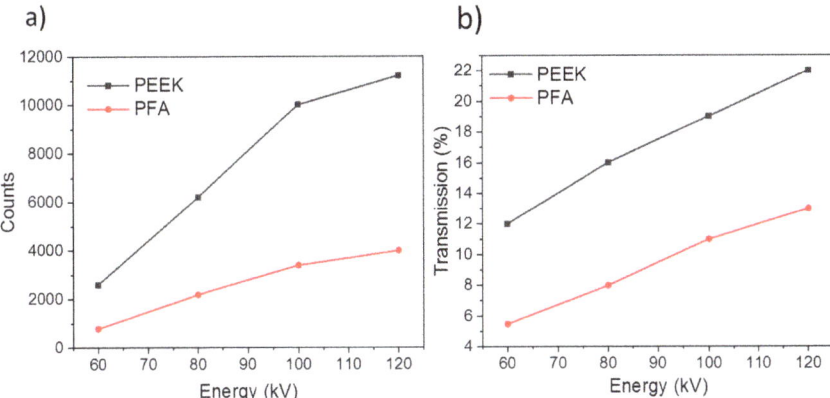

Figure 1. (**a**) Detector counts and (**b**) transmission as a function of source voltage for a 60 s exposure time through an NMC111 electrode in the PFA and PEEK cell.

2.3. Number of Projections

The Nyquist-Shannon theorem can be applied to determine the angular resolution or an equivalent number of projections, which is theoretically required for reconstruction. The theorem states that an object has to be sampled with a frequency greater than twice the highest frequency of the features within the object. For a comprehensive mathematical treatment of sampling conditions for various beam geometries, the reader is directed to texts such as those by Natterer [23], Epstein [24], or Herman [25]. Zhao et al. suggested a general rule of thumb for cone beam CT where the number of projections, N_{proj}, should be spaced to ensure the angular separation between each projection at the edge of the field of view (FOV) is equivalent to the voxel size, b_{vox}.

$$N_{proj} \geq \frac{2\pi}{\arctan(2b_{vox}/FOV)} \qquad (4)$$

As described earlier, the sample diameter is likely to exceed the detector size for in-situ cells and the theoretical number of projections becomes a function of the detector size: a 2K detector with 2048 pixels will require the acquisition of ca. 6400 projections. In reality, it is unlikely for the spatial frequency of features within the sample to exceed the sampling frequency and fewer projections may be acquired without compromising image quality [26].

2.4. Magnification and Resolution

There are two sources of magnification in a conventional laboratory micro-CT without X-ray optics: geometric and optical magnification. Micro-focus sources with a small spot size produce a divergent cone beam that provides geometric magnification. The image formed on the scintillator can also be magnified through the use of objective lenses. On the detector, counts can be improved at the expense of resolution by combining neighboring pixels in a process called binning.

The pixel size achieved after pixel binning and magnification must be capable of capturing the phenomena of interest at a representative spatial resolution. For example, if the cell geometry and the alignment of its components are to be observed, a lower magnification and a higher binning may be appropriate while, for electrode-level phenomena and detailed microstructure, a higher magnification and the lowest possible binning should be used.

2.5. Optimization of Battery Electrodes Used for In-Situ Cells

Most commercial Li-ion batteries utilize a 'full-cell' configuration consisting of a transition metal oxide positive electrode paired with a graphite negative electrode [27] and containing porous networks where Li-ions can reversibly intercalate during charge and discharge processes. Commercially relevant positive electrode materials include Lithium Cobalt Oxide (LiCoO$_2$–LCO), Lithium Manganese Oxide (LiMnO$_4$–LMO), and Lithium Nickel Manganese Cobalt Oxide (LiNi$_{0.33}$Mn$_{0.33}$Co$_{0.33}$O$_2$, NMC111), which is a potential complicating factor for in-situ cells due to the relatively dense and attenuating nature of these materials.

The in-situ cells presented in this paper were mainly prepared in a 'half-cell' arrangement where the electrode of interest is paired with a lithium metal negative electrode instead. This arrangement is commonly used in materials research since the specific electrochemical performance and practical voltage range of the electrodes of interest can be decoupled. Electrochemical parameters include the voltage range and C-rate that a battery is subject to and are important factors that influence the rate of degradation. A C-rate of 1C is equivalent to the current that will charge or discharge the entire capacity of a battery in an hour and can be calculated from the specific capacity, active material loading, and the mass or diameter of an electrode.

The electrodes, electrode arrangement, and cell environment have to be carefully optimized to ensure performance comparable to larger format cells, which is a non-trivial task given the conflicting need to minimize cell dimensions for micro-CT. Some factors that have to be considered include: electrode alignment and distance between electrodes that determines the ionic resistance across the electrolyte, compressive forces within the cell that controls the contact resistance between the electrode current collectors and plungers, and dead volume within the cell that influences electrode wetting due to gas evolution. Thus, the iterative approach, as presented in this work, was necessary to optimize electrochemical performance.

2.6. Evolution of Cell Designs

In most cases, region of interest (ROI) tomographies are carried out since the sample size is much larger than the detector size. Thus, grayscale data obtained in the reconstruction process does not correspond directly with linear attenuation coefficients even though discontinuities are captured. In this section, we explore the evolution of in-situ cell designs for electrochemical control.

2.6.1. In-Situ Coin Cell

The initial iteration of in-situ tomography cells we developed consisted of a modified coin cell with a Kapton window, which is illustrated in Figure 2. In this geometry, angular projections were acquired with a planar scan trajectory (i.e., with centre of rotation parallel to the current collector of the electrode). The Kapton window provided an angular range of 147° through which the X-ray beam could pass through unobstructed by the coin cell casing.

The X-ray transparent window significantly minimizes attenuation of the X-ray beam by the stainless steel casing, which reduces imaging artefacts such as beam hardening observed in Figure 2d. In-situ cells in this geometry were tested with both laboratory and synchrotron micro-CT instruments and appear to yield better imaging results with the latter. With synchrotron micro-CT, local tomograms with a voxel size of ca. 0.365 µm were obtained as opposed to larger voxel sizes (up to 2 µm) with laboratory micro-CT. This is due to limitations in achievable geometric magnification caused by the size of the in-situ cell. Whilst some reconstruction artefacts are to be expected due to the truncated angular range of the projections, which is shown in Figure 2c, image quality obtained with synchrotron micro-CT is remarkably comparable to ex-situ micro-CT scans of the same electrode material where the full angular range was captured.

The main drawback experienced with the in-situ coin cell design was regarding the electrochemical performance and stability shown in Figure 3. The cell was cycled in constant current-constant voltage mode at a rate of C/20 and achievable capacity is comparable to standard coin cells for the first ca. 50 h beyond which cell capacity rapidly deteriorates. As outlined in Table 1, moisture impermeability is crucial for electrochemical performance and stability because Li-ion electrolytes are highly sensitive to moisture. Although materials such as Kapton and Mylar have been used extensively as X-ray transparent materials, we hypothesize that the poor stability of the cell was due to moisture absorption through the Kapton window. This was evident in white deposits that formed in the cell over time after exposure to ambient conditions.

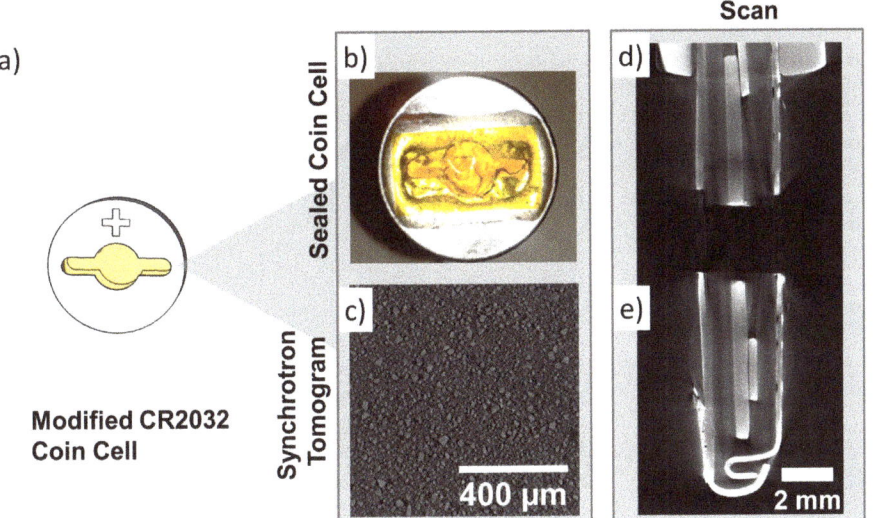

Figure 2. (a) Modified CR2032 coin cell rendering and (b) image of the Kapton window attached to the cell. (c) Reconstructed slice of a LiMnO$_2$ electrode acquired at a synchrotron facility and (d,e) post-mortem CT scan of the entire cell showing its components.

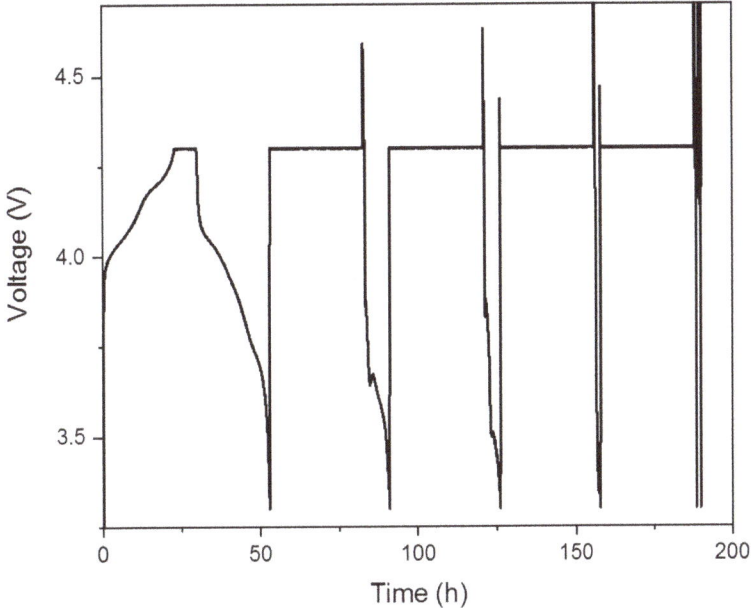

Figure 3. Voltage profile of in-situ windowed coin cell cycled at C/20 in constant current-constant voltage mode.

2.6.2. In-Situ Swagelok-Type PFA Cell

The electrochemical stability issues faced in the in-situ coin cell design resulted in the need to develop a cell capable of extended cycling and yet remaining easy to assemble and suitable for X-ray imaging. Inspired by larger Swagelok-type cells described by previous authors [28], we have modified 1/8" PFA Swagelok straight unions to be used for in-situ X-ray characterization, which is illustrated in Figure 4a. Consisting of stainless steel plungers on both sides in contact with the positive and negative electrodes, the cell is mounted upright and imaged with a cylindrical scan trajectory. By virtue of this design, the cell body in the beam path is thinned down and no highly attenuating phases (such as stainless steel) enter the field of view of the entire cell stack during tomography, which reduces undesirable artefacts significantly. Furthermore, the rotational symmetry of the cell ensures compatibility with laboratory micro-CT where projections have to be acquired for the full 360° due to the cone beam nature of the X-ray source [29].

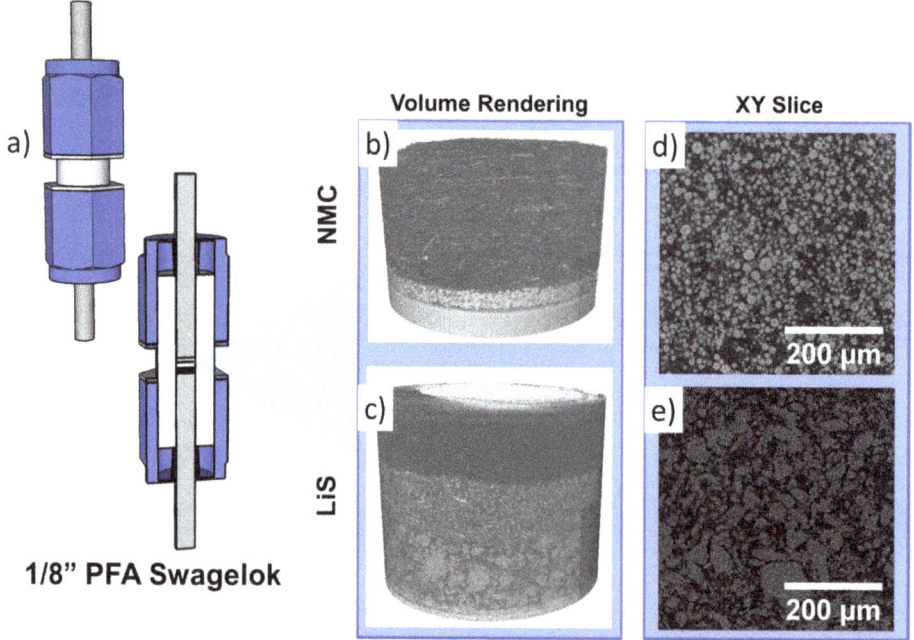

Figure 4. (a) Rendering of the 1/8" PFA Swagelok in-situ cell. (b) Volume rendering of NMC and (c) Li-S electrodes and the (d,e) their respective virtual slices.

Li-ion half-cells with NMC as a positive electrode were constructed in the PFA in-situ cell and exhibited excellent electrochemical stability over numerous cycles, as presented in Figure 5, with virtually no capacity degradation across 10 cycles and achieved an areal capacity of ca. 1.25 mAh cm^{-2}.

While the PFA cell design exhibits excellent electrochemical stability, the diameter of the resulting cell (ca. 10 mm) is relatively large and, therefore, more suited to synchrotron micro-CT where beam brilliance is not a limiting factor. As shown in Figure 4e, where we have previously reported an in-situ study of Li-S cells [13] with laboratory micro-CT, tomograms are relatively noisy. Acquisition times in laboratory micro-CT may be about ca. 48 h at high magnifications since long exposure times are required to achieve an adequate signal-to-noise ratio. Furthermore, polymers such as PFA and PTFE are also known to degrade from exposure to radiation, which turns brittle as a function of the total X-ray dose.

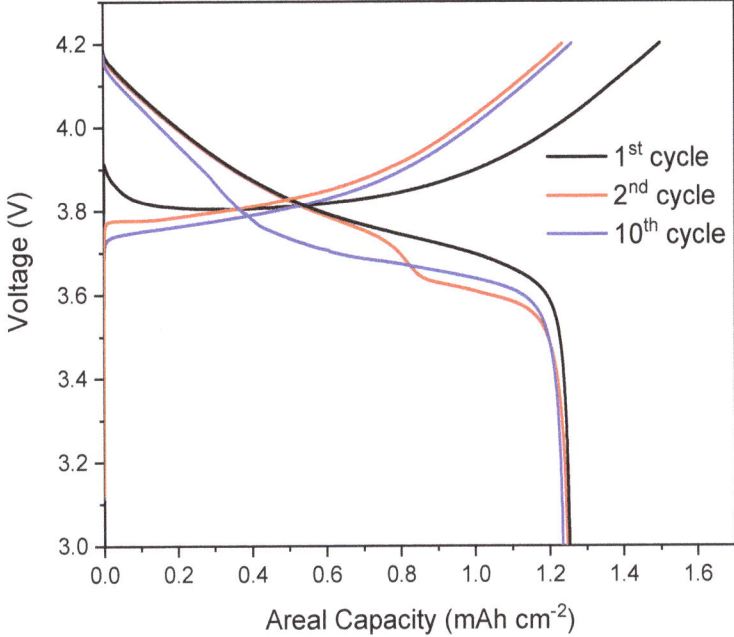

Figure 5. Charge and discharge curves for the 1st, 2nd, and 10th cycles for the NMC cell.

2.6.3. In-Situ PEEK Cell

To improve radiation resistance and further reduce sample size to improve image quality and acquisition times in laboratory micro-CT, a bespoke PEEK cell was developed. PEEK was selected since it is compatible with common Li-ion electrolytes and is non-reactive with metallic lithium. The PEEK cell has a comparable geometry to the PFA cell with a factor of 4 reduction in electrode diameter from ca. 3.2 mm to ca. 0.8 mm. For comparison, the same NMC material was cycled in the PEEK cell and the electrochemical performance, shown in Figure 6a, is similar to the PFA cell even though some capacity loss occurs after 10 cycles. A Li-S cell was also constructed with the same PEEK cell with electrochemical data presented in Figure 6a. Despite the marginally poorer electrochemical performance and stability of the PEEK cell compared to the PFA cell due to the smaller electrode diameter, this is a reasonable compromise considering the improvements in spatial and temporal resolution achievable with this cell design.

The total attenuation due to the cell body and electrode is decreased with a reduction in the diameter of the in-situ cell, which leads to shorter acquisition times in laboratory micro-CT. Furthermore, it is desirable to minimize the source-sample and sample-detector distances in laboratory micro-CT to reduce attenuation by air and the profile of the PEEK cell is highly suited to this. Relevant information is shown in Figure 7a.

A tomogram of the Li-S PEEK cell was acquired at 40× magnification with a laboratory micro-CT instrument and a virtual slice of the sulfur electrode is shown in Figure 7e along with the associated volume rendering in Figure 7c. The volume rendering in Figure 7b shows the different layers present within the cell with lithium metal on top, glass fibre separator in the middle and NMC electrode below, which is sandwiched between two current collector plungers. In Figure 7c, the higher magnification enables the electrode to be visualized in greater detail at the expense of FOV. Most significantly, a voxel size of ca. 230 nm was achieved, which permits the identification of sub-micron features within the electrode, as shown in Figure 7e, with an adequate signal-to-noise ratio and a relatively fast acquisition time due mainly to the diminutive electrode size of the in-situ PEEK cell. In addition to the relatively featureless and X-ray transparent PEEK cell body, the smaller electrode diameter leads to a reduction of complex geometries external to the field of view, and, in turn, fewer artefacts after reconstruction. Furthermore, the radiation resistance of PEEK also expands the possibilities for long-term micro-CT studies on battery degradation to be carried out in-situ. For instance, this includes the investigation of particle cracking within Li-ion positive electrode materials after a large number of cycles.

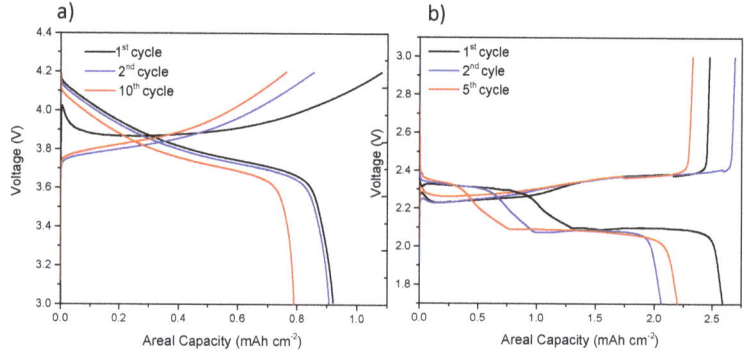

Figure 6. Electrochemical cycling data of (**a**) Li-ion half-cell with NMC positive electrode and (**b**) Li-S cell with elemental sulfur electrode.

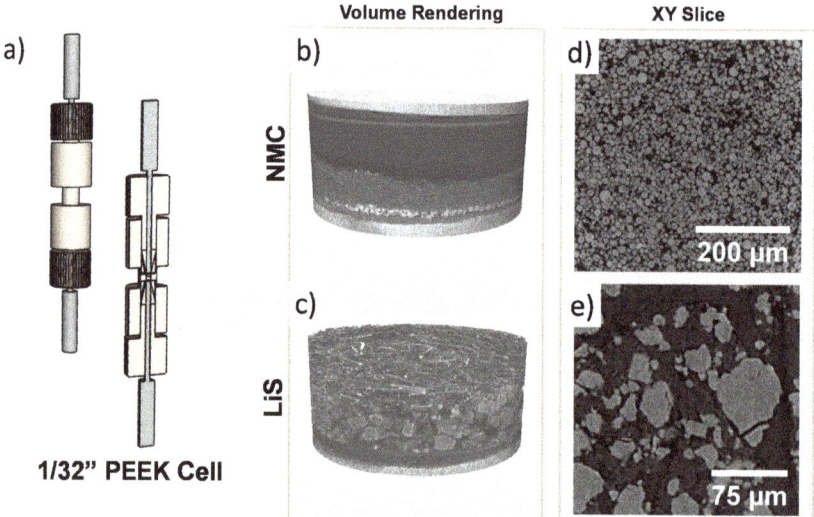

Figure 7. (**a**) Rendering of the 1/32" PEEK cell. (**b**) Volume rendering of NMC acquired at 20× magnification and (**c**) Li-S electrodes acquired at 40× magnification and the (**d**,**e**) their respective virtual slices.

2.7. *Improvement in Image Quality and Electrochemical Performance through In-Situ Cell Optimization*

Improvements in image quality over the iterations of in-situ cell design are demonstrated in Figure 8 and, while these improvements cannot be quantified due to the combination of synchrotron and laboratory CT used, in-situ studies that were once exclusive to synchrotron sources (Figure 8a) are now within the realm of capability of laboratory micro-CT instruments. It is expected that the marked improvement in image quality between the PFA cell shown in Figure 8b and PEEK cell in Figure 8c will also translate to synchrotron studies due to the reduction of material external to the FOV. This is advantageous due to the limited availability and greater competition for synchrotron time compared to the wider availability of lab-based instruments.

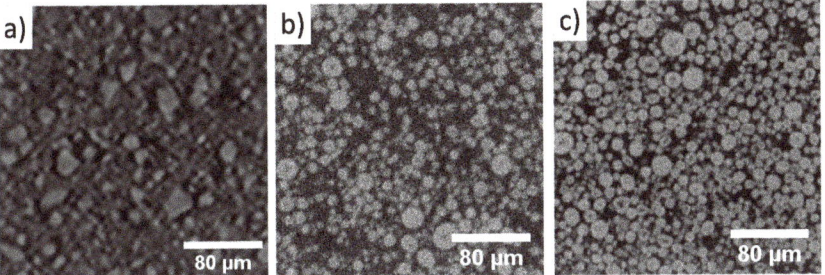

Figure 8. Improvement in image quality between (**a**) coin cell scan performed at a synchrotron facility for an LMO electrode, (**b**) Swagelok-type PFA cell, and (**c**) PEEK cell scans performed with a laboratory micro-CT instrument, Zeiss Xradia Versa 520 for an NMC electrode.

Improvements in image quality not only provide better statistics when analyzing electrode level degradation but also enable higher quality segmentation with the possibility of identifying individual particles. Image analysis techniques that were once largely within the domain of ex-situ electrode scans can now be conducted on the same sample volume as a function of variables including SoC and cycle number. Thus, parameters that can be quantified in-situ include active material loading, active material distribution within the electrode, contact area between the solid phases, and pore and particle size distributions. Furthermore, degradation phenomena occurring at the electrodes can be investigated at various length scales, which improves the understanding of their influence on electrochemical performance and lifetime.

The considerations presented throughout and summarized in Figure 9 indicate that reducing the size of the cell is paramount to decreasing the total scan time while considerably improving both spatial and temporal resolutions and image quality.

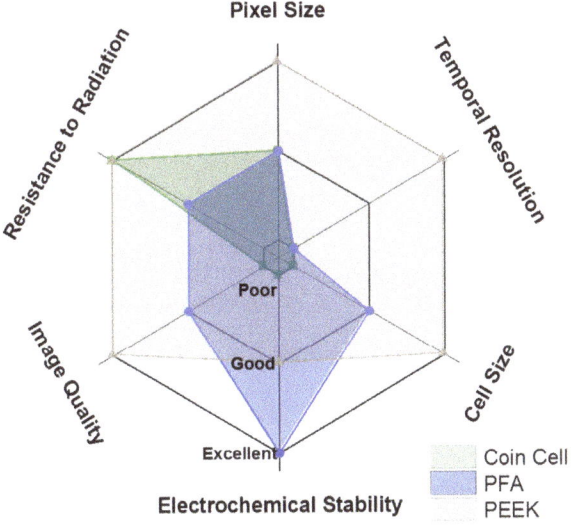

Figure 9. Comparison of the various iterations of cell designs.

2.8. Other Avenues for Improvement and Future Work

In addition to the parameters considered here, other avenues for improvement include maintaining constant compression within the cell and controlling the electrolyte volume during initial cell filling. It has been reported extensively that cell compression [30] and insufficient electrolyte can influence the electrochemical performance of a cell [31]. It is, therefore, desirable to establish a method that allows constant contact pressure between the electrodes and the current collectors and sufficient electrolyte volume. Future designs may look to incorporate spring loading into the cell assembly and vacuum filling of electrolyte to further optimize the in-situ cell.

Another avenue for improvement may come during tomographic reconstruction since the widely used filtered back-projection algorithm requires a high angular resolution for successful reconstruction. Iterative reconstruction methods that employ intelligent algorithms to reduce the number of projections required for an equivalent reconstruction may be beneficial in the study of electrochemical devices. Many iterative reconstruction algorithms exist but they are typically based around the same method. A forward projection of an estimate creates artificial data which is compared to the real raw data that is collected from the sample. From this comparison, a correction term is computed that is back-projected onto the estimate. These iterative reconstruction computations conclude after a certain number of iterations or once the difference between the estimate and the raw data (known as the 'update') converges to a sufficiently small value.

Iterative reconstruction methods based upon the improvement of either individual pixels, the entire projection, or a subset are known as the Algebraic Reconstruction Technique (ART), the Simultaneous Algebraic Reconstruction Technique (SART), or the Ordered Subset (OS) method, respectively. All methods are additive in nature, based on the addition of the update onto the current solution, and OS-based methods are typically the fastest even though they may be susceptible to artefacts. Multiplicative methods also exist such as the Multiplicative Algebraic Reconstruction Technique (MART).

The use of iterative reconstruction techniques may permit higher temporal resolution studies, which are particularly of interest for studies involving highly dynamic mechanisms such as thermal runaway [32] and rapid charging of batteries [33]. For more information on iterative reconstruction techniques, the reader is directed to work by Beister et al. [34].

3. Conclusions

As research into novel cell chemistries has expanded over the past few decades, it is clearly of great scientific interest to visualize the electrodes of electrochemical cells at all stages in their developmental lifecycle to extract performance gains and maximize their capacity, power, and lifetime. Through in-situ and operando X-ray tomography, electrode microstructures can be visualized in 3D as a function of variables such as state of charge and cell age and conditions such as thermal abuse or overcharging. We believe that in-situ and operando tomography will ultimately achieve a similar impact to X-ray diffraction and spectroscopy in the design and engineering of new battery materials.

In addition to qualitative observations on electrode degradation, techniques such as digital volume correlation can be used to track and quantify microstructure evolution chronologically within the electrodes. With a better understanding of the fundamental processes occurring at the electrodes and how cell conditions and configurations affect these processes, electrode materials can be optimized to improve their lifetime and performance. The improvements in spatial and temporal resolutions gained from the use of optimized cell designs for micro-CT far outweigh the impact on electrochemical performance these designs may have.

4. Methodology

4.1. Electrode Preparation

For the Li-ion cell electrodes, LiNi$_{0.33}$Mn$_{0.33}$Co$_{0.33}$O$_2$ (NCM111, MTI Corp., St. Louis, MO, USA) or LiMn$_2$O$_4$ (LMO, MTI Corp.), conductive carbon black (Super C65, Imerys, Paris, France), and polyvinylidenefluoride (PVDF) (Solef 5130, Solvay, Brussels, Belgium) in a 90:5:5 mass ratio were homogenized with a planetary centrifugal mixer (ARE-250, Thinky Corporation, Tokyo, Japan) using anhydrous N-methyl-2pyrrolidinone (NMP, Sigma Aldrich, Saint Louis, MO, USA) as solvent. The slurry was cast on a 15 μm thick aluminium foil (MTI Corp.) using a micrometre adjustable film applicator set to a blade gap of 150 μm.

For the Li-S cell, elemental sulfur (325 mesh, Alfa Aesar, Haverhill, MA, USA), conductive carbon black (Super C65, Timcal, Bodio, Switzerland), Ketjenblack (EC600-JD, Akzo Nobel, Amsterdam, The Netherlands), and polyvinylidene fluoride binder (Solef 5130, Solvay) in a 75:12:3:10 mass ratio were homogenized with a high shear laboratory mixer (L5M, Silverson, Buckinghamshire, UK) to form an ink with 20% total solids content with anhydrous NMP as solvent. The ink was cast onto 15 μm thick aluminium foil using a micrometre adjustable film applicator set to a blade gap of 400 μm.

The electrode sheets were initially dried on a hot plate at 80 °C and subsequently dried overnight under vacuum at 120 °C (for Li-ion electrodes) or 60 °C (for S electrode). Additionally, 3.15 mm and 0.77 mm disks were cut from the sheets by using a laser micro-machining instrument (A Series Compact Micromachining System, Oxford Lasers Ltd., Didcot, UK).

4.2. Coin Cell with Kapton Window

CR2032-type coin cell (CR2032, MTI) were cleaned in isopropanol ($\geq$99.5% purity) and dried overnight under vacuum at 60 °C prior to use. A 16 mm wide × 3 mm high letterbox-shaped aperture with a 6 mm diameter circular hole in the centre was drilled into the can, cap, and spacer components of the coin cells. Rectangular strips (ca. 19 mm × 5.5 mm) of 50 μm thick adhesive Kapton tape were applied on both the internal and external surfaces of the coin cell cap and can components to create an X-ray transparent window. Epoxy adhesive (Araldite) was then applied over the edges of the external Kapton strips to create a hermetic seal around both windows. The circular hole in the centre of the coin cell was designed to aid in sample alignment during tomography scans and the letterbox shaped portion of the X-ray window was designed to provide a sufficiently large angular range for tomographic acquisition while maintaining the mechanical stability of the coin cell.

This window design meant that tomographic acquisition performed with the in-situ coin cells were limited angle scans with the angular range dependent on the field of view. During coin cell assembly, the X-ray windows on the cell casings and spacer components were carefully aligned in order to avoid beam attenuation by the dense metal components, which ensures a clear 'line-of-sight' for the X-ray beam being transmitted through the electrode material. All coin cells were assembled with the LMO electrode as a working electrode, glass fiber separator soaked in LiPF$_6$-based electrolyte and a metallic lithium counter electrode.

4.3. 1/8″ PFA Swagelok

The 3.15 mm electrode disks were dried in a transferrable vacuum oven (Glass Oven B-585 Drying, Buchi, Flawil, Switzerland) at 120 °C (NMC electrode) or 60 °C (S electrode) overnight and transferred to an argon filled glovebox (MBraun, LABstar, Garching, Germany) where both O$_2$ and H$_2$O levels were maintained below 0.5 ppm. Customized 1/8″ PFA Swagelok unions (PFA-220-6, Swagelok, Soren, OH, USA) were used as cell bodies and these were assembled using 1/8″ 316L stainless steel plungers as the current collector. Excess material was removed from the centre of the PFA union to reduce the X-ray attenuation. Lithium metal punched to 1/8″ was used as the counter electrode with glass fiber punched to 4 mm (GF/D, Whatman, Maidstone District, UK) as a separator. For Li-ion cells, 1.2 M lithium hexafluorophosphase (LiPF$_6$) in ethylene carbonate and ethyl methyl carbonate

(EC:EMC, 1:2 v/v, Soulbrain, Northville Township, MI, USA) was used as an electrolyte. For Li-S cells, 1 M lithium bis(trifluoromethane) sulfonimide (LiTFSI) in 1,3-dioxolane and 1,2-dimethoxyethane (DOL:DME, 1:1 v/v) with 0.3 M lithium nitrate as an additive (Soulbrain, Northville Township, MI, USA) was used as an electrolyte.

4.4. 1/32" PEEK Union

The 0.77 mm electrode disks were dried in a transferrable vacuum oven (Glass Oven B-585 Drying, Buchi) at 120 °C (NMC electrode) or 60 °C (S electrode) overnight and transferred to an argon filled glovebox (MBraun, LABstar, Garching, Germany) where both O_2 and H_2O levels were maintained below 0.5 ppm. Bespoke 1/32" polyether ether ketone (PEEK) unions were used as cell bodies for the miniature tomography cells with 316L stainless steel plungers as the current collector.

Lithium metal punched to 0.8 mm was used as the counter electrode with glass fiber punched to 1 mm (GF/D, Whatman) as a separator. For Li-ion cells, 1.2 M lithium hexafluorophosphate ($LiPF_6$) in ethylene carbonate and ethyl methyl carbonate (EC:EMC, 1:2 v/v, Soulbrain, Northville Township, MI, USA) was used as an electrolyte. For Li-S cells, 1 M lithium bis(trifluoromethane) sulfonimide (LiTFSI) in 1,3-dioxolane and 1,2-dimethoxyethane (DOL:DME, 1:1 v/v) with 0.3 M lithium nitrate as an additive (Soulbrain, Northville Township, MI, USA) was used as an electrolyte.

4.5. Synchrotron Micro-CT Acquisition and Reconstruction

Synchrotron micro-CT was performed at the i13-2 beamline at Diamond Light Source (Harwell, UK) in the absorption contrast imaging mode. A parallel beam was used for the interior tomography of an LMO electrode sample assembled within the in-situ coin cell. The incident X-ray beam was monochromatized to 16 keV by a water-cooled double crystal Si <111> monochromator. The sample to detector distance was set to 25 mm and an average useful rotation range of 147° was achieved through the Kapton window. Projection images were acquired when the sample was rotated through angular steps of 0.1° about its long axis with a 6 s exposure time per projection. A 9.6 µm thick GGG:Eu scintillator was coupled to a 10× objective lens and projections were captured with a 2000 × 2000 pixel pco4000 CCD detector, which resulted in an effective pixel size of 0.365 µm.

4.6. Laboratory Micro-CT Acquisition and Reconstruction

X-ray micro-CT was performed on the PFA and PEEK in-situ cells with a lab-based micro-CT instrument (Zeiss Xradia Versa 520, Carl Zeiss Inc., Oberkochen, Germany). The instrument consisted of a polychromatic micro-focus sealed source set to an accelerating voltage of 80 kV on a tungsten target at a maximum power of 7 W. The scintillator was coupled to either a 20× or 40× objective lens and 2048 × 2048 pixel CCD detector with a pixel binning of 1, which results in a pixel size of ca. 460 nm and a field of view of ca. 940 µm for the 20× objective and ca. 230 nm and a field of view of ca. 470 µm for the 40× objective. There was no significant geometric magnification since the sample was set close to the detector to reduce the effects of penumbral blurring arising from the cone beam nature of the source. The sample was rotated through 360° with radiographs collected at discrete angular intervals amounting to a total of 1601 projections. The radiographic projections were then reconstructed with proprietary reconstruction software (Version 11.1.8043, XMReconstructor, Carl Zeiss Inc.) by using a modified Feldkamp-David-Kress (FDK) algorithm for cone beam geometry.

Author Contributions: C.T. and S.R.D. contributed equally in writing the manuscript and all experimental work. O.O.T provided data for the synchrotron study and coin cell design and T.M.M.H. contributed to the theoretical descriptions of X-ray C.T., D.J.L.B. and P.R.S. directed research. All authors discussed the results and contributed to the manuscript.

Funding: This research was funded by the EPSRC under grants EP/R020973/1, EP/N032888/1 and through the Faraday Institution, the Royal Academy of Engineering under grant CiET1718/59, and Diamond Light Source for beamtime under MT11539.

Acknowledgments: The authors would like to acknowledge the EPSRC for funding under grants EP/R020973/1, EP/N032888/1, and the Faraday Institution. PRS acknowledges funding from the Royal Academy of Engineering for financial support under the Chair in Emerging Technologies scheme. The authors acknowledge Diamond Light Source for synchrotron beam time on the Diamond-Manchester Branchline (I13-2) of the I13 imaging and coherence beamline.

Conflicts of Interest: The authors declare no conflict of interest.

References

1. Tarascon, J.M.; Armand, M. Issues and challenges facing rechargeable lithium batteries. *Nature* **2001**, *414*, 359–367. [CrossRef] [PubMed]
2. Bruce, P.G.; Freunberger, S.A.; Hardwick, L.J.; Tarascon, J.M. Li-O_2 and Li-S batteries with high energy storage. *Nat. Mater.* **2012**, *11*, 19–29. [CrossRef] [PubMed]
3. Shearing, P.R.; Howard, L.E.; Jørgensen, P.S.; Brandon, N.P.; Harris, S.J. Characterization of the 3-dimensional microstructure of a graphite negative electrode from a li-ion battery. *Electrochem. Commun.* **2010**, *12*, 374–377. [CrossRef]
4. McBreen, J.; Balasubramanian, M. Rechargeable lithium-ion battery cathodes: In-situ xas. *J. Miner. Met. Mater. Soc.* **2002**, *54*, 25–28. [CrossRef]
5. Buchberger, I.; Seidlmayer, S.; Pokharel, A.; Piana, M.; Hattendorff, J.; Kudejova, P.; Gilles, R.; Gasteiger, H.A. Aging analysis of graphite/LiNi$_{1/3}$Mn$_{1/3}$Co$_{1/3}$ O_2 cells using XRD, PGAA, and AC impedance. *J. Electrochem. Soc.* **2015**, *162*, A2737–A2746. [CrossRef]
6. Meirer, F.; Cabana, J.; Liu, Y.; Mehta, A.; Andrews, J.C.; Pianetta, P. Three-dimensional imaging of chemical phase transformations at the nanoscale with full-field transmission X-ray microscopy. *J. Synchrotron Radiat.* **2011**, *18*, 773–781. [CrossRef] [PubMed]
7. Chen-Wiegart, Y.-C.K.; Liu, Z.; Faber, K.T.; Barnett, S.A.; Wang, J. 3D analysis of a LiCoO$_2$–Li(Ni$_{1/3}$Mn$_{1/3}$Co$_{1/3}$)O_2 Li-ion battery positive electrode using X-ray nano-tomography. *Electrochem. Commun.* **2013**, *28*, 127–130. [CrossRef]
8. Shearing, P.R.; Brandon, N.P.; Gelb, J.; Bradley, R.; Withers, P.J.; Marquis, A.J.; Cooper, S.; Harris, S.J. Multi length scale microstructural investigations of a commercially available li-ion battery electrode. *J. Electrochem. Soc.* **2012**, *159*, A1023–A1027. [CrossRef]
9. Randjbar Daemi, S.; Brett, D.J.L.; Shearing, P.R. A lab-based multi-length scale approach to characterize lithium-ion cathode materials. *ECS Trans.* **2017**, *77*, 1119–1124. [CrossRef]
10. Eastwood, D.S.; Bradley, R.S.; Tariq, F.; Cooper, S.J.; Taiwo, O.O.; Gelb, J.; Merkle, A.; Brett, D.J.L.; Brandon, N.P.; Withers, P.J.; et al. The application of phase contrast X-ray techniques for imaging li-ion battery electrodes. *Nucl. Inst. Methods Phys. Res. B* **2014**, *324*, 118–123. [CrossRef]
11. Jiao, L.A.; Li, X.; Ren, L.L.; Kong, L.Y.; Hong, Y.L.; Li, Z.W.; Huang, X.B.; Tao, X.F. 3D structural properties study on compact LiFePO$_4$s based on X-ray computed tomography technique. *Powder Technol.* **2015**, *281*, 1–6. [CrossRef]
12. Casimir, A.; Zhang, H.G.; Ogoke, O.; Amine, J.C.; Lu, J.; Wu, G. Silicon-based anodes for lithium-ion batteries: Effectiveness of materials synthesis and electrode preparation. *Nano Energy* **2016**, *27*, 359–376. [CrossRef]
13. Yermukhambetova, A.; Tan, C.; Daemi, S.R.; Bakenov, Z.; Darr, J.A.; Brett, D.J.; Shearing, P.R. Exploring 3D microstructural evolution in Li-sulfur battery electrodes using in-situ X-ray tomography. *Sci. Rep.* **2016**, *6*, 35291. [CrossRef] [PubMed]
14. Tan, C.; Heenan, T.M.M.; Ziesche, R.F.; Daemi, S.R.; Hack, J.; Maier, M.; Marathe, S.; Rau, C.; Brett, D.J.L.; Shearing, P.R. Four-dimensional studies of morphology evolution in lithium–sulfur batteries. *ACS Appl. Energy Mater.* **2018**. [CrossRef]
15. Shearing, P.; Wu, Y.; Harris, S.J.; Brandon, N. In situ X-ray spectroscopy and imaging of battery materials. *Electrochem. Soc. Interface* **2011**, *20*, 43–47. [CrossRef]
16. Wang, J.; Chen-Wiegart, Y.C.; Wang, J. In operando tracking phase transformation evolution of lithium iron phosphate with hard X-ray microscopy. *Nat. Commun.* **2014**, *5*, 4570. [CrossRef] [PubMed]
17. Zhou, Y.N.; Ma, J.; Hu, E.; Yu, X.; Gu, L.; Nam, K.W.; Chen, L.; Wang, Z.; Yang, X.Q. Tuning charge-discharge induced unit cell breathing in layer-structured cathode materials for lithium-ion batteries. *Nat. Commun.* **2014**, *5*, 5381. [CrossRef] [PubMed]

18. Zhou, Y.-N.; Yue, J.-L.; Hu, E.; Li, H.; Gu, L.; Nam, K.-W.; Bak, S.-M.; Yu, X.; Liu, J.; Bai, J.; et al. High-rate charging induced intermediate phases and structural changes of layer-structured cathode for lithium-ion batteries. *Adv. Energy Mater.* **2016**, *6*, 1600597. [CrossRef]
19. Hubbell, J.H. Photon mass attenuation and energy-absorption coefficients from 1 keV to 20 MeV. *Int. J. Appl. Radiat. Isot.* **1982**, *33*, 1269–1290. [CrossRef]
20. Tables of X-ray Mass Attenuation Coefficients and Mass Energy-Absorption Coefficients from 1 keV to 20 MeV for Elements Z = 1 to 92 and 48 Additional Substances of Dosimetric Interest. Available online: https://nvlpubs.nist.gov/nistpubs/Legacy/IR/nistir5632.pdf (accessed on 28 September 2018).
21. Evaluation of Transmission Based Image Quality Optimisation for X-ray Computed Tomography. Available online: https://www.ndt.net/article/ctc2012/papers/255.pdf (accessed on 28 September 2018).
22. Daemi, S.R.; Tan, C.; Volkenandt, T.; Cooper, S.J.; Palacios-Padros, A.; Cookson, J.; Brett, D.J.L.; Shearing, P.R. Visualizing the carbon binder phase of battery electrodes in three dimensions. *ACS Appl. Energy Mater.* **2018**, *1*, 3702–3710. [CrossRef]
23. Natterer, F. *The Mathematics of Computerized Tomography*; Society for Industrial and Applied Mathematics: Philadelphia, PA, USA, 2001.
24. Epstein, C.L. *Introduction to the Mathematics of Medical Imaging*, 2nd ed.; Society for Industrial and Applied Mathematics: Philadelphia, PA, USA, 2007.
25. Herman, G.T. *Fundamentals of Computerized Tomography: Image Reconstruction from Projections*; Springer Science & Business Media: Berlin, Germany, 2009.
26. Mathematics and Physics of Computed Tomography (CT): Demonstrations and Practical Examples. Available online: http://cdn.intechopen.com/pdfs/43595/InTech-Mathematics_and_physics_of_computed_tomography_ct_demonstrations_and_practical_examples.pdf (accessed on 28 September 2018).
27. Scrosati, B.; Garche, J. Lithium batteries: Status, prospects and future. *J. Power Sources* **2010**, *195*, 2419–2430. [CrossRef]
28. Tarascon, J.M.; Disalvo, F.J.; Murphy, D.W.; Hull, G.W.; Rietman, E.A.; Waszczak, J.V. Stoichiometry and physical properties of ternary molybdenum chalcogenides $M_xMo_6X_8$ (X = S, Se; M = Li, Sn, Pb). *J. Solid State Chem.* **1984**, *54*, 204–212. [CrossRef]
29. Johnson, R.H.; Hu, H.; Haworth, S.T.; Cho, P.S.; Dawson, C.A.; Linehan, J.H. Feldkamp and circle-and-line cone-beam reconstruction for 3D micro-CT of vascular networks. *Phys. Med. Biol.* **1998**, *43*, 929–940. [CrossRef] [PubMed]
30. Cannarella, J.; Arnold, C.B. Stress evolution and capacity fade in constrained lithium-ion pouch cells. *J. Power Sources* **2014**, *245*, 745–751. [CrossRef]
31. Long, B.R.; Rinaldo, S.G.; Gallagher, K.G.; Dees, D.W.; Trask, S.E.; Polzin, B.J.; Jansen, A.N.; Abraham, D.P.; Bloom, I.; Bareno, J.; et al. Enabling high-energy, high-voltage lithium-ion cells: Standardization of coin-cell assembly, electrochemical testing, and evaluation of full cells. *J. Electrochem. Soc.* **2016**, *163*, A2999–A3009. [CrossRef]
32. Finegan, D.P.; Scheel, M.; Robinson, J.B.; Tjaden, B.; Hunt, I.; Mason, T.J.; Millichamp, J.; Di Michiel, M.; Offer, G.J.; Hinds, G.; et al. In-operando high-speed tomography of lithium-ion batteries during thermal runaway. *Nat. Commun.* **2015**, *16*, 58–59. [CrossRef] [PubMed]
33. Ahmed, S.; Bloom, I.; Jansen, A.N.; Tanim, T.; Dufek, E.J.; Pesaran, A.; Burnham, A.; Carlson, R.B.; Dias, F.; Hardy, K.; et al. Enabling fast charging—A battery technology gap assessment. *J. Power Sources* **2017**, *367*, 250–262. [CrossRef]
34. Beister, M.; Kolditz, D.; Kalender, W.A. Iterative reconstruction methods in X-ray CT. *Phys. Med.* **2012**, *28*, 94–108. [CrossRef] [PubMed]

© 2018 by the authors. Licensee MDPI, Basel, Switzerland. This article is an open access article distributed under the terms and conditions of the Creative Commons Attribution (CC BY) license (http://creativecommons.org/licenses/by/4.0/).

Article

An X-ray Tomographic Study of Rechargeable Zn/MnO₂ Batteries

Markus Osenberg [1,2,*], Ingo Manke [2], André Hilger [1,2], Nikolay Kardjilov [2] and John Banhart [1,2]

[1] Institute of Material Science and Technologies, Technical University Berlin, Hardenbergstraße 36, 10623 Berlin, Germany; hilger@helmholtz-berlin.de (A.H.); banhart@helmholtz-berlin.de (J.B.)
[2] Helmholtz-Centre Berlin for Materials and Energy GmbH, Hahn-Meitner-Platz 1, 14109 Berlin, Germany; manke@helmholtz-berlin.de (I.M.); kardjilov@helmholtz-berlin.de (N.K.)
* Correspondence: markus.osenberg@tu-berlin.de

Received: 28 July 2018; Accepted: 17 August 2018; Published: 21 August 2018

Abstract: We present non-destructive and non-invasive in operando X-ray tomographic investigations of the charge and discharge behavior of rechargeable alkaline-manganese (RAM) batteries (Zn-MnO₂ batteries). Changes in the three-dimensional structure of the zinc anode and the MnO₂ cathode material after several charge/discharge cycles were analyzed. Battery discharge leads to a decrease in the zinc particle sizes, revealing a layer-by-layer dissolving behavior. During charging, the particles grow again to almost their initial size and shape. After several cycles, the particles sizes slowly decrease until most of the particles become smaller than the spatial resolution of the tomography. Furthermore, the number of cracks in the MnO₂ bulk continuously increases and the separator changes its shape. The results are compared to the behavior of a conventional primary cell that was also charged and discharged several times.

Keywords: alkaline manganese batteries; X-ray tomography; in operando; in situ; zinc powder

1. Introduction

The development of new energy storage materials and systems is currently one of the most important challenges in materials research. Batteries play a crucial role in the future replacement of conventional mobile or stationary energy sources based on fossil fuels. However, batteries with high storage capacities and low weights are still by far too expensive. Furthermore, the general shortage in various resources puts constraints on the development of many battery types. Therefore, the development of cost-efficient production methods and use of easily accessible raw materials are key issues in battery research.

Zinc is a widely available and inexpensive material, and it is a candidate for future use in rechargeable batteries for mobile and stationary applications [1–7]. The well-known alkaline-manganese battery is still one of the most common types in use [1]. Reasons include their low self-discharge and environmental friendliness compared to other battery types. Such batteries are cheap to produce, maintenance-free, and safe compared to lithium-based batteries. In addition, in the charged state they provide a voltage of 1.5 V, which is higher than many other (e.g., nickel-metal hydride) batteries. Their main disadvantage is that they are normally designed as primary cells, i.e., they are not rechargeable.

Because of the many fundamental advantages of alkaline-manganese batteries, much effort has been put into developing and optimizing primary cells and, even more important for future applications, developing rechargeable alkaline-manganese batteries (RAM) [8–13]. Up to now, RAM still suffer from an unreliable cyclic behavior. Some individual batteries can be recharged up to 500 times, while others last only a few cycles.

In the past, various methods have been applied to study alkaline primary cells. For the investigation of the zinc particles, electron microscopy and optical microscopy have been used [14–16]. Preparation of the samples is very difficult because the oxidation and corrosion of Zn, and the carbonation of ZnO alter the structure of the material. Horn et al. have developed a dedicated preparation technique [14]. However, all these measurement techniques do not allow for an in situ study of the material inside the entire volume of the battery. Only the sectioned material is accessible.

Imaging techniques based on X-rays have been successfully used to study battery materials [17–29]. Since these techniques are non-destructive and non-invasive, they are especially suited for in situ or in operando measurements [30–33]. X-ray tomography using both table-top and synchrotron radiation sources was used to investigate alkaline primary cells and zinc-air batteries in three dimensions [34–37]. Moreover, neutron imaging has been used to investigate alkaline primary cells [34,38].

In this paper, structural changes in RAM cells were examined in situ and non-destructively by X-ray tomography.

2. Experimental Set-Up and Data Processing

2.1. The Alkaline-Manganese Battery

2.1.1. Set-Up

The alkaline-manganese battery consists of a steel shell into which the hollow cylinder of the cathode material—consisting of manganese dioxide and an electrolyte—was inserted by the manufacturer. The anode was made of a mixture of zinc powder and an electrolyte, and it was injected into the shell. Between the anode and the cathode, a separator is located. A metallic nail at the bottom of the battery acts as the negative pole of the battery. It protrudes into the anode and acts as a charge collector. Between the bottom and the cathode, a seal prevents leakage of the cell.

2.1.2. Chemical Processes in an Alkaline-Manganese Battery

During the initial discharge, a reduction reaction takes place at the cathode; see Equations (1) and (2) [1]:

$$MnO_2 + H_2O + e^- \rightarrow MnOOH + OH^-, \tag{1}$$

$$3MnOOH_2 + e^- \rightarrow Mn_3O_4 + OH^- + H_2O \tag{2}$$

Due to the formation of MnOOH, the cathode expands in volume by about 17%. At the anode, as given in Equation (3), zinc initially forms zincate. After the electrolyte is supersaturated with zincate, the reaction product changes to zinc hydroxide, see Equation (4), which is then slowly dehydrated to zinc oxide, see Equation (5):

$$Zn + 4OH^- \rightarrow [Zn(OH)_4]^{2-} + 2e^- \tag{3}$$

$$Zn + 2OH^- \rightarrow Zn(OH)_2 + 2e^- \tag{4}$$

$$Zn(OH)_2 \rightarrow ZnO + H_2O \tag{5}$$

The overall discharge redox reaction is shown in Equation (6) [1]:

$$2MnO_2 + Zn + 2H_2O \rightarrow 2MnOOH + Zn(OH)_2 \tag{6}$$

For a small to medium discharge, the reaction in Equation (7) predominantly takes place in alkaline-manganese batteries [8]:

$$3MnO_2 + 2Zn \rightarrow 2Mn_3O_4 + 2ZnO + Zn(OH)_2 \tag{7}$$

2.1.3. Setup for Charge and Discharge of Alkaline-Manganese Batteries

The batteries were discharged with a VOLTCRAFT Multicharger VC 1506 that was connected to a computer. The batteries were charged using the charger type ACP62 PowerSet AA. For measuring the behavior of the charger, an oscilloscope was connected in parallel to a battery. To avoid damage, the conventional primary cells were charged in a pulsed mode and not to above 1.72 V. RAM were discharged to 0.9 V at currents of 100 mA, 200 mA, and 400 mA. The primary cell was discharged at 200 mA current.

2.1.4. X-ray Tomography System and Measurement Procedure

The setup consisted of a fixed Hamamatsu microfocus X-ray tube (L8121-3) with a stable spot size of 7 μm and a Hamamatsu flat panel detector (C7942SK-05). The X-ray tube had an operational voltage range of 40 kV to 150 kV, and the target spot had a diameter of 7 μm [39]. The detector comprised a gadolinium oxysulfide (Gadox)-based scintillator on a 2316 × 2316 pixel detector array with a pixel size of 50 μm. On a goniometer, a sample can be mounted and moved with 5 degrees of freedom.

To avoid image analysis on large zinc-free spaces, cells with a homogenous zinc distribution inside the field of view were preferred. For that, multiple RAM and multiple primary cells were radiographed. Batteries were discarded if air inclusions were visible. Eventually, one primary cell and three RAM were selected. Before and after the first, second, third, fifth, 10th and if possible, 15th, charging step a tomogram was recorded.

For tomography, a source–object distance of 58 mm, and a source–detector distance of 350 mm were selected, which resulted in an effective pixel size of 8.3 μm, and thus a special resolution of 16.6 μm. The magnification chosen in this way was the largest that projected the entire image onto the detector, and not just a part of it. To achieve maximum contrast and the best signal-to-noise ratio, the X-ray tube was operated at 130 kV and 76 μA with a 0.5 mm copper filter. Furthermore, an exposure time of 1.6 s for each of the in total 1500 projections over 360° was applied, resulting in a total scanning time of 1.8 hr per tomogram. After image acquisition, the images were reconstructed using the software 'Octopus' (version 8, XRE, Gent, Belgium).

2.2. Data Processing

For particle analysis, it is usually necessary to filter the data, because otherwise individual particles touching each other would be counted as a single particle, or noise artefacts would be interpreted as small particles. The choice of the filter thus had a strong impact on the significance of particle analysis. The reconstructed 3D data were filtered with the Software 'Fiji' (version 1.52a) [40,41], and then analyzed with the software 'Avizo' (version 8.1, Thermo Fisher Scientific, Waltham, MA, USA). Figure 1 demonstrates some of the main steps of the measurement and image analysis procedure schematically.

2.2.1. Median Filter

Median filtering consists of first sorting all voxels to be analyzed, and their neighbors with respect to their grey values, and then lining them up in a list. The voxel to be analyzed then receives the grey level located in the middle of this list (i.e., the median value of all voxels in the neighborhood). With this method, noise is partially eliminated.

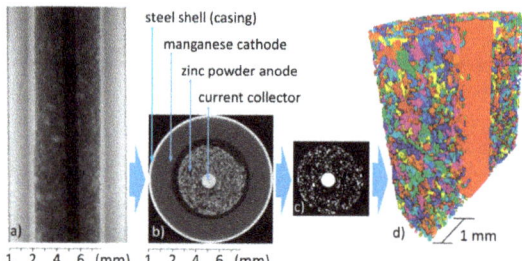

Figure 1. Sketch illustrating the reconstruction and data preparation process. After capturing all 1500 radiographic projection images—one shown in (**a**)—a tomographic 3D data set is reconstructed in (**b**). After binarization (**c**), the individual zinc particles are labelled (and, for example, color-coded as in (**d**)), which allows for a shape analysis of each individual particle.

2.2.2. Threshold Filter/Binarization

The 3D data sets were binarized with a threshold filter [42]. If a voxel of the dataset belonged to a zinc particle, the value 1 was assigned to it, whereas all other voxels received the value 0. After setting a threshold, all voxels above this value were set to 1 and all others to 0. The choice of the threshold value was crucial in this stage. The larger this threshold was chosen, the higher the X-ray absorption of a voxel had to be, to qualify as belonging to a particle.

2.2.3. Erosion/Dilation

Eroding removes the edge/outer shell of a particle [42]. Mainly 'noise dots' (single voxels), but also very small particles that result from the recording process—for example, intersecting streak artefacts—and that do not belong to particles, disappear completely. Subsequent dilatation then adds the missing edges to the particles, but not to the now missing noise dots. In this way, the particles largely regain their previous volume, but the noise dots are removed. However, previously sharp particle edges tend to be smoothened.

2.2.4. Watershed Transformation

To carry out a watershed transformation, a function is initially used to assign a distance from the particle edge to each voxel of a particle. If one now interprets the equidistance lines as water level lines after fictitious flooding with water, the volume is divided into several pools. If one fills them up with virtual water, one first obtains several smaller lakes that are merged to a larger lake at a watershed. Along this watershed, the particle is divided into two [42].

2.2.5. Location Retrieval

In order to follow the dissolution process of individual particles, it was necessary to retrieve the location of the particles after each discharging or charging step. Therefore, a distinctive particle was chosen for each battery in the corresponding data set. After each cycle, this particle was located to ensure that the same areas were examined in the batteries. These particles were selected from areas neither directly next to the nail, nor to the separator, as it has been suggested that in these regions that a typical particle disintegration may occur.

In the image corresponding to the RAM discharged at 100 mA current for 10 cycles, it was no longer possible to find the initially selected particle. However, by comparing a number of other areas in this data set (around the initial particles) the original location of the particle was successfully reconstructed.

3. Experimental Results and Discussion

3.1. Properties of the Cells at Different Discharge Rates and Numbers of Cycles

Figure 2a shows measured capacities of the individual cells per discharge/charge cycle. The final discharge voltage was 0.9 V. According to the manufacturer, the RAM should deliver up to 800 mAh, while the primary cell should deliver up to 1220 mAh at a discharge current of 30 mA.

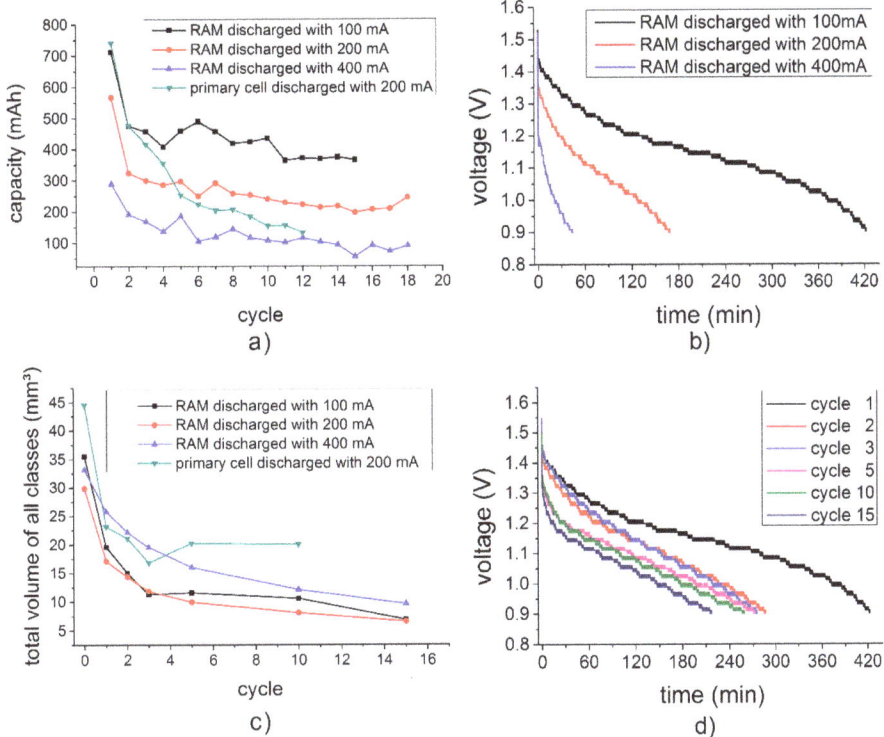

Figure 2. (**a**) Battery capacities remaining after a given number of discharge cycles, (**b**) discharge curves of pristine rechargeable alkaline-manganese (RAM) batteries at different discharge currents, (**c**) total volume of all segmented zinc particles and (**d**) discharge curves of a RAM battery (100 mA) during different cycles.

It can be seen in Figure 2b that a higher discharge current results in a faster drop of voltage, which is associated with a lower discharge capacity after the cycle, and an increased internal resistance. The voltage also decreases with an increasing amount of cycles (Figure 2d). Of course discharging with lower currents also takes longer, as shown in the discharge curves in Figure 2b for the three different discharge currents applied in our work, namely 100 mA, 200 mA, and 400 mA.

Additionally a difference in the dissolving behavior of the zinc particles can be seen in Figure 2c. While the zinc particles in the RAM cells dissolve continuously during cycling, the zinc particles in the primary cell stopped dissolving after the third cycle.

3.2. 3D Structural Analysis

Figure 3 shows tomographic cross sections through samples in different charging stages. The cross sections of the fully charged and partially discharged batteries were always taken at the same locations (Section 2.2.5). After a first discharge, many of the smaller zinc particles were dissolved in the electrolyte (compare Figure 3, first and second column). Dissolution continuously progresses until most of the zinc particles have been dissolved into small particles that form a homogeneous gel in the electrolyte (i.e., particles are smaller than the spatial resolution of the tomography setup).

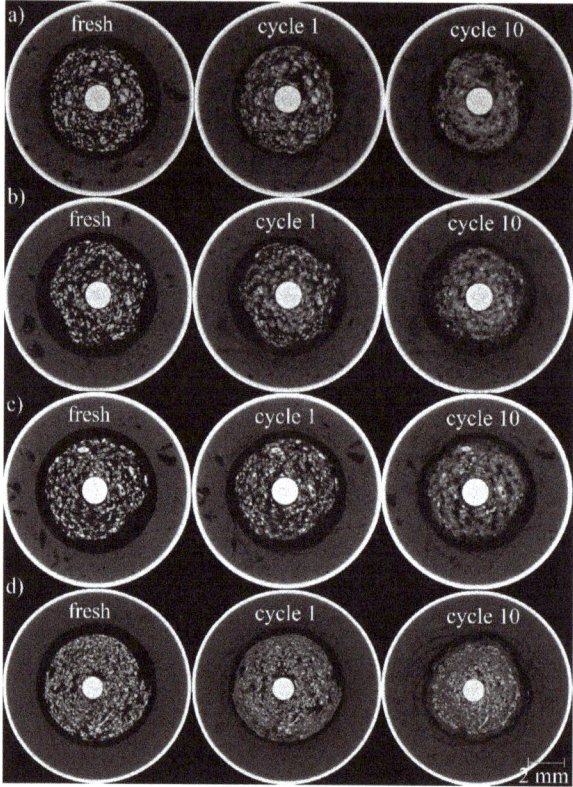

Figure 3. Tomographic cross sections showing three different cycle states for each of the four batteries studied. (**a**) RAM battery discharged at a current of 100 mA, (**b**) 200 mA, (**c**) 400 mA, and (**d**) alkaline manganese primary cell discharged at a current of 200 mA.

However, in the images describing the RAM cells (Figure 3a–c), larger particles, especially at the outer ring area, still remained even after 10 cycles (see also Figure 4). Their size decreased, but their inner core remained almost unchanged. This reveals a layer-by-layer dissolving behavior (see also Figure 4). On the (outer) cathode side, the cylinder comprising the MnO_2 and electrolyte slowly swelled and moves inwards, while at the same time cracks formed. Especially in the primary cell, the deformation resulted in the development of pointy structures in the separator region between the electrodes. The gap between both electrodes that was filled by the separator became smaller.

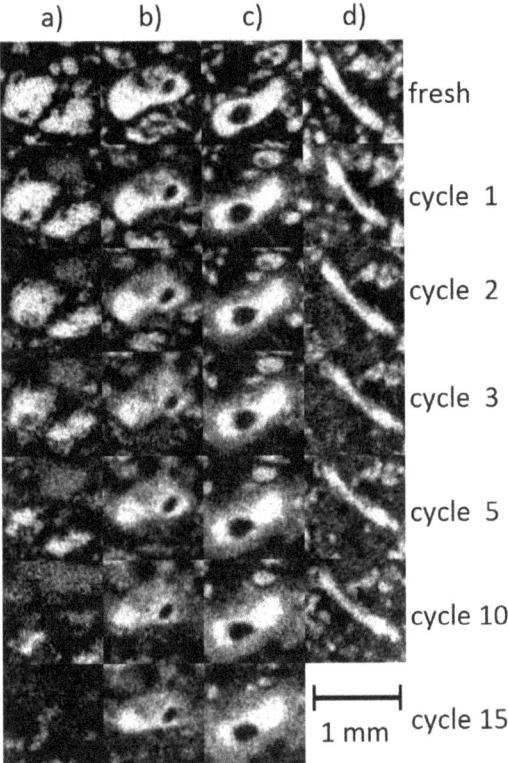

Figure 4. Overview over four selected particles in (**a**) RAM battery discharged with 100 mA, (**b**) 200 mA, (**c**) 400 mA, and (**d**) primary alkaline-manganese cell discharged with 200 mA.

3.3. Dissolution of Individual Particles During Cycling

Cross sections through individual large particles occurring in each of the four tomographic measurement series showed the shapes of the particles after several cycling steps (Figure 4). Due to their size and good recognizability, even after several cycles, these particles were also used as "markers" for the location adjustments made when recording the data for Figure 3, i.e., they were used to find the same locations in the cell after different cycling steps. Obviously, the particles maintained their shapes over several cycling steps. However, the particle size slowly decreased and an area around the particles occurred that appeared fuzzy and in intermediate (grey) contrast, possibly containing small zinc particles that were no longer resolved by our measurement technique.

3.4. Quantitative 3D Analysis

For a quantitative analysis of the zinc particles in the batteries, the air surrounding the battery, the steel casing, and the manganese dioxide cathode were "cut off", as shown in Figure 1c, so that the respective image data contained only the anode material and the nail in the center. To this data set, the threshold filter was applied, after which the particles were white, and particle-free areas were left black. The same threshold was applied to all batteries and for all cycles. To reduce the noise in both the white and black areas, a median-3D filter was applied to the data sets (Section 2.2.3). Using the distinctive particles shown in Figure 4 as a starting point, the locations of 600 layers in each battery tomogram were adjusted to the same position, and the resulting data sets were quantitatively analyzed.

This procedure ensures that exactly the same volume range for all cycling states of the battery were analyzed and any possible drift was eliminated After this, a watershed transformation was applied to improve particle separation (Section 2.2.4) using the program 'Avizo 7.0'. For the radius-dependent analysis, each particle coordinate was converted from Cartesian to cylindrical coordinates, with the nail as a center. It was taken into account that the nails stuck askew in the batteries. The graphs in Figure 5 present the results of this particle size analysis. The particle diameter represents the corresponding spherical diameter. Figure 5a,c show the results for the RAM cell discharged at either 100 mA or 400 mA, while Figure 5b,d show the corresponding results for the RAM cell and the primary cell after discharging at 200 mA. Typical particle diameters range from 100 µm to 200 µm for both battery cell types. In all four cases, a rapid drop in the overall particles volume was found after the first discharge/charge cycle. This indicates that large zinc amounts were dissolved in the electrolyte, and that many particles reached a size where they could no longer be resolved by tomography and therefore did not contribute to the overall volume in the graphs shown. After each cycle, the volume found in the analyzed (larger) particles decreases in accordance to the findings in Figure 3. The size distributions of the particles remained almost unchanged.

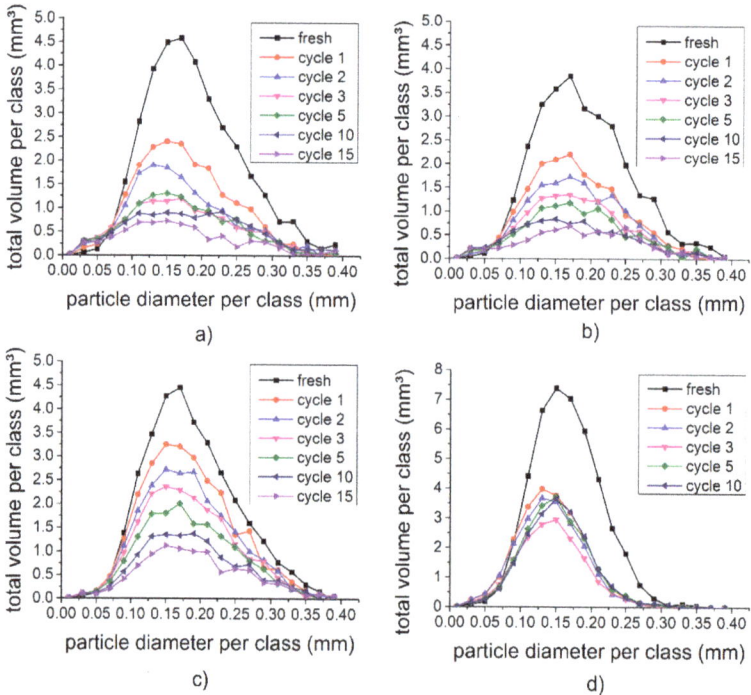

Figure 5. Particle size distributions in RAMs discharged at (**a**) 100 mA, (**b**) 200 mA, and (**c**) 400 mA and (**d**) primary cell discharged at 200 mA.

One might expect that all particles shrink continuously. However, Figure 3 demonstrates that many particles dissolved quickly and "disappeared", the smaller particles did so even faster than the larger ones. On the other hand, many larger particles became smaller and/or broke up, and they contributed to the amount of smaller particles while at the same time, several larger particles seemed to be largely unaffected by the cycling process. Eventually, the remaining particles seemed to have a size distribution that was very similar to the initial one, although many individual particles changed their size.

Figures 4 and 5 reveal that particles dissolved more rapidly at lower than at higher discharge currents. Especially, the first discharge had a large effect. From the second discharge, the dissolution process was much slower and decreased continuously with increasing cycle numbers. During the discharge, the edges of large particles, dissolved and a shell with a lower absorption coefficient was formed around the particle. During charging, these clouds became smaller again. Very small particles were completely dissolved. This process was more noticeable near the separator.

Furthermore, during charging some very small new particles are formed, indicating the development of seeds for the growth of zinc crystals (see Supplementary Video 1).

3.5. Local Effects

The changes in particle sizes were not uniformly distributed over the entire cell. With progressing cycling, increased particle migration was observed, the distance and direction of which depended on the distance of the individual particle to the current collector nail. The particles in the RAM cell appeared to drift away from the collector nail towards the separator. This was observed only in the fifth cycle or later.

In the primary cell, in contrast, it seemed that starting from the first cycle, particles close to the nail migrated towards the center, while particles located close to the separator moved outwards, thus forming a ring with a lower particle density. The particle analyses of the pristine RAM cell and of the RAM cell after the 10th cycle, both discharged at 200 mA, as displayed in Figure 6. Here, the data set was divided into two annular disks (or in 3D hollow cylinders) of equal areas. The inner ring had a radius of 0.70 mm to 2.20 mm and the outer ring was 2.20 mm to 3.03 mm. The graphs show that with progressive cycling, the particles at a larger distance to the collector nail had less total volume and a smaller average particle diameter compared to other particles in the cell. Particle migration seemed to cause or at least contribute to a separation between the larger and smaller particles.

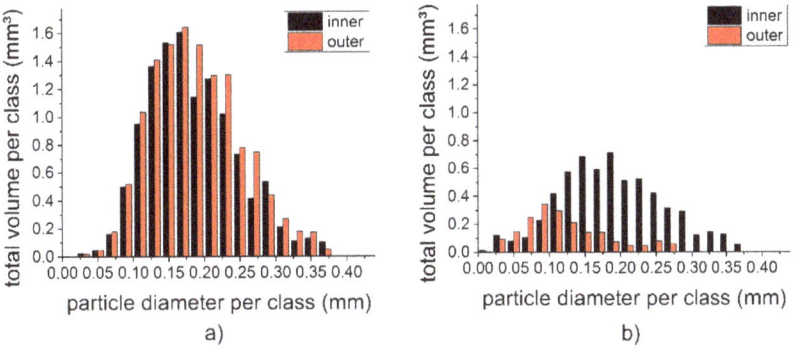

Figure 6. Size distributions of the zinc particles in the discharged RAM cell. (**a**) Pristine and (**b**) after the 10th cycle (discharge current 200 mA).

3.6. Comparison Between the Primary Cell and the RAM

As can be seen in Figure 5, a much larger volume of zinc particles was present in the primary cell (compare Figure 5d to Figure 5a–c). In addition, the maximum total volume per diameter class in the primary cell was observed for particles with 0.15 mm ± 0.02 mm diameter, which was slightly smaller compared to the 0.17 mm ± 0.02 mm for the RAM. Unlike the large particles in the RAM, the large particles of the primary cell dissolved faster than the smaller particles, thus leading to a shift of the average particle diameter upon cycling (Figure 5d).

Direct comparison of slices taken from the RAM (Figure 7 bottom) and from the primary cell (Figure 7 top) revealed the different design of the two battery types. The graph in Figure 2c, representing the total volume of all particles, shows that the particle volumes declined sharply in both types of batteries after the first cycle. In the alkaline primary battery, the particle volume remained roughly constant in the ensuing cycles, whereas in the RAM it continued to decrease. This can be seen as well in Figure 5.

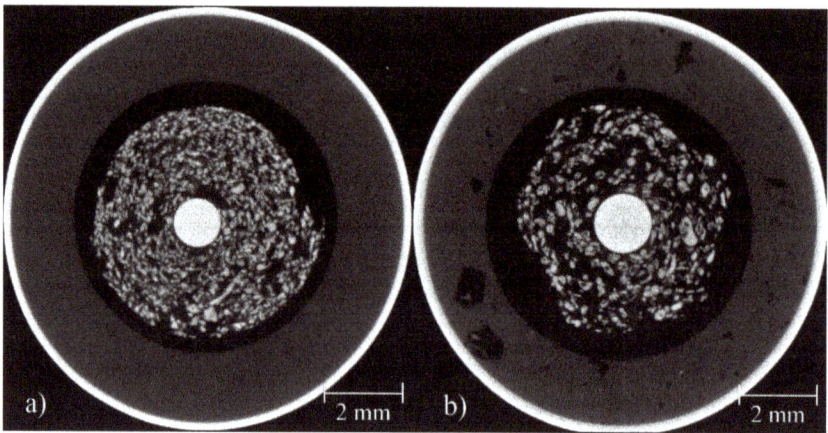

Figure 7. Comparative cut of the primary cell (**a**) and the RAM battery (**b**). Both cells are shown in their pristine state.

During the first discharge, the primary cell had the largest capacity (Figure 2a), which could be explained by the amount of zinc used. After their first cycle, all batteries lost a large part of their capacities. After 10 cycles, the capacities of the RAMs were still about 50% of their corresponding original capacity, while the primary cell reached only about 25% of the initial level. Furthermore, with an increasing number of cycles, the volume of the manganese dioxide cathode also increased, and the anode as well as the separator was increasingly compressed. This did not happen uniformly, but it created splinter-shaped structures that penetrated the separator. In the primary cell, this effect was much more evident. After the 12th cycle, the primary cell was no longer rechargeable and it began to leak.

After the failure of the primary alkaline battery, the tomogram exhibited some irregularities. Figure 8 demonstrates that many manganese dioxide spikes formed that pierced the separator (red circle). In addition, the manganese dioxide in the area around the separator was slightly brighter (blue arrow). This can be explained by a damaged separator so that zinc-enriched electrolyte could freely pass through to the cathode side. Therefore, the battery discharged itself and the manganese dioxide layer continued to expand, the internal pressure increased, and the battery started to leak. Presumably, in addition to the process of discharge, gas was formed, which may also have contributed to the increasing pressure [43].

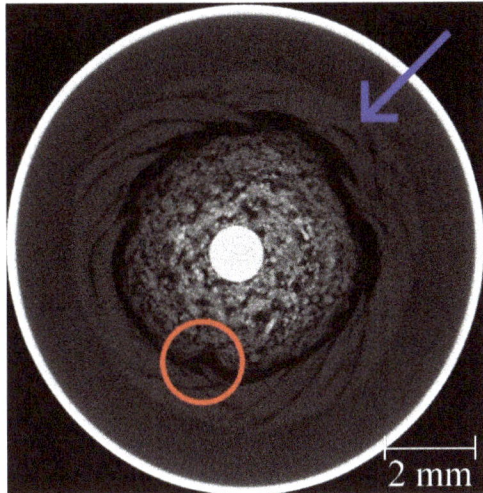

Figure 8. Tomographic cross section of the primary cell after 12 cycles.

4. Conclusions

By X-ray tomography, we have analyzed structural and morphological changes in rechargeable alkaline-manganese batteries, and in non-rechargeable primary cells, during repeated charge and discharge. We applied three different discharging currents (100 mA, 200 mA, and 400 mA) to the RAM cells. The size distributions of the zinc particles were calculated and compared (Figure 5). We found that first the smaller zinc particles disappear and after about 10 cycles also the larger ones dissolve. The degree and pace of dissolution differs between various locations in the cell. The zinc particles dissolve layer by layer and become increasingly smaller.

The structural changes in the primary alkaline cell are different than in the RAM cell. After the first discharge of the primary alkaline cell the overall zinc particle volume remains almost constant during cycling. Furthermore we found that new zinc particles are formed after cycling (see Supplementary Video 1). For the first few cycles, the primary alkaline battery performed well, and provided more capacity than the RAM battery. The total failure of the primary cell came suddenly and unexpectedly without any signs (see Figure 2a), may have been caused by separator penetration from one of the needle-like structures formed at the electrodes.

We think that the analysis and the corresponding results presented here can significantly contribute to the fundamental understanding and development of rechargeable alkaline batteries and of zinc-based batteries in general. In the future, similar studies might be done on other zinc-based systems such as zinc-air batteries, and they may contribute to the development not only of RAM cells with increased durability, but also of rechargeable zinc-air batteries.

Supplementary Materials: The following are available online at http://www.mdpi.com/1996-1944/11/9/1486/s1, Video S1: Tomographic cross sections of the evolution of the RAM cell that was cycled with 200 mA.

Author Contributions: Conceptualization, M.O. and I.M.; Methodology, M.O. and A.H.; Software, M.O. and A.H.; Validation, M.O., A.H., N.K. and I.M.; Formal Analysis, M.O.; Investigation, M.O.; Resources, A.H. and I.M.; Data Curation, M.O.; Writing-Original Draft Preparation, M.O. and I.M.; Writing-Review & Editing, N.K., A.H. and J.B.; Visualization, M.O.; Supervision, A.H. and I.M.; Project Administration, I.M. and J.B.; Funding Acquisition, J.B.

Funding: We acknowledge support by the German Research Foundation (DFG) and the Open Access Publication Funds of Technische Universität Berlin.

Conflicts of Interest: The authors declare no conflicts of interest.

References

1. Linden, D.; Reddy, T.B. *Handbook of Batteries*, 3rd ed.; McGraw-Hill: New York, NY, USA, 2002.
2. Harting, K.; Kunz, U.; Turek, T. Zinc-Air Batteries: Prospects and Challenges for Future Improvement. *Z. Phys. Chem.* **2012**, *226*, 151–166. [CrossRef]
3. Li, Y.G.; Gong, M.; Liang, Y.Y.; Feng, J.; Kim, J.E.; Wang, H.L.; Hong, G.S.; Zhang, B.; Dai, H.J. Advanced zinc-air batteries based on high-performance hybrid electrocatalysts. *Nat. Commun.* **2013**, *4*. [CrossRef] [PubMed]
4. Lee, D.U.; Park, H.W.; Higgins, D.; Nazar, L.; Chen, Z.W. Highly Active Graphene Nanosheets Prepared via Extremely Rapid Heating as Efficient Zinc-Air Battery Electrode Material. *J. Electrochem. Soc.* **2013**, *160*, F910–F915. [CrossRef]
5. Chen, Z.; Choi, J.Y.; Wang, H.J.; Li, H.; Chen, Z.W. Highly durable and active non-precious air cathode catalyst for zinc air battery. *J. Power Sources* **2011**, *196*, 3673–3677. [CrossRef]
6. Kordesch, K.; Daniel-Ivad, J. Advances in battery systems for energy storage. In *Proceedings of the Second International Symposium on New Materials for Fuel Cell and Modern Battery Systems, Montréal, QC, Canada, 6–10 July 1997*; Ecole polytechnique de Montreal: Montreal, QC, Canada, 1997; pp. 2–13.
7. Daniel-Ivad, J.; Kordesch, K. The status of the rechargeable alkaline manganese dioxide zinc battery. In *Proceedings of the Symposium on Aqueous Batteries*; Bennett, P.D., Gross, S., Eds.; The Electrochemical Society, Inc.: Pennington, NJ, USA, 1997; Volume 96, pp. 11–22.
8. Kordesch, K.; Binder, L.; Taucher, W.; Faistauer, C.; Daniel-Ivad, J. The Rechargeable Alkaline MnO_2-Zn System (New Theoretical and Technological Aspects). *J. Power Sources* **1993**, *14*, 193–216.
9. Binder, L.; Odar, W.; Kordesch, K. A Study of Rechargeable Zinc Electrodes for Alkaline Cells Requiring Anodic Limitation. *J. Power Sources* **1981**, *6*, 271–289. [CrossRef]
10. Binder, L.; Kordesch, K.; Urdl, P. Improvements of the rechargeable alkaline MnO_2-Zn cell. *J. Electrochem. Soc.* **1996**, *143*, 13–17. [CrossRef]
11. Stani, A.; Taucher-Mautner, W.; Kordesch, K.; Daniel-Ivad, J. Development of flat plate rechargeable alkaline manganese dioxide-zinc cells. *J. Power Sources* **2006**, *153*, 405–412. [CrossRef]
12. Cheng, F.Y.; Chen, J.; Gou, X.L.; Shen, P.W. High-Power Alkaline Zn–MnO_2 Batteries Using γ-MnO_2 Nanowires/Nanotubes and Electrolytic Zinc Powder. *Adv. Mater.* **2005**, *17*, 2753–2756. [CrossRef]
13. Im, D.; Manthiram, A. Role of bismuth and factors influencing the formation of Mn_3O_4 in rechargeable alkaline batteries based on bismuth-containing manganese oxides. *J. Electrochem. Soc.* **2003**, *150*, A68–A73. [CrossRef]
14. Horn, Q.C.; Shao-Horn, Y. Morphology and spatial distribution of ZnO formed in discharged alkaline Zn/MnO_2 AA cells. *J. Electrochem. Soc.* **2003**, *150*, A652–A658. [CrossRef]
15. Turner, S.; Buseck, P.R. Defects in nsutite ([gamma]-MnO_2) and dry-cell battery efficiency. *Nature* **1983**, *304*, 143–146. [CrossRef]
16. Powers, R.W.; Breiter, M.W. The Anodic Dissolution and Passivation of Zinc in Concentrated Potassium Hydroxide Solutions. *J. Electrochem. Soc.* **1969**, *116*, 719–729. [CrossRef]
17. Zielke, L.; Hutzenlaub, T.; Wheeler, D.R.; Chao, C.-W.; Manke, I.; Hilger, A.; Paust, N.; Zengerle, R.; Thiele, S. Three-Phase Multiscale Modeling of a $LiCoO_2$ Cathode: Combining the Advantages of FIB-SEM Imaging and X-Ray Tomography. *Adv. Energy Mater.* **2015**, *5*, 1401612. [CrossRef]
18. Ebner, M.; Geldmacher, F.; Marone, F.; Stampanoni, M.; Wood, V. X-Ray Tomography of Porous, Transition Metal Oxide Based Lithium Ion Battery Electrodes. *Adv. Energy Mater.* **2013**, *3*, 845–850. [CrossRef]
19. Shearing, P.R.; Howard, L.E.; Jorgensen, P.S.; Brandon, N.P.; Harris, S.J. Characterization of the 3-dimensional microstructure of a graphite negative electrode from a Li-ion battery. *Electrochem. Commun.* **2010**, *12*, 374–377. [CrossRef]
20. Stenzel, O.; Westhoff, D.; Manke, I.; Kasper, M.; Kroese, D.P.; Schmidt, V. Graph-based simulated annealing: A hybrid approach to stochastic modeling of complex microstructures. *Model. Simul. Mater. Sci. Eng.* **2013**, *21*, 055004. [CrossRef]
21. Ebner, M.; Marone, F.; Stampanoni, M.; Wood, V. Visualization and Quantification of Electrochemical and Mechanical Degradation in Li Ion Batteries. *Science* **2013**, *342*, 716–720. [CrossRef] [PubMed]

22. Mitsch, T.; Kraemer, Y.; Feinauer, J.; Gaiselmann, G.; Markoetter, H.; Manke, I.; Hintennach, A.; Schmidt, V. Preparation and Characterization of Li-Ion Graphite Anodes Using Synchrotron Tomography. *Materials* **2014**, *7*, 4455–4472. [CrossRef] [PubMed]
23. Zielke, L.; Hutzenlaub, T.; Wheeler, D.R.; Manke, I.; Arlt, T.; Paust, N.; Zengerle, R.; Thiele, S. A Combination of X-ray Tomography and Carbon Binder Modeling: Reconstructing the Three Phases of $LiCoO_2$ Li-Ion Battery Cathodes. *Adv. Energy Mater.* **2014**, *4*, 1301617. [CrossRef]
24. Zielke, L.; Barchasz, C.; Waluś, S.; Alloin, F.; Leprêtre, J.C.; Spettl, A.; Schmidt, V.; Hilger, A.; Manke, I.; Banhart, J.; et al. Degradation of Li/S Battery Electrodes On 3D Current Collectors Studied Using X-ray Phase Contrast Tomography. *Sci. Rep.* **2015**, *5*, 10921. [CrossRef] [PubMed]
25. Eastwood, D.S.; Yufit, V.; Gelb, J.; Gu, A.; Bradley, R.S.; Harris, S.J.; Brett, D.J.L.; Brandon, N.P.; Lee, P.D.; Withers, P.J.; et al. Lithiation- Induced Dilation Mapping in a Lithium- Ion Battery Electrode by 3D X-Ray Microscopy and Digital Volume Correlation. *Adv. Energy Mater.* **2014**, *4*, 1300506. [CrossRef]
26. Weker, J.N.; Liu, N.; Misra, S.; Andrews, J.C.; Cui, Y.; Toney, M.F. In situ nanotomography and operando transmission X-ray microscopy of micron-sized Ge particles. *Energy Environ. Sci.* **2014**, *7*, 2771–2777. [CrossRef]
27. Steinbock, L.; Dustmann, C.H. Investigation of the inner structures of ZEBRA cells with a microtomograph. *J. Electrochem. Soc.* **2001**, *148*, A132–A136. [CrossRef]
28. Gonzalez, J.; Sun, K.; Huang, M.; Dillon, S.; Chasiotis, I.; Lambros, J. X-ray microtomography characterization of Sn particle evolution during lithiation/delithiation in lithium ion batteries. *J. Power Sources* **2015**, *285*, 205–209. [CrossRef]
29. Gonzalez, J.; Sun, K.; Huang, M.; Lambros, J.; Dillon, S.; Chasiotis, I. Three dimensional studies of particle failure in silicon based composite electrodes for lithium ion batteries. *J. Power Sources* **2014**, *269*, 334–343. [CrossRef]
30. Hoch, C.; Schier, H.; Kallfass, C.; Totzke, C.; Hilger, A.; Manke, I. Electrode deterioration processes in lithium ion capacitors monitored by in situ X-ray radiography on micrometre scale. *Micro-Nano Lett.* **2012**, *7*, 262–264. [CrossRef]
31. Tariq, F.; Yufit, V.; Eastwood, D.S.; Merla, Y.; Biton, M.; Wu, B.; Chen, Z.; Freedman, K.; Offer, G.; Peled, E.; et al. In-Operando X-ray Tomography Study of Lithiation Induced Delamination of Si Based Anodes for Lithium-Ion Batteries. *ECS Electrochem. Lett.* **2014**, *3*, A76–A78. [CrossRef]
32. Sun, F.; Markötter, H.; Dong, K.; Manke, I.; Hilger, A.; Kardjilov, N.; Banhart, J. Investigation of failure mechanisms in silicon based half cells during the first cycle by micro X-ray tomography and radiography. *J. Power Sources* **2016**, *321*, 174–184. [CrossRef]
33. Sun, F.; Markötter, H.; Zhou, D.; Alrwashdeh, S.S.S.; Hilger, A.; Kardjilov, N.; Manke, I.; Banhart, J. In Situ Radiographic Investigation of (De)Lithiation Mechanisms in a Tin-Electrode Lithium-Ion Battery. *ChemSusChem* **2016**, *9*, 946–950. [CrossRef] [PubMed]
34. Manke, I.; Banhart, J.; Haibel, A.; Rack, A.; Zabler, S.; Kardjilov, N.; Hilger, A.; Melzer, A.; Riesemeier, H. In situ investigation of the discharge of alkaline $Zn–MnO_2$ batteries with synchrotron X-ray and neutron tomographies. *Appl. Phys. Lett.* **2007**, *90*, 214102. [CrossRef]
35. Haibel, A.; Manke, I.; Melzer, A.; Banhart, J. In Situ Microtomographic Monitoring of Discharging Processes in Alkaline Cells. *J. Electrochem. Soc.* **2010**, *157*, A387–A391. [CrossRef]
36. Schröder, D.; Arlt, T.; Krewer, U.; Manke, I. Analyzing transport paths in the air electrode of a zinc air battery using X-ray tomography. *Electrochem. Commun.* **2014**, *40*, 88–91. [CrossRef]
37. Arlt, T.; Schroeder, D.; Krewer, U.; Manke, I. In operando monitoring of the state of charge and species distribution in zinc air batteries using X-ray tomography and model-based simulations. *Phys. Chem. Chem. Phys.* **2014**, *16*, 22273–22280. [CrossRef] [PubMed]
38. Riley, G.V.; Hussey, D.S.; Jacobson, D. In Situ Neutron Imaging of Alkaline and Lithium Batteries. *ECS Trans.* **2010**, *25*, 75–83.
39. Kim, F.H.; Penumadu, D.; Gregor, J.; Kardjilov, N.; Manke, I. High-Resolution Neutron and X-Ray Imaging of Granular Materials. *J. Geotech. Geoenviron. Eng.* **2013**, *139*, 715–723. [CrossRef]
40. Schindelin, J.; Arganda-Carreras, I.; Frise, E.; Kaynig, V.; Longair, M.; Pietzsch, T.; Preibisch, S.; Rueden, C.; Saalfeld, S.; Schmid, B.; et al. Fiji: An open-source platform for biological-image analysis. *Nat. Methods* **2012**, *9*, 676–682. [CrossRef] [PubMed]

41. Rueden, C.T.; Schindelin, J.; Hiner, M.C.; DeZonia, B.E.; Walter, A.E.; Arena, E.T.; Eliceiri, K.W. ImageJ2: ImageJ for the next generation of scientific image data. *BMC Bioinform.* **2017**, *18*, 529. [CrossRef] [PubMed]
42. Banhart, J. *Advanced Tomographic Methods in Materials Research and Engineering*; Oxford University Press: Oxford, UK, 2008.
43. Sun, F.; Markoetter, H.; Manke, I.; Hilger, A.; Kardjilov, N.; Banhart, J. Three-Dimensional Visualization of Gas Evolution and Channel Formation inside a Lithium-Ion Battery. *ACS Appl. Mater. Interfaces* **2016**, *8*, 7156–7164. [CrossRef] [PubMed]

© 2018 by the authors. Licensee MDPI, Basel, Switzerland. This article is an open access article distributed under the terms and conditions of the Creative Commons Attribution (CC BY) license (http://creativecommons.org/licenses/by/4.0/).

Review

Correlation of Materials Property and Performance with Internal Structures Evolvement Revealed by Laboratory X-ray Tomography

Lei Zhang * and Shaogang Wang *

Shenyang National Laboratory for Materials Science, Institute of Metal Research, Chinese Academy of Sciences, Shenyang 110016, China
* Correspondence: lzhang@imr.ac.cn (L.Z.); wangshaogang@imr.ac.cn (S.W.);
  Tel.: +86-24-2397-1551 (L.Z.); +86-24-2397-1823 (S.W.)

Received: 17 July 2018; Accepted: 19 September 2018; Published: 21 September 2018

Abstract: Although X-rays generated from a laboratory-based tube cannot be compared with synchrotron radiation in brilliance and monochromaticity, they are still viable and accessible in-house for ex situ or interrupted in situ X-ray tomography. This review mainly demonstrates recent works using laboratory X-ray tomography coupled with the measurements of properties or performance testing under various conditions, such as thermal, stress, or electric fields. Evolvements of correlated internal structures for some typical materials were uncovered. The damage features in a graded metallic 3D mesh and a metallic glass under mechanical loading were revealed and investigated. Micro-voids with thermal treatment and void healing phenomenon with electropulsing were clearly demonstrated and quantitatively analyzed. The substance transfer around an electrode of a Li-S battery and the protective performance of a Fe-based metallic glass coating on stainless steel were monitored through electrochemical processes. It was shown that in situ studies of the laboratory X-ray tomography were suitable for the investigation of structure change under controlled conditions and environments. An extension of the research for in situ laboratory X-ray tomography can be expected with supplementary novel techniques for internal strain, global 3D grain orientation, and a fast tomography strategy.

Keywords: X-ray tomography; in situ; mechanics; corrosion; biomaterial; battery

1. Introduction

It is of common interest to understand the properties or performance of materials with their underlying structures in the corresponding length scales from atomic to macro in materials research. In practice, the structures of materials are examined by sampling before or after the measurements of the properties or performance testing. Samples need to be cut and carefully prepared to fulfill various analytical instruments by which the structural information can be obtained. Then, the researcher explains or deduces the material properties from the data collected statically. Technical obstacles make it difficult to understand the properties of the materials or the performances involved in changes in internal structures during their measurement or testing in real conditions and living states.

Images, patterns, and spectroscopy are commonly employed to show the material's structure through morphology, crystallography, or chemistry. These techniques of microanalysis originate from the reflection, diffraction, and interaction of materials with physical probes, such as light, lasers, X-rays, electrons, neutrons, or ions, etc. By manipulating these probes on a micro and nano scale, microanalysis can provide two-dimensional (2D) data of the material's structure with spatial resolution in terms of the interaction volume. Researchers are likely to use this individual cross-section of the material structure to relate to the property of the bulk material. This works reasonably well for properties related to homogenous or uniformly distributed structure, for example, the elastic modulus of structural materials. However, the structure of the sliced material may miss some critical information on local or heterogeneous structural variation on which the property of the material relies. Crack growth in composites and damage induced by void accumulation in plastic deformation are two obvious cases [1]. Tomography is a state-of-the-art technique to complement the ordinary characterization of the planar structure with the spatial structure in three dimensions (3D). With the increases in computer power and software resources, it is now feasible to reconstruct a 3D digital volume of a sample with an enormous amount of planar structural data. One can use either stacked tomography from sliced structural data produced layer by layer or computed tomography (CT) from projections with the rotation angles downstream to the probe beam transmission of the sample.

The stacked tomography by optical microscopy (OM) and scanning electron microscopy (SEM) need a number of consecutive sections from a sample surface, called serial sections, which can be performed by microtomy, ion milling, or consecutive polishing steps [2]. When accompanied with a focused ion beam (FIB) or a plasma FIB [3], recently-developed SEM tomography can be made more automatic while the data collection is interlaced with removing of material layer-by-layer on the sample surface. Practically, the spatial resolution can be as high as 10 nm with FIB serial sectioning. So far atom probe tomography (APT) is the most advanced technique producing a 3D volume composed of atom species in stacked tomography [4]. In practice, the outmost atomic layer at the tip of a needle shaped sample can be ionized and kicked off layer by layer in a pulsed high-voltage field. A flat panel time-of-flight (TOF) detector continuously collects the atomic-layer signals in sequence, from which the tip of the sample can be reconstructed [5,6]. However, the preparation of the needle sample with tens of nanometers in diameter needs skillful thinning and sharpening techniques. These stacked tomography techniques are inevitably destructive. In computed tomography, transmission electron microscopy (TEM) combined with CT technique, 3D-TEM, can reconstruct in 3D the volume with a resolution down to the nanometer range [7,8]. However, the sample has to be thinned down to tens of nanometers for the penetration of the high-voltage accelerated electrons. X-ray tomography (XRT) is another popular class in computed tomography. The obvious feature of XRT is that it is nondestructive. The XRT technique is, nowadays, commonly employed for clinical diagnosis as X-rays can easily penetrate biological tissues and bones. It can also be used for materials analysis through optional higher energy X-rays suitable for the penetration of specified materials and sample thicknesses.

1.1. XRT and In Situ Experiment

The invention and development of XRT for materials research since the last century has been reviewed in detail [9,10]. XRT has been widely used in the characterization of internal structures for various materials. It covers traditional materials including polymers, ceramics, metals and alloys, and also advanced materials with complex structures, such as composites [11–13], foams and cellular structures [14–16], biomaterials [17–19], etc. The recent development of XRT techniques have been reviewed comprehensively [20] and reach specified characterizations not only for internal damage [1], but also grain orientation [21–24], chemical information [25], and even internal strain [26] with a global 3D view.

In addition to the ability to show the internal structure of a wide variety of materials in static states, the capability of XRT to set test conditions facilitates two paralleled processes of both 3D imaging of internal structures and measurement of properties in dynamic states. Their relationship can be correlated more easily and directly. Pioneering works of the in situ XRT in materials science can be traced back to the 1990s. Guvenilir et al. investigated the crack evolution in an aluminum–lithium alloy employing XRT combined with in situ loading experiments [27]. Since then, the damage inside materials induced by processing, cycling, or deformation have been investigated from 3D viewpoint. Vivid 3D structures of materials and quantitative analysis of the temporal state provide valuable insights as material properties rely on microstructure evolvement. Research works are still very active and versatile in the experimental mechanics field. Deformation and fracture are still common interest for metals and alloys [28–42], foams and porous materials [11,14,43–46], and composites [13,47–49]. XRT has also been considerably extended with other in situ measurements including materials processing [50–55], materials interaction with specified conditions involving extreme temperature [56,57], corrosion [58–60], electrochemical environments, as in batteries [61–63], etc. The finite element (FE) method has also been employed to analyze the dynamic processes with the change of applied fields based on 3D models from XRT [16,64–66]. Detailed reviews have also been published on in situ XRT for materials science [20,67], which clearly shows the fast growth of the technique in the past two decades. Such in situ experiments usually provide comprehensive 3D quantitative data and related properties or the performance of materials.

1.2. Laboratory-Based and Synchrotron Radiation XRT

X-rays from either a laboratory-based (LB) tube or synchrotron radiation (SR) are available to carry out XRT with in situ experiments. Due to the X-ray characteristics from different sources, the instrumentation for a specified experiment has to be considered beforehand.

On one hand, X-ray source and imaging methods affect the quality of XRT for fine structures. LB-XRT usually uses radiography or cone beam imaging using a micro-focus source, which is usually a feature of a divergent and polychromatic X-ray source. The spatial resolution from a point source combined with the geometric magnification has routinely reached the micrometer range. Tuning the focus of the electron beam on a metallic thin film can produce an X-ray source with finer spot size down to hundreds of nanometers range. Such nano CT systems are commercially available, including the GE Nanotom and the Skyscan Ultra-High Resolution Nano-CT, with a resolution of around 200~400 nm [10]. A nano CT system based on the SEM was also developed accordingly [68]. Another type of the LB-XRT approaching 50 nm in spatial resolution was produced by Xradia (now merged into Zeiss). X-ray optics, such as condenser and Fresnel zone plates (FZP), were employed [69]. The other popular X-ray source is from synchrotron radiation due to magnetic field bending charged particles with relativistic velocity. Brilliance of the third generation SR might be ten orders of magnitude greater than that emitted from the LB X-ray tube [70]. The electromagnetic wave from SR has a wide range of spectra, covering hard X-ray to micro waves. X-ray beams used for imaging or computed tomography can be focused or parallel, polychromatic or monochromatic by tuning optics like a mirror and monochromator on the beamline. Characteristics of higher brightness and tunable wavelength bring significant improvement and possibility to X-ray imaging and computed tomography. The resolution can be below 50 nm for full field imaging or tomography by using optics like a Kirkpatrick-Baez (KB) focus mirror and FZP for the X-ray source in SR [20,71,72].

On the other hand, post processing of the collected XRT data is also relevant to the scanning configuration. Before the reconstruction of the collected 3D imaging dataset, wobble and shift during specimen rotation, as well as image registration, needed to be corrected first with post-processing software. Due to the differences between the LB and SR X-ray source, it has to consider a suitable reconstruction process. For LB XRT using a cone beam, a weighted back projection that considers the cone shape of the X-ray beam must be used instead of the simple back projection processing used for a parallel beam [10,73]. For this purpose, the filtered back projection (FBP) algorithm is generally used for reconstruction in the cone beam tomographs. Monochromatic X-ray beams from SR can simplify the tomographic reconstruction algorithm. The absorption of polychromatic X-ray is also an issue of concern. As low energy X-rays are easily absorbed when incoherent X-rays pass through the materials, polychromatic X-rays from the LB source gives rise to the effect of beam hardening [73,74]. This means an uneven transmission from the edge and internal regions of the sample due to the broad X-ray energy range. The penetration of higher-energy X-rays is more at the edge than the interior, so the rim looks brighter than the center of the sample, which is an artifact on the reconstructed 3D images. This artifact is generally corrected in a commercial laboratory tomography system with bundling hardware and software for optimization and correction.

1.3. Timing for XRT

For LB cone beam system, one absorption contrast XRT needs to take a series of projections with a full 360 degrees of sample rotation. Hundreds or thousands of images with a specified voxel size according to the resolution requirements are recorded within a durable time, in which the sample structure is expected to be unchanged. Therefore, 3D imaging data collection by XRT is time consuming. The typical acquisition time is approximately in hours to obtain a 3D volume of $1024 \times 1024 \times 1024$ voxels with a pixel size down to 1 µm for a LB XRT. A brighter X-ray source is more effective in resolving features with very small differences of absorptivity with a better relative signal to noise ratio (SNR). For comparison of cone beam system on the third SR, new optics and detectors used facilitated the spatial resolution reaching below 100 nm and the acquisition time in minutes to obtain a volume of $1024 \times 1024 \times 1024$ voxels. There is no technical problem to combine XRT with an in situ experiment. However, the temporal resolution of XRT must be in the same time scale so that the structure does not change when the specified condition is applied with the measurement of property or performance testing. A higher contrast images are more easily acquired when using a brighter source. It can easily infer that the XRT by the SR source is much faster than that by the LB source in terms of the imaging speed with a proper contrast. Increase exposure time helps to improve the contrast of the imaging.

In addition to XRT with absorption contrast, which related to local density variation of the interior of materials, the chemical components of internal structure can be revealed by XRT with different approaches. Based on X-ray absorption near edge spectroscopy (XANES), a specific element or a chemical state can be distinguished by a full field imaging near its absorption edge. With the complementation of XRT, it is also feasible to show the elemental distribution not only in a 2D imaging but also in a volume of materials [75]. The 3D XANES microscopy was used to study electrochemical reduction and re-oxidation of the NiO electrode at a voxel size down to 30 nm × 30 nm × 30 nm in a FOV of 15 µm × 15 µm × 15 µm. The XANES XRT at one energy out of 13 distinct energy points spent about 1.4 h [76]. Due to the polychromatic X-ray source of the LB tube, alternative approaches for an elemental contrast tomography were developed. Dual-energy CT systems can differentiate component in materials by X-ray absorption contrast at low and high tube energies. However, extra data processing steps and calibration are required. Instead of changing tube energy, Ross-pair filters can also be utilized to produce tunable X-rays with defined energy bandwidths [77]. In terms of the K-edge energy of Rh, a combination of two filter pairs of Nb/Mo and Pd/Ag was selected for the LB XRT to identify the Rh in the Al foam. The exposure time of 15 seconds for one projection is acceptable for the LB-XRT. The other approach by employing a hyperspectral detector has much more potential applications for 3D chemical imaging. Every pixel of the newly developed detector is able to detect individual photons and extract quantitative hard X-ray spectra [78], and obtain higher spectral statistics with optimized retrieval method [79]. Such spectroscopy LB-XRT shows advantages of discrimination of the chemical components with different X-ray energies [80]. It was demonstrated in the characterization of the metallic catalyst in a porous structure and in the identification of the inclusion phases in an ore sample [81]. Another advantage of the spectroscope LB-XRT is that only one scan is needed to acquire chemical information in a 3D volume, and the scanning time is similar to normal LB-XRT.

1.4. Effective Factors of LB-XRT

The brightness of the X-ray sources is not the only factor for in situ XRT. The quality of the imaged projections also relies on the penetration of the X-rays through the material by considering the absorption for a proper sample thickness. In principle, hard X-rays can transmit most materials with the energy spectrum from keV to a hundred keV. However, limitations exist for in situ experiments with the LB or SR-XRT, especially for the samples with high density such as ferrous metals and alloys, copper alloys and refractory alloys, etc. To image millimeter sized samples of such kinds of materials, the X-ray energy needs to be more than a hundred keV. In addition to the attenuation coefficient of materials at different X-ray energy, resolution and the full pixels of the detector also constrain the choice of sample size. 50 µm is estimated to be the largest dimension for a full view of the detector with 1000–2000 pixels at the resolution of 50 nm [82]. When the available sample thickness is reduced down to hundreds or tens of micrometers, however, the in situ experiment is difficult due to sample preparation and manipulation.

To acquire magnified image with a resolution in micrometers, X-ray microscopy with cone beam projection is available for LB-XRT by using a focused point X-ray source and geometric magnification (GM). To further enlarge the image, optical magnification (OM) of visible light converted by a scintillator was feasible in the Xradia XRT systems [83]. In a SEM, cone beam X-ray can also be generated from finely focused electrons on a thin foil target. The X-ray source size can be limited by smaller interaction volume between the electrons and the thinner target, and produce the XRT with a resolution better than 100 nm [84,85]. Higher resolution in tens of nanometers are routinely available by employing X-ray objective lens as FZP to magnify full field imaging for both LB and SR-XRT. Associated with the focus of a line emission X-ray source, hard X-rays with energy at 8 keV can resolve 30 nm lines of a gold spoke pattern by using SR X-ray imaging [86] and 50 nm thick Cu tracks on a Si substrate by using LB X-ray imaging [87]. However, it takes minutes for a single projection and days for a 3D tomography with such LB systems. Table 1 shows the capability of full-field imaging and the characteristics of XRT on some typical SR beamlines and LR facilities. A comprehensive review summarized and compared these arts on X-ray nanotomography [82]. The key limitation of LB-XRT is the brightness of the available X-ray source. Recently, increment of nearly two orders of magnitude in brightness is demonstrated by using a liquid metal target in LB X-ray source. This emerging technique can load more power on the target with better heat dissipation and without melting anymore [88]. It is expected to enhance the capability of LB-XRT for faster tomography with higher resolution.

Table 1. Typical SR beamlines and LB facilities for X-ray nano tomography with full field view.

Source	Facility	Beamline/Model	Energy keV	Flux phs/s	Res. nm	Ref.	XRT
SR	ESRF	ID11	18–140	~10^{14}	100	[89]	Monochromatic/Pink
	APS	34 ID-C	5–15	5×10^9	100	[90]	Tunable energy Minutes Tomo.
	Spring-8	BL29XUL	4.4–37.8	6×10^{13}	50	[91]	Time-lapse Tomo.
	SSRF	B13W1	8–72.5	3×10^{10}	100	[92]	Absorption contrast Phase contrast
LB *	MXIF	Gatan XuM	9.7		200	Web.	Polychromatic/Monochromatic
	IMR	Xradia Versa 500	30–160	5×10^8	700	Spec.	Hours Tomo. Interrupted Tomo.
	IMR	Xradia 810 Ultra	8		50	Spec.	Absorption contrast Phase contrast

* Data from the website (Web.) and the specifications (Spec.) measured with the facility.

One also needs to take the X-ray absorption differences of the material components into consideration. On one hand, large absorption differences may result in the over exposure due to high density phases. Suitable energy selection for the SR-XRT can be utilized to resolve small concentrations of the specified element in 3D volume. Dual or more energies of the SR X-ray can be used to increase sensitivity to tell the features in different phases. With the help of the sharp variation at the elemental absorption edge, the enhanced contrast provides information on the chemical difference. On the other hand, if the X-ray attenuation of different phases is similar, the absorption contrast is difficult to tell any significant differences of the phases in the grey level. The techniques of X-ray phase contrast imaging (PCI) can be applied to emphasize the appearance of the refraction at the phase interfaces. Coherent X-ray beams from SR can resolve the phase information in a straightforward approach with several mature techniques such as interferometry, propagation, edge-illumination, and grating-analyzer, etc. The propagation method of the PCI is much more easily and commonly applied to SR-XRT. Phase ring [87,93,94] and increment of propagation distance [80] are usually employed for phase contrast imaging for LB-XRT with a polychromatic cone-beam based on absorption contrast. The methods for the reconstruction of phase contrast tomography have two classes, phase retrieval and direct methods. The phase retrieval is to derive the refractive part of the X-ray through a sample and produce differential phase contrast imaging [95]. The direct method directly obtain the refractive index from the in-line plane intensity without intermediate step of phase retrieval [96]. The filtered back projection (FBP) algorithm is also commonly used for the reconstruction of such phase contrast images, by which reconstruction is processed in one step. Alternatively, an iterative algorithm can produce good reconstruction by using fewer and noisy images. A fast 3D reconstruction strategy was recently developed to perform reconstruction in 3D rather than slice-by-slice [97].

The unique characteristics of synchrotron radiation X-rays play a very important role in the advancement of in situ XRT for materials research. Environmental fixtures can be settled in the SR imaging beamline, and are employed to simulate a similar condition under service circumstances including mechanical, thermal, electrical, and electrochemical environments. Most recently, SR-XRT is able to monitor dynamic processes at high temperature. Powder sintering or liquid droplet nucleation can be recorded in 3D real time [71]. One XRT with a voxel size about 100 nm^3 could be completed within 20 s. The temporal and spatial resolution of high speed SR-XRT has opened new windows for in situ experiments to investigate the dynamic evolution of the structures in a nanoscale range.

Although there are many advantages of SR-XRT, access to the facilities is limited by the availabilities of scheduled beam time. LB-XRT systems will help to fulfill the growth in usage for specified applications with in situ experiments. LB-XRT is currently undergoing fast development and growth with various applications. Some companies produce commercial LB-XRT systems with continuous improvement for conventional X-ray sources and optics. Routinely, available spatial resolutions have fallen into the sub-micrometer and even nanometer range. For in situ experiments, the add-on accessories can be self-made in terms of the specified object for loading, heating, charging, and so on.

1.5. Studied Cases with In Situ XRT

Most in situ experiments investigate the performance of materials, whether in a modality of a time-lapse or an interrupted in situ, can be carried out with LB-XRT and SR-XRT [98,99]. Heating and cooling apparatus were used for the materials processing studies. The in situ XRT clearly exhibited the evolution of solid/liquid mixtures and dendritic growth, and can produce real digital structures for the simulation of phase-field methods. The dendritic coarsening was mimicked in 3D view, and the speed of the interface movement could be calculated [100]. The deformation response of highly porous materials during loading was studied by in situ XRT [15,16]. The evolution of the cellular microstructure was characterized quantitatively, and the deformation mechanisms deduced by finite element simulation with the meshed model from the XRT. Damage in materials due to loading or fatigue has also been a favorable subject for in situ XRT studies [5]. The mechanisms of damage formation such as cavitation, fracture, micro-cracking, fatigue cracking, and stress corrosion cracking are proposed from a 3D viewpoint. Failure parameters of fracture and damage have been extracted and examined with the quantitative analysis of tomography images. The internal deformation of different types of materials during loading and heating were quantified and analyzed with the implementation of self-designed loading and heating equipment for the XRT [67]. SR-XRT is favorable for these in situ experiments with its advantages in brightness, monochrome, or energy options, and even space for the accessories of the in situ instrumentation. However, in situ LB-XRT experiments need to consider more limitations, e.g., the contrast from the polychromatic source, and the spaces for the sample coupled with the process and the condition control for a static and stable state during XRT acquisition.

The following review will focus on our recent works with LB-XRT exploring the mechanical, physical, chemical properties, and the correlated response of microstructures with the change of environments including mechanical loading, heat treatment, electrical treatment, corrosion, and electrochemical reactions for some typical materials. In addition, the latest advancements of in situ experiments with nano scale LB-XRT from the work of Patterson et al. will also be introduced [31].

2. Experimental Method

All XRT works were carried out on our LB system, an Xradia Versa XRM-500 system (Carl Zeiss X-ray Microscopy Inc., Pleasanton, California, United States). This cone beam system was operated with the voltage in a range of 50–160 keV. The generated point X-ray source utilized Bremsstrahlung continuum spectrum. Transmitted X-ray through the sample with the geometric magnified projections traveled down to detector. The X-ray was then converted to visible light by a scintillator. The image signal magnified by optical lens was recorded by a 2000 × 2000 CCD camera (Andor Technology Ltd., Belfast, Northern Ireland, UK). The absorption contrast was imaging for the internal structure investigation in terms of the density difference from compositions, phases, or defects like voids, cracks, etc. Generally, 1600~2000 projections with a 360° rotation were taken for one 3D volume with a pixel size in the range of 0.5~40 µm. All of the datasets were performed a correction of beam hardening and then reconstructed by the bundled software kit with the FBP algorithm. Reconstructed 3D tomography data were visualized and processed with Avizo software (V7.1, Visualization Sciences Group, Bordeaux, France). through which segmentation and quantitative analysis can provide clear and solid information of the internal structure as the concern of the investigated properties. The thickness of the samples used in the following cases was carefully determined with the consideration of the attenuation of the investigated materials and checked by real XRT prior to subsequent in situ study. The properties and performances varied case by case. The experimental details are briefly described in each case. One can find more information in the referred literature of the specified cases.

3. LB-XRT Examples

In this section, several typical examples were selected and introduced to show what role in situ LB-XRT can play in the understanding of the correlation of materials property and performance with internal structures evolvement. The focus was the new insights obtained from in situ LB-XRT compared to the traditional characterization techniques. These examples covered some developing advanced materials such as additive manufacturing titanium alloys, zirconium-based bulk metallic glasses, third-generation single-crystal nickel-based superalloy, iron-based amorphous coatings, and new degradable biomaterials, and their key properties and performance under service environments including mechanical loading, high temperature, corrosion, electrochemical conditions, etc.

3.1. Mechanical Loading

Metallic cellular structures have potential applications in the human body if porous architecture, improved fatigue properties, and high energy absorption capability can be simultaneously satisfied. However, homogeneous cellular structures always own mutually opposing properties between porous architecture and mechanical strength. The topological design of the porous material may solve this problem. Zhao et al. illustrated that functionally graded Ti-6Al-4V interconnected meshes fabricated through additive manufacturing not only manifested high fatigue strength, but also integrated low density and high energy absorption, which could not be achieved by the ordinary uniform meshes [101]. Considering the coarse surface roughness of the mesh struts and requirements to see failure evolution, two kinds of in situ XRT experiments were used to elucidate the basic principles responsible for the unique mechanical behavior of the graded meshes.

The first in situ experiment was a uniaxial compression test on the specimens with a dimension of 15 × 30 × 30 mm^3 at a displacement rate of 1 × 10^{-3} mm/s [101]. The loading direction was perpendicular to the graded direction. The test was stopped at 1000 N, and the specimen was scanned using XRT. Then, the specimen was continually loaded to 2000 N. Such an interruption scanning was adopted at 1000 N, 2000 N, and 3000 N, respectively, in Figure 1a. The digital volume correlations (DVC) technique running on Davis platform offered by LaVision (LaVision GmbH, Göttingen, Germany) was used to analyze the displacement and strain field of the graded meshes based on the XRT results [101]. The color in Figure 1c–e represents the glyph vector magnitude in ParaView software. The 2D XRT slices in Figure 1f–h were used to track the sites of cracks generated at 1000 N, 2000 N, and 3000 N, respectively. The results indicated that the G1 mesh endured a process of the highest stress (Figure 1b), the highest strain (Figure 1c), cracks formed (Figure 1g), and the modulus decreased. Then, the G2 mesh successively exhibited a similar process followed by the G3 mesh. Under compressive load, such a non-uniform deformation behavior in the entire meshes originated from the graded mechanical properties of the G1, G2, and G3 meshes. The G1 part had the highest modulus and strength, so it could support the highest stress. Once cracks nucleated in the G1 part, the redistribution of the stress occurred. While many studies captured the outer or surface deformation features of graded meshes during compression test by a camera or scanning electron microscopy [102,103], the results here showed the interior deformation characteristics of graded meshes during compression by in situ XRT. This internal information gave new insights into how the substructure of graded meshes deformed and the crack nucleated in microscale while the entire sample was under elastic deformation in macroscale.

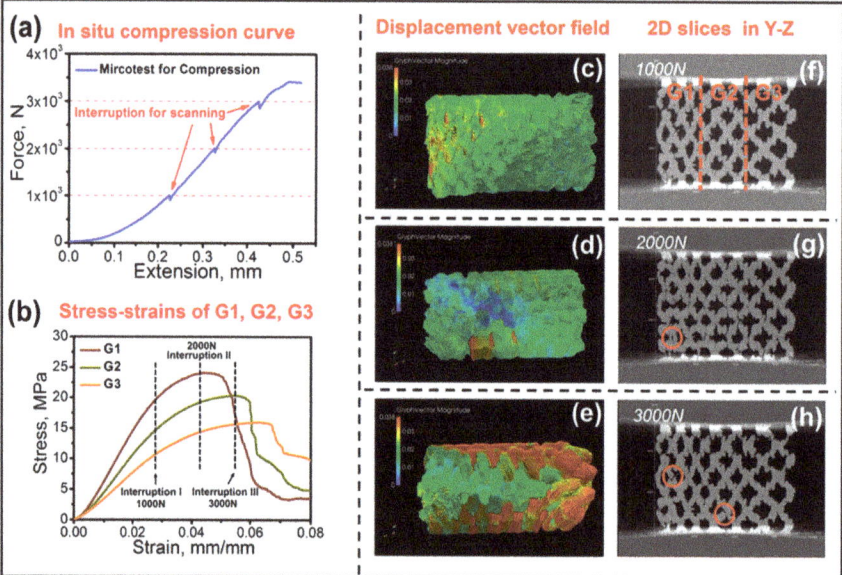

Figure 1. (a) Force-extension result acquired from in situ compression, (b) stress-strain behaviors of individual uniform meshes, (f–h) 2D slices showing the site of cracks marked in red circles, and (c–e) displacement vector fields of the graded structures corresponding to three interruptions during in situ compression. From the displacement maps, a phenomenon of the transition of the maximum deformation sites could be observed [101].

The second in situ experiment was a high cycle compressive fatigue test on the specimens with a dimension of 2 × 6 × 10 mm at a stress ratio, R, of 0.1 and a frequency of 10 Hz [101]. The loading direction was also perpendicular to the graded direction. The strain-cycle curves (ε-N) of the graded cellular structures in Figure 2 could be divided into three stages according to the slope of the curves in Figure 2a. In situ XRT was employed to find the sites of the cracks during cyclic compressive fatigue deformation in Figure 2b. In stage I, the cracks first formed in the G1 part. In stage II, the old cracks in the G1 part propagated at a slow speed while new cracks were detected in both the G1 and G2 parts. In stage III, additional new cracks were detected in the G3 mesh when the fatigue cycles achieved the sudden increase strain zone. The behavior of nucleation and propagation showed that the cracks were prone to initiate in the G1, G2, and G3 parts than near the transition zone, even though the distribution of the stress at the transition zone was not continuous. The progressive initiation of the cracks in the order of G1, G2, and G3 made a big difference from the crack behavior of the uniform meshes where the cracks propagated quickly until the whole mesh sample fractured once the cracks initiated. It was implied that the graded meshes with continuous stress redistribution could retard the abrupt collapse during fatigue, which might also be the origin of the unique cyclic ratcheting rate for the graded meshes in Figure 2. It should be noted that the theoretical analyses also showed that the fatigue behavior of the graded meshes was mainly determined by the stress distribution in the constituent meshes during cyclic compressive fatigue deformation [101]. Such a stress distribution phenomenon for graded meshes was the origin of the combination of low density, high fatigue strength, and high energy absorption, which could not be simultaneously achieved according to the reported metallic cellular structures with uniform density [104–106].

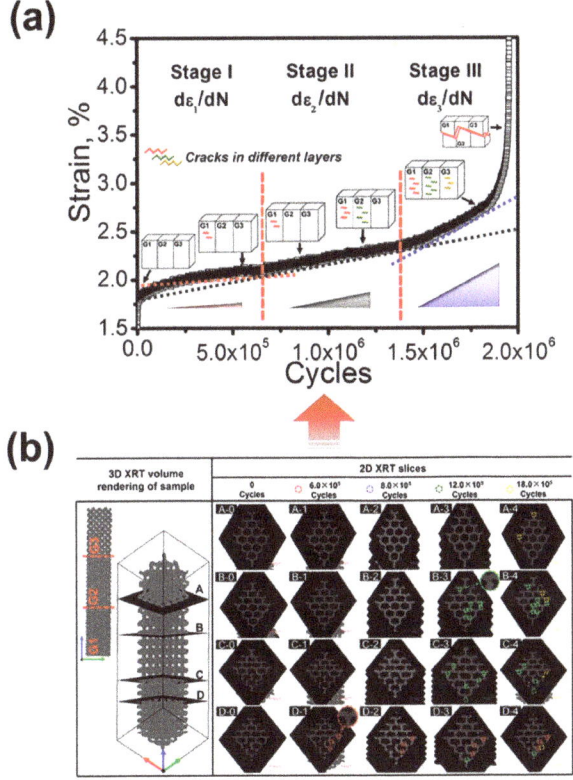

Figure 2. (a) strain-cycle curves of the graded mesh, (b) XRT images of the graded meshes which are corresponding to the arrow points shown in (a). 2D slices A, B, C, and D are located at the G3, G2, and G1 part is shown specifically sitting in the volume rendering of the sample. The numbers 0, 1, 2, 3, and 4 are in accordance with the 5 arrow point, respectively. The colored dash cycles in (b) indicate the sites of the cracks observed at each arrow point [101].

The above two experiments displayed the internal deformation and fracture behavior of graded meshes under different loading mode. One was loaded at a constant displacement rate; the other was loaded and unloaded under compression cyclic mode with a stress smaller than 3.6 MPa. These in situ XRT results would enrich and deepen the understanding of the relationship between topological designs of graded structure and different properties.

For metallic glass (MG), how the shear band (SB) cracks if there is no negative pressure is still an open question. To answer this question, three uniaxial compression tests at a strain rate of 10^{-4} s^{-1} using the same sample with dimensions of 2 × 2 × 4 mm for the ductile $Zr_{65}Fe_5Al_{10}Cu_{20}$ (at %) MG was conducted [107]. After the first compression, the specimen was taken away to be scanned using XRT. Then, the specimen performed the second compression, followed by the second XRT scanning and the third compression and XRT scanning. XRT was used to reveal the evolution behavior of the interior SB cracks during in situ compression in Figure 3. To our knowledge, this XRT work was the pioneering study of 3D imaging on internal shear-banding cracks and their evolution. Several key findings were obtained from the longitudinal 2D slices and 3D extracted cracks. A phenomenon of discontinuous nucleation, linkage, and propagation of the crack, as well as shrinkage or closure was seen along the major SB. The long-narrow-thin 3D crack with a thickness of 16–27 μm can be regarded as the affected zone of shear-banding, which was much larger than the initial SB (~10 nm). Some unique features such as non-coplanar behavior and the largest crack with a curved plane were also captured. These in situ XRT results accompanied by SEM results verified that SB cracking may be traced back to one or a combination of the three sources including the excess free volume [108–110], shearing of non-planar SB [111,112], and SB interaction. The reported experiments in the literatures usually obtained limited SB cracking results from either fracture surface morphology or polished sections [111,113,114]. This in situ XRT results contributed new findings on how SB evolves into crack under compression loading mode.

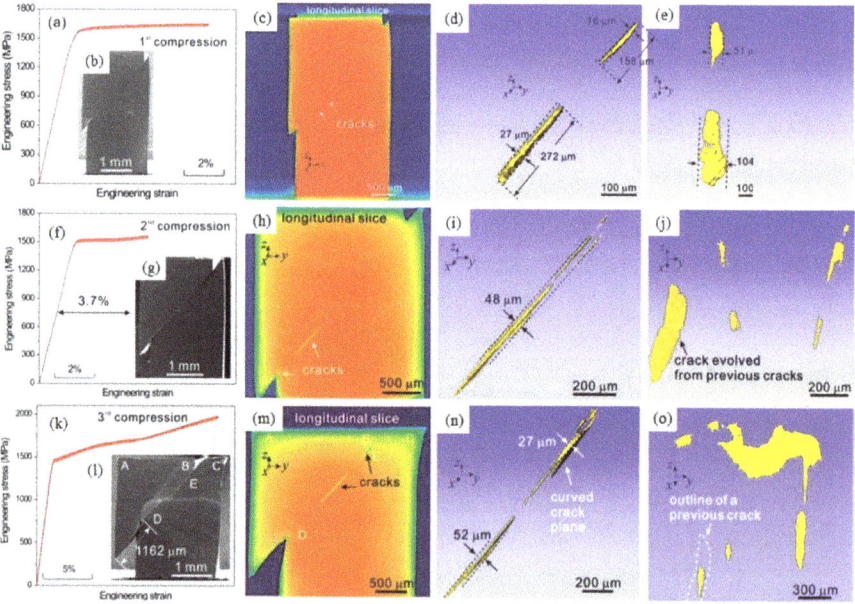

Figure 3. Shear band cracking behavior in an in situ compression test. The compression, SEM, and XRT results in the rows after the first stop (**a–e**), after the second stop (**f–j**), and after the third stop (**k–o**), respectively [107].

Generally, most materials will have a more brittle behavior at low temperatures when compared with a high temperature. Surprisingly, when a relatively brittle $Ti_{32.8}Zr_{30.2}Ni_{5.3}Cu_9Be_{22.7}$ (at %) MG was tested under compression at a strain rate of 10^{-4} s^{-1} using the sample with dimensions of 2 × 2 × 4 mm, both the plasticity and the yield strength were improved while the apparent softening rate decreased with the decrease of the temperature from room temperature to 173 K as shown in Figure 4 [115]. The XRT was used to extract the 3D internal cracks to understand the shear band cracking behavior of the MG. Two samples deformed at 198 K and 173 K unloaded at different periods of compressive deformation showed interesting XRT images in Figure 4. Scattered and small cracks were found in the shear band plane for the sample with a 68 µm shear offset tested at 198 K, while long and large cracks were observed for the sample with a 701 µm shear offset tested at 173 K. This illustrated that the testing temperature accompanied by the amount of shear plastic deformation influenced the crack evolution. In addition, a split crack was detected in the sample at 173 K in Figure 4c2–d2, which may have originated from the local bending moment. The 2D slices and 3D XRT images also showed many uncracked areas which carried the load during the test. These uncracked areas would decrease, while total stress reduced when the plastic deformation increased accordingly. As a consequence, the apparent softening behavior occurred. Within several samples instead of one of the same samples, these XRT results clearly demonstrated that the apparent softening was more likely to come from SB cracking rather than the commonly accepted view of SB dilation.

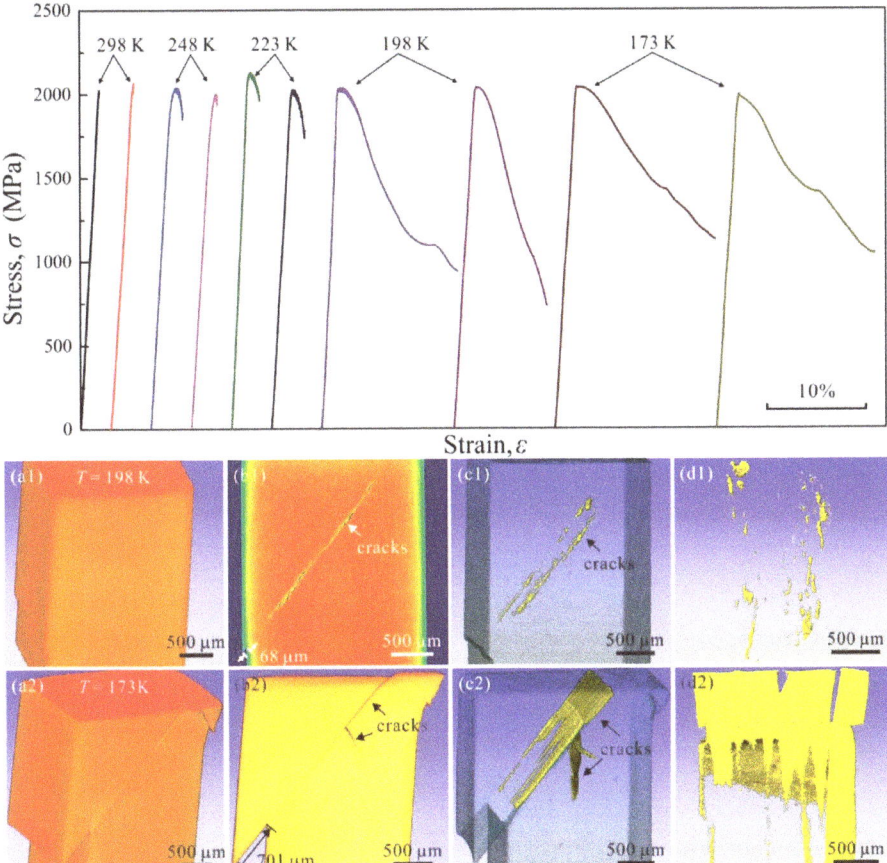

Figure 4. Typical engineering stress–strain curves under compression at temperatures from 298 K to 173 K showing the influence of testing temperature on the stress-strain behavior of Ti32.8Zr30.2Ni5.3Cu9Be22.7 metallic glass and the XRT results corresponding to the samples at 198 K (**a1–d1**) and 173 K (**a2–d2**). (**a1–a2**) the global 3D volume renderings of the samples. (**b1–b2**) 2D slice or cross-section showing the internal cracks well located in the shear band plane. (**c1–c2**) and (**d1–d2**) 3D extracted internal cracks observed from different perspectives. The XRT volume renderings together with the 3D internal cracks show the temperature effect on the deformation of the MG sample [115].

3.2. Heat Treatment

The high-temperature capabilities make single-crystal (SX) nickel-based superalloys one of the material choices in turbine blades. Key properties such as the fatigue and creep properties of SX superalloys are critically affected by micro-pores. Micro-pores are associated with solidification and heat treatment. In the final stage of solidification, the liquid metal contracts, and gas solubility sharply decreases. The pores formed in this process are called S-pores. The solution heat treatment time and temperature can also generate micro-pores named H-pores. Although many studies have reported the growth of micro-pores at high temperatures [116,117], the S-pores and H-pores in these studies were treated as the same micro-pores. How the S-pore and H-pore evolve and their respective fundamental mechanisms are still unsolved issues [118].

In situ XRT was performed on the same sample with dimensions of 10 mm in length and 1 mm × 1 mm in cross-section to characterize the S-pores and H-pores and quantitatively study the evolution behavior of each pore during solution heat treatment at 1603 K for 1, 4, 7, 12, and 20 h, respectively. The samples were first scanned using XRT, then sealed in vacuum tubes and exposed at 1603 K for 1 h for the solution heat treatment. After that, the samples were taken out to be re-scanned using XRT. This process was repeated several times with the only difference of the solution heat treatment was time, by substituting 4, 7, 12, and 20 h for one hour [118]. The XRT volume renderings of the micro-pores in the same sample showed the characteristics during the solution heat treatment in Figure 5. It is quite interesting to observe that new H-pores (indicated by the arrows) formed and grew during the solution heat treatment. For the S-pores, the volume fraction gradually decreased at 1603 K at 1 and 4 h, then increased at 7, 12, and 20 h. For the H-pores, the volume fraction increased during the entire solution process. Considering that the cross-diffusion of elements was imbalanced, a vacancy would form and diffuse during heat treatment. Such vacancy behavior was believed to be related to the evolution of both S-pores and H-pores. The micro-pores may have a maximum volume fraction during the solution heat treatment according to the experimental results [116] and theoretical study [119]. Such a maximum was not observed in the present solution heat treatment at 1603 K for less than 20 h. If the solution heat treatment time was further increased, the maximum would eventually occur.

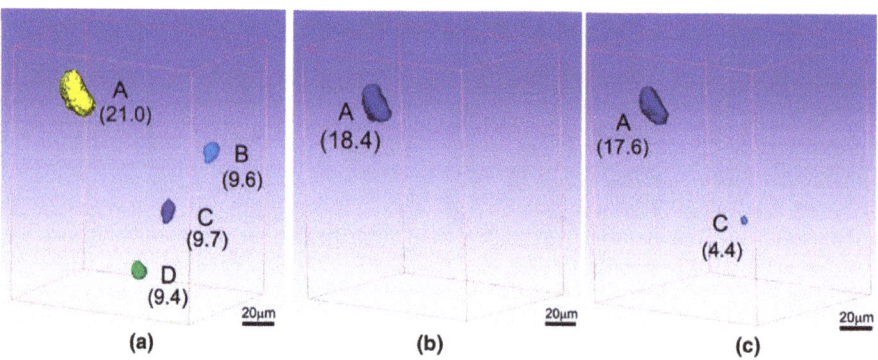

Figure 5. *Cont.*

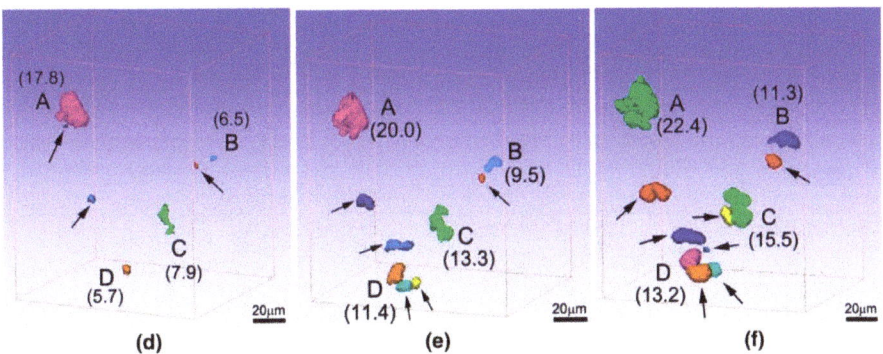

Figure 5. XRT volume renderings of micro-pores in a same sample with a heat treatment at 1603 K for (**a**) 0 h, (**b**) 1 h, (**c**) 4 h, (**d**) 7 h, (**e**) 12 h, and (**f**) 20 h. The equivalent diameter (μm) of the S-pores is also given [118].

3.3. Electropulsing Treatment

Twinning-induced plasticity (TWIP) steel, as a future candidate for the aerospace industry, greatly combines high tensile strength and uniform elongation by means of the high strain-hardening capacity. If TWIP steels are subjected to plastic deformation, the initiation of micro-voids could undermine the strain-hardening capability and result in premature failure. Improving the mechanical properties is expected by eliminating the micro voids using electropulsing treatment (EPT) [120]. Figure 6a–c shows an interesting phenomenon where EPT could recover the strength and plasticity, but annealing treatment had no such ability. The evidence on the sample surface as shown in Figure 6d–g proved that the voids could indeed be healing as well as that the inclusions could be eliminated by EPT. This led to the increase of the strain-hardening capability and resulted in the recovery of strength and plasticity of the TWIP steels samples when compared with those annealed counterparts. One may doubt that the void healing effect may be a pseudomorph on the sample surface due to the quasi in situ processing.

To clarify the internal structure change, in situ XRT was used to verify the healing effect of EPT. A specimen with dimensions of 1.8 × 1.8 × 50 mm was scanned using XRT before EPT. Then, the specimen was electropulse treated using self-made equipment with a capacitor bank discharge circuit with the discharge voltage of 250 V and current pulse duration of 400 ns at room temperature. After EPT, the specimen was re-scanned using XRT to detect the evolution of the macro voids and cracks.

Figure 6h,i show the XRT images of a long 3D crack composed of several short cracks in a strained TWIP steel specimen before and after EPT, respectively. The overall length of the crack decreased from about 400 μm before EPT to 300 μm after EPT. In addition, a macro-void at the bottom in Figure 6h,i became much smaller after EPT when compared with that before EPT. Therefore, the XRT results proved that macro voids in the interior of the material could indeed be healed by EPT in the TWIP steel. The healing effect of EPT may originate from two aspects: one is the elevated temperature near the crack, and the other is the thermal compressive stress around the crack [121]. The material near the crack may be molten or expand, while that far away from the crack remains solid or did not expand so much [122]. Thus, the melted material may endure compression and be gradually compressed into the crack [123]. Thus, the cracks could be healed after cooling. This healing process may also be used to explain the healing of the macro voids by EPT. This damage-healing method could be expanded to heal the damage in those engineering alloys under cyclic deformation and prolong their service life.

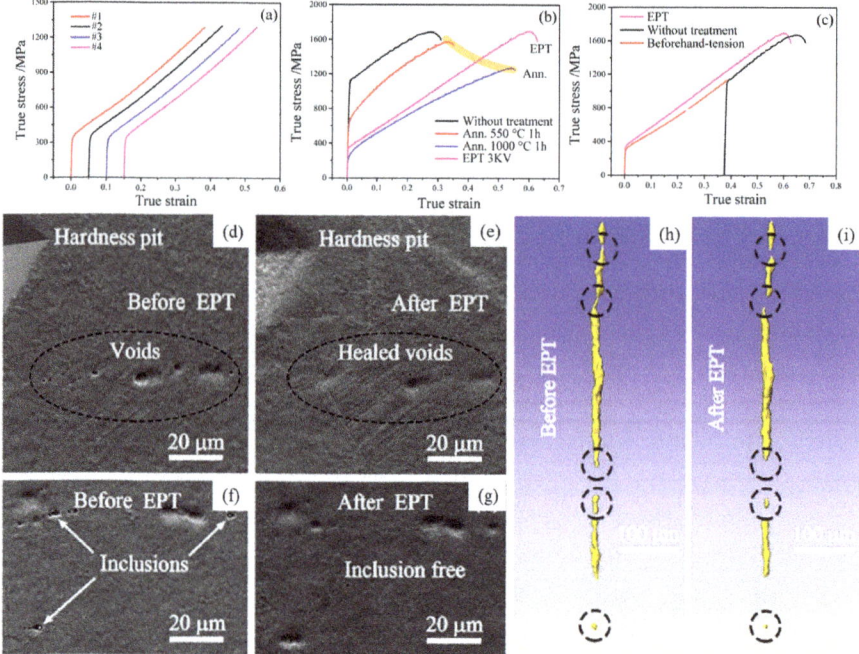

Figure 6. (a–c) tensile results, (d–g) SEM images, and (h–i) XRT volume renderings of a crack for Fe–22Mn–0.9C TWIP steel specimens. These results show the healing effect with EPT, which is superior to the annealing technique [120].

3.4. Protective Coating from Corrosion

Thermal sprayed coatings have been widely used in industry to protect the surfaces of metals and alloys against corrosion and wear. The corrosion resistance of the coatings is related to the coating porosity, which can be categorized as through-porosity and non-through porosity on the basis of its behavior in the corrosion of the coated materials. According to the potentiodynamic polarization curves of the $Fe_{49.7}Cr_{18}Mn_{1.9}Mo_{7.4}W_{1.6}B_{15.2}C_{3.8}Si_{2.4}$ amorphous coating with different thicknesses, the anodic current density increased when the coating thickness decreased from 700 μm to 60 μm when the potential was higher than 0.3 V_{SCE} [124]. This phenomenon was inferred to be related with the through-porosity, which could lead to direct paths between the substrate and corrosive environment. The electrolyte easily penetrated the substrate from through-porosity, and caused the substrate to dissolve once the potential was above the pitting potential of the substrate. However, direct evidence was needed to confirm the substrate corrosion beneath the coating.

In situ XRT technique was used to observe the substrate corrosion evolution beneath the coating before and after electrochemical measurements. The initial specimen with a coating thickness of 150 μm and coating surface of 1 mm × 1 mm was scanned using XRT, and then taken out to be dynamically polarized in the anodic direction at a rate of 0.333 mV/s and interrupted at 0.7 V_{SCE} for the following XRT measurement. To clearly demonstrate substrate corrosion, a further potentiostatic polarization test was performed on the specimen at 0.7 V_{SCE} for 10,000 s, and then scanned by XRT. All XRT scanning was configured with the same parameters.

The 3D XRT volume renderings are given in Figure 7a–g for the same coated sample acquired before and after the potentiodynamic polarization test and subsequent potentiostatic polarization. A preferential site in the substrate for corrosion at the interface between the coating and substrate was found and tracked. A large corrosion pit was verified at the last stage. Focusing the view on the coating surface, we could see that the corrosion products accumulated and piled, which may be attributed to the continuous flow of corrosion products from the through-porosity (Figure 7h). This evidence helps confirm that the abrupt variation of the anodic current of a thinner coating at the potentials above 0.3 V_{SCE} originated from the through-porosity. Compared with non-through porosity, through-porosity could cause increased detriment to the coated material [125,126]. Until this work, direct evidence of substrate corrosion beneath the coating was captured. If the presence and the role of through-porosity is known, the critical coating thickness can be estimated, which is beneficial for the development of advanced metallic coatings for corrosion resistance.

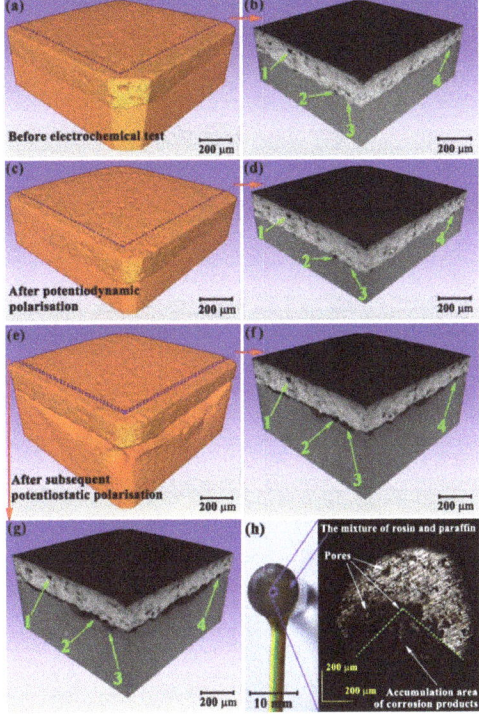

Figure 7. (a–g) 3D XRT images showing the substrate corrosion evolution beneath the coating during three stages, and (h) coating surface morphology showing the accumulation of corrosion products [124].

Magnesium-based biodegradable implants can be used in the fields of orthopedics because of their safe degradation behavior and no need to remove after bone healing. The application limitation mainly focuses on the initial rapid degradation, which will lead to hydrogen bubbling and loss of mechanical support, and then, decrease the bone growth. Surface modification may solve this potential issue for future applications. Han et al. [89] fabricated an Mg–1.5 wt % Sr alloy with Sr–Ca–P containing a micro-arc oxidation (MAO) coating to evaluate the role of the coating on the degradation behavior of the implants.

Since in vitro degradation measurements including mass loss, hydrogen evolution, and electrochemical behavior cannot draw complete maps for the in vivo status, 3D XRT with a pixel size of 25.94 μm was applied to visualize the in vivo degradation amount of the distal femora of rabbit at the implantation periods of eight weeks post-surgery. The dimensions of the rods for the MAO-coated Mg–Sr alloy and the control Mg–Sr alloy without coating were Φ 1.5 mm × 20 mm.

Figure 8 displays the 2D slices and volume renderings of the remaining Mg–Sr alloy with and without Sr–Ca–P coating implants as well as the surrounding bone tissue in the rabbit distal femur.

For the Mg–Sr alloy, pitting corrosion occurred around the entire rod after eight weeks, and only the central area remained in the original state. In comparison, uniform corrosion was observed on the Mg–Sr alloy with MAO coating except for some pitting areas. The lower part of the rod still maintained its integrity while the upper part in the proximal site showed severe degradation. Similar phenomena were also reported by Gu et al. [127]. After the calculations according to the uncorroded parts using the quantitative XRT results, the Mg–Sr alloy with MAO coating showed a degradation speed of 0.75 mm/year while the speed of the Mg–Sr alloy without coating was about 1.3 mm/year. These results manifested that the MAO coating can play a role as a corrosion barrier [128] and help to keep the integrity of the Mg–Sr alloy in vivo [129].

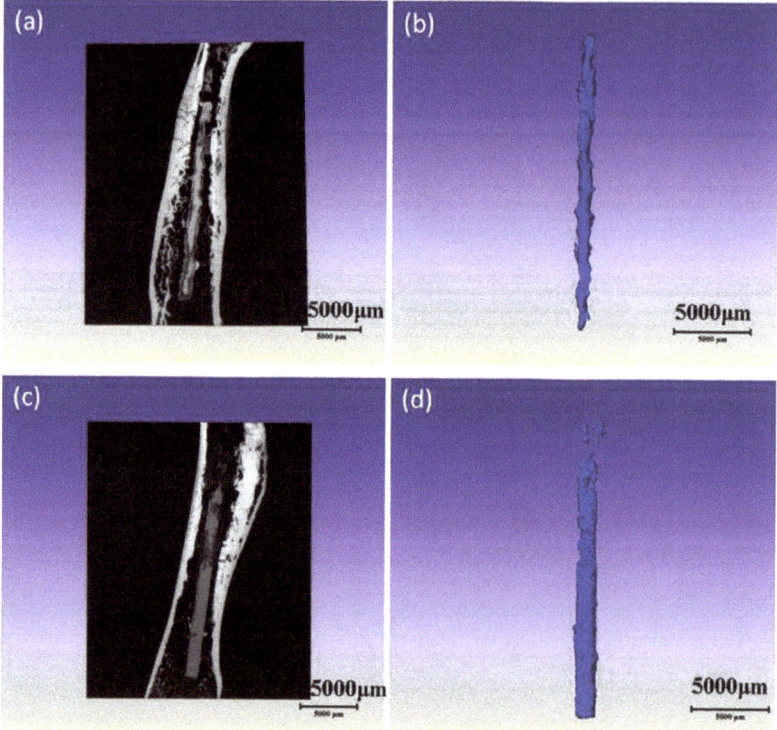

Figure 8. 2D slices and 3D volume renderings of (**a**,**b**) Mg–Sr alloy and (**c**,**d**) Mg–Sr alloy with Sr–CaP coating visualized in vivo using the XRT technique after implantation of 8 weeks in rabbit femur [129].

3.5. Electrochemical Reaction

The Li-S battery is regarded as an excellent potential candidate for high capacity energy storage devices [130,131]. The limitations of advanced high energy density Li-S batteries in real use mainly focus on two factors: low sulfur loading and low sulfur content when the electrode is only considered. This problem impedes the commercial application of Li-S batteries. In the common sense of the community of Li-S batteries, it is hard to simultaneously increase the electrochemical performance and sulfur loading. Therefore, it is still an unresolved key issue on how to increase the sulfur loading and sulfur content without the sacrifice of the electrochemical performance. Design cathodes with a 3D structure may be a solution to this problem [132–134].

3D graphene foam electrodes were subtly designed to achieve high sulfur loading for high energy density Li-S batteries. The highest sulfur loading was about 10.1 mg cm^{-2}. During the XRT tests, two samples were used. One sample of 3D graphene foam electrodes before cycling was cut into pieces and scanned using XRT. The other sample was assembled into 2025-type stainless steel coin cells. After 1000 cycles, the graphene foam electrodes were taken out to perform XRT scanning at different length scales.

Before cycling, we could visualize the 3D structure of graphene foam using the XRT technique with a pixel size of 0.68 µm in Figure 9a,b. It was quite easy to distinguish the sulfur particles from the carbon black and graphene. The sulfur particles were located at the network of the 3D graphene foam. The areal capacity of the electrode with the highest sulfur loading could achieve 13.4 mA h cm^{-2}. Such an excellent value of the electrochemical performance was superior to the published results for the Li-S electrodes. After 1000 cycles, there was still a large amount of sulfur in the active material layer in the electrode. Both the shape and size of the sulfur greatly changed in Figure 9c,d. The pixel size of 0.68 µm was not enough to provide clear details of the sulfur, so an Xradia 800 ultra was used for ultra-high spatial resolution. Figure 9e,f present the XRT results with a pixel size of 64 nm. It can be clearly seen that the size of the large sulfur with a needle-like shape was in a range of 150 nm to 20 µm while the size of small sulfur is less than 150 nm. All the sulfur distributed on the outer surface of the slurry. This work displayed the sulfur morphology and distribution change before cycling and after 1000 cycles. Such a 3D XRT characterization on the electrodes of Li-S batteries after 1000 cycles has never been done before. Researchers have been more prone to capture the changes during different stages of one cycle using the 2D synchrotron XRT method [135] or have investigated the sulfur degradation behavior during the initial 10 cycles using X-ray phase contrast tomography [136]. Nevertheless, the above results proved that the design of a 3D graphene foam based flexible electrode can satisfy the demand for the Li-S battery with high energy density, high power density, and long cyclic life to some extent.

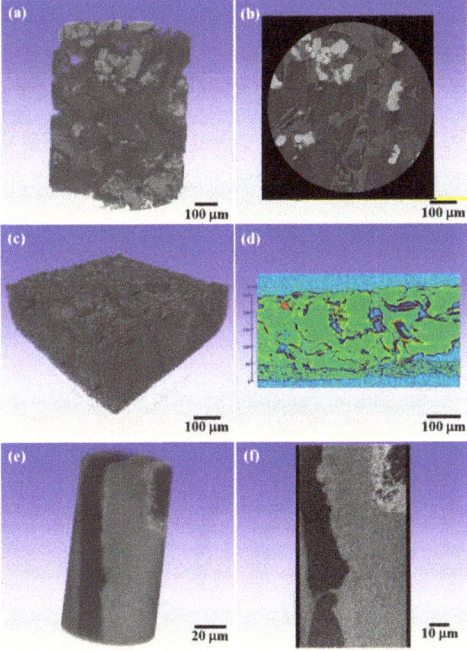

Figure 9. 3D XRT volume renderings and 2D slices of the 3D grapheme foam electrodes with 10.1 mg cm^{-2} sulfur loading: (**a,b**) original state; (**c–f**) after 1000 charge/discharge cycles; (**a–d**) the pixel size is about 0.7 µm; and (**e,f**) the pixel size is about 64 nm.

3.6. Nano XRT In Situ Experiment

Most works of in situ LB-XRT under mechanical loading, heating, or harsh environment have a limited resolution at the micrometer scale. Recently, the spatial resolution of the in situ LB-XRT has been greatly improved. It was claimed that the initiation and propagation of the crack could be accurately probed at the scale of 200 nm [31]. This progress was achieved by using an in situ mechanical stage where a load arm was driven by a piezo motor. This stage had been integrated into the nano X-ray tomography systems. Three operating modes including nanoindentation, compression, and tension were supported on this assembly. Data could be recorded in a quasistatic, interrupted in situ manner while one 3D tomographic data acquisition needed several hours.

The crack growth in the dentin of elephant tusk under nanoindentation was investigated. It provided insights into anisotropic fracture behavior and crack-shielding mechanisms. In this in situ experiment, a cone indenter with a tip radius of 1 µm and tip angle of 90° was used to initiate and propagate cracks on a cone shape dentin sample with the height of 1 mm and top surface diameter of 50 µm. The load was applied incrementally and held in a manner of displacement control at different stages (Figure 10a–c). Zernike phase contrast mode was used with 721 projections. Each projection was exposed for 60 s with 64 nm pixel size. Figure 10a–c present continuous crack propagation as the indentation was increased. Further information was obtained from the 3D segmented image (Figure 10e). The extension of the crack was mainly in a radial direction from the indenter, which seemed to be a crack path with low energy. A deflecting crack path could also be deduced from the cracks growing in other directions. Therefore, cracking and bridging seem to be the main fracture toughening mechanisms of tubules for the tusk. This example demonstrated that enhanced insight on material performance was no longer limited in synchrotron-based X-ray tomography, and could be explored through lab-based X-ray nano-tomography.

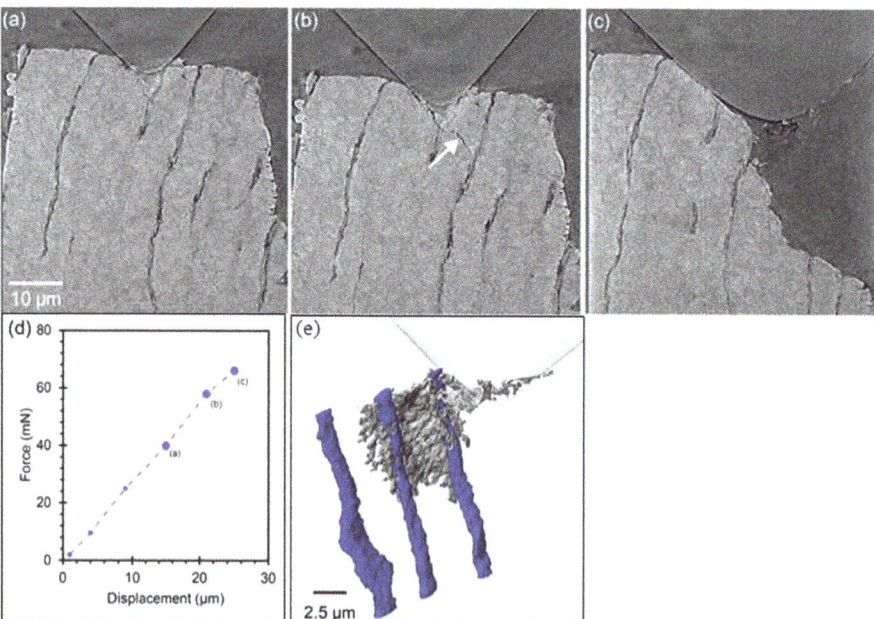

Figure 10. (**a–c**) 2D slices showing crack evolution corresponding to the different stages of indenter loading. (**d**) The force-displacement curve during the nano-indentation test. (**e**) 3D volume rendering of an extracted crack at the loading point (**b**) on the force-displacement curve showing the relationship of the crack with selected dentin tubules [31].

4. Conclusions and Perspective

With high brilliance and monochromatic X-rays, the advantages of the SR-XRT boosted the in situ experiments for the investigation of the temporal evolution of the 3D structure coupled with the measurements of the properties. The researchers now are devoted to develop 3D tomography system with the temporal resolution stepped into seconds, the voxel size down to a few tens of nanometers, or even the combination. Meanwhile, the brightness of the X-ray tube in the laboratory constrains the advancement towards real-time in situ XRT. However, LB-XRT is also viable and easily accessible for specified research interests covering a controllable condition or environment where the structure studied can be stable and imaged by 3D tomography statically. As in the cases we showed in this review, LB-XRT is feasible for some common properties or performances of materials to correlate with their 3D structures. It can focus on, but is not limited to, research fields such as crack or damage growth during loading or electric stressing, structure and void evolution with thermal processing, composition change or transport in controllable chemical or electrochemical environment, etc.

Recently, the diffraction contrast tomography (DCT) solution was developed for laboratory-based Xradia Versa 520 system. Crystallographic information of the individual grains from polycrystalline samples can be identified and demonstrated in 3D space with colorful marks for different orientations [137–139]. The DCT developed in the past decade have been based on synchrotron monochromatic X-rays. It is now available in laboratories as a routine tool for non-destructive 3D grain mapping. As a complementary to the XRT in-house from absorption or phase contrast tomography, it is possible to carry out characterization and also study of the mechanisms of damage, deformation, and growth related to grains, rather than the sole phase or morphology structures.

Expanded application fields for LB-XRT need to be explored by a coordination of novel technologies. One of them nowadays attracts fast growing interest. The 3D digital volume from a real structure can be meshed and simulated by the Finite Elements tool kit to see the local strain distribution with the corresponding structure variation in the applied field. Another technique is the so-called digital volume correlation (DVC), which can be performed to measure the local displacement inside the bulk and reveal the strain field in a 3D global manner [35,140–143]. The combination of these novel techniques will play a critical role in investigating materials with heterogeneous structures such as various composites, architecture materials made by additive manufacturing, biomaterials with hierarchical structures, etc. To understand the mechanism of the initiation and extension of cracks or damage in such materials and improve the resistivity with structure tuning might be fascinating research for the LB-XCT to be used.

According to the present modality of LB-XRT, the limited brightness of the X-ray source is an obstacle that is difficult to cross for the real time 3D characterization of structure for dynamic processes in a few minutes or seconds. Instead of the hardware improvement, the optimization of data redundancy in the tomography by software development can be another route for fast tomography. This new strategy takes only effective information or compresses information with little loss. In practical operation, a new approach named the projection-based digital volume correlation (P-DVC) was proposed to make accurate reconstructions with fewer projections [144,145]. A fast tomography of a tensile test on a cast iron sample with radiography and LB-XRT realized 127 loading steps in a total time of 10–15 min, a time saving of at least two orders of magnitude. This emerging method shows a very bright future for fast tomography in-house with time resolution in seconds.

Acknowledgments: The authors acknowledge all of the professional cooperators and continuous support from the Shenyang National Laboratory for Materials Science in the past six years.

Conflicts of Interest: The authors declare that they have no conflicts of interest.

References

1. Withers, P.J.; Preuss, M. Fatigue and damage in structural materials studied by X-ray tomography. *Annu. Rev. Mater. Res.* **2012**, *42*, 81–103. [CrossRef]
2. Zankel, A.; Wagner, J.; Poelt, P. Serial sectioning methods for 3D investigations in materials science. *Micron* **2014**, *62*, 66–78. [CrossRef] [PubMed]
3. Burnett, T.L.; Kelley, R.; Winiarski, B.; Contreras, L.; Daly, M.; Gholinia, A.; Burke, M.G.; Withers, P.J. Large volume serial section tomography by Xe plasma FIB dual beam microscopy. *Ultramicroscopy* **2016**, *161*, 119–129. [CrossRef] [PubMed]
4. Gault, B. Atom probe microscopy. In *Springer Series in Materials Science*; Springer: New York, NY, USA, 2012; p. 396.
5. Cerezo, A.; Clifton, P.H.; Lozano-Perez, S.; Panayi, P.; Sha, G.; Smith, G.D.W. Overview: Recent progress in three-dimensional atom probe instruments and applications. *Microsc. Microanal.* **2007**, *13*, 408–417. [CrossRef] [PubMed]
6. Blavette, D.; Cadel, E.; Pareige, C.; Deconihout, B.; Caron, P. Phase transformation and segregation to lattice defects in Ni-base superalloys. *Microsc. Microanal.* **2007**, *13*, 464–483. [CrossRef] [PubMed]
7. Saghi, Z.; Midgley, P.A. Electron tomography in the (S)TEM: From nanoscale morphological analysis to 3D atomic imaging. *Annu. Rev. Mater. Res.* **2012**, *42*, 59–79. [CrossRef]
8. Midgley, P.A.; Weyland, M. 3D electron microscopy in the physical sciences: The development of Z-contrast and EFTEM tomography. *Ultramicroscopy* **2003**, *96*, 413–431. [CrossRef]
9. Stock, S.R. X-ray microtomography of materials. *Int. Mater. Rev.* **1999**, *44*, 141–164. [CrossRef]
10. Stock, S.R. Recent advances in X-ray microtomography applied to materials. *Int. Mater. Rev.* **2008**, *53*, 129–181. [CrossRef]
11. Atturan, U.A.; Nandam, S.H.; Murty, B.S.; Sankaran, S. Deformation behaviour of in-situ TiB2 reinforced A357 aluminium alloy composite foams under compressive and impact loading. *Mater. Sci. Eng. A* **2017**, *684*, 178–185. [CrossRef]
12. Ferre, A.; Dancette, S.; Maire, E. Damage characterisation in aluminium matrix composites reinforced with amorphous metal inclusions. *Mater. Sci. Technol. Lond.* **2015**, *31*, 579–586. [CrossRef]
13. Nikishkov, Y.; Seon, G.; Makeev, A.; Shonkwiler, B. In-situ measurements of fracture toughness properties in composite laminates. *Mater. Des.* **2016**, *94*, 303–313. [CrossRef]
14. Dillard, T.; Guyen, F.; Maire, E.; Salvo, L.; Forest, S.; Bienvenu, Y.; Bartout, J.D.; Croset, M.; Dendievel, R.; Cloetens, P. 3D quantitative image analysis of open-cell nickel foams under tension and compression loading using X-ray microtomography. *Philos. Mag.* **2005**, *85*, 2147–2175. [CrossRef]
15. Maire, E. X-ray tomography applied to the characterization of highly porous materials. *Annu. Rev. Mater. Res.* **2012**, *42*, 163–178. [CrossRef]
16. Petit, C.; Meille, S.; Maire, E. Cellular solids studied by X-ray tomography and finite element modeling—A review. *J. Mater. Res.* **2013**, *28*, 2191–2201. [CrossRef]
17. Hornberger, B.; Bale, H.; Merkle, A.; Feser, M.; Harris, W.; Etchin, S.; Leibowitz, M.; Qiu, W.; Tkachuk, A.; Gu, A.; et al. X-ray microscopy for in situ characterization of 3D nanostructural evolution in the laboratory. In *X-ray Nanoimaging: Instruments and Methods II*; SPIE: Bellingham WA, USA, 2015; Volume 9592.
18. Fratini, M.; Campi, G.; Bukreeva, I.; Pelliccia, D.; Burghammer, M.; Tromba, G.; Cancedda, R.; Mastrogiacomo, M.; Cedola, A. X-ray micro-beam techniques and phase contrast tomography applied to biomaterials. *Nucl. Instrum. Methods B* **2015**, *364*, 93–97. [CrossRef]
19. Bradley, R.S.; Withers, P.J. Correlative multiscale tomography of biological materials. *Mrs Bull.* **2016**, *41*, 549–554. [CrossRef]
20. Maire, E.; Withers, P.J. Quantitative X-ray tomography. *Int. Mater. Rev.* **2014**, *59*, 1–43. [CrossRef]
21. Ludwig, W.; King, A.; Herbig, M.; Reischig, P.; Marrow, J.; Babout, L.; Lauridsen, E.M.; Proudhon, H.; Buffiere, J.Y. Characterization of polycrystalline materials using synchrotron X-ray imaging and diffraction techniques. *JOM* **2010**, *62*, 22–28. [CrossRef]
22. Ludwig, W.; Schmidt, S.; Lauridsen, E.M.; Poulsen, H.F. X-ray diffraction contrast tomography: A novel technique for three-dimensional grain mapping of polycrystals. I. Direct beam case. *J. Appl. Crystallogr.* **2008**, *41*, 302–309. [CrossRef]

23. Poulsen, H.F. An introduction to three-dimensional X-ray diffraction microscopy. *J. Appl. Crystallogr.* **2012**, *45*, 1084–1097. [CrossRef]
24. Poulsen, H.F.; Jensen, D.J.; Vaughan, G.B.M. Three-dimensional X-ray diffraction microscopy using high-energy X-rays. *MRS Bull.* **2004**, *29*, 166–169. [CrossRef]
25. Requena, G.; Degischer, H.P. Three-dimensional architecture of engineering multiphase metals. *Annu. Rev. Mater. Res.* **2012**, *42*, 145–161. [CrossRef]
26. Toda, H.; Maire, E.; Aoki, Y.; Kobayashi, M. Three-dimensional strain mapping using in situ X-ray synchrotron microtomography. *J. Strain. Anal. Eng.* **2011**, *46*, 549–561. [CrossRef]
27. Guvenilir, A.; Breunig, T.M.; Kinney, J.H.; Stock, S.R. Direct observation of crack opening as a function of applied load in the interior of a notched tensile sample of Al-Li 2090. *Acta Mater.* **1997**, *45*, 1977–1987. [CrossRef]
28. Babout, L.; Maire, E.; Fougeres, R. Damage initiation in model metallic materials: X-ray tomography and modelling. *Acta Mater.* **2004**, *52*, 2475–2487. [CrossRef]
29. Hannard, F.; Pardoen, T.; Maire, E.; Le Bourlot, C.; Mokso, R.; Simar, A. Characterization and micromechanical modelling of microstructural heterogeneity effects on ductile fracture of 6xxx aluminium alloys. *Acta Mater.* **2016**, *103*, 558–572. [CrossRef]
30. Maire, E.; Zhou, S.X.; Adrien, J.; Dimichiel, M. Damage quantification in aluminium alloys using in situ tensile tests in X-ray tomography. *Eng. Fract. Mech.* **2011**, *78*, 2679–2690. [CrossRef]
31. Patterson, B.M.; Cordes, N.L.; Henderson, K.; Mertens, J.C.E.; Clarke, A.J.; Hornberger, B.; Merkle, A.; Etchin, S.; Tkachuk, A.; Leibowitz, M.; et al. In situ laboratory-based transmission X-ray microscopy and tomography of material deformation at the nanoscale. *Exp. Mech.* **2016**, *56*, 1585–1597. [CrossRef]
32. Pottmeyer, F.; Bittner, J.; Pinter, P.; Weidenmann, K.A. In-situ CT damage analysis of metal inserts embedded in carbon fiber-reinforced plastics. *Exp. Mech.* **2017**, *57*, 1411–1422. [CrossRef]
33. Revil-Baudard, B.; Cazacu, O.; Flater, P.; Chandola, N.; Alves, J.L. Unusual plastic deformation and damage features in titanium: Experimental tests and constitutive modeling. *J. Mech. Phys. Solids* **2016**, *88*, 100–122. [CrossRef]
34. Marrow, J.; Reinhard, C.; Vertyagina, Y.; Saucedo-Mora, L.; Collins, D.; Mostafavi, M. 3D studies of damage by combined X-ray tomography and digital volume correlation. *Proc. Mater. Sci.* **2014**, *3*, 1554–1559. [CrossRef]
35. Mostafavi, M.; Collins, D.M.; Cai, B.; Bradley, R.; Atwood, R.C.; Reinhard, C.; Jiang, X.; Galano, M.; Lee, P.D.; Marrow, T.J. Yield behavior beneath hardness indentations in ductile metals, measured by three-dimensional computed X-ray tomography and digital volume correlation. *Acta Mater.* **2015**, *82*, 468–482. [CrossRef]
36. Mostafavi, M.; Vertyagina, Y.; Reinhard, C.; Bradley, R.; Jiang, X.; Galano, M.; Marrow, J. 3D studies of indentation by combined X-ray tomography and digital volume correlation. *Key Eng. Mater.* **2014**, *3*, 592–593. [CrossRef]
37. Girault, B.; Vidal, V.; Thilly, L.; Renault, P.O.; Goudeau, P.; Le Bourhis, E.; Villain-Valat, P.; Geandier, G.; Tranchant, J.; Landesman, J.P.; et al. Small scale mechanical properties of polycrystalline materials: In situ diffraction studies. *Int. J. Nanotechnol.* **2008**, *5*, 609–630. [CrossRef]
38. Salvo, L.; Belestin, P.; Maire, E.; Jacquesson, M.; Vecchionacci, C.; Boller, E.; Bornert, M.; Doumalin, P. Structure and mechanical properties of AFS sandwiches studied by in-situ compression tests in X-ray microtomography. *Adv. Eng. Mater.* **2004**, *6*, 411–415. [CrossRef]
39. Lin, J.L.; Mo, K.; Yun, D.; Miao, Y.B.; Liu, X.; Zhao, H.J.; Hoelzer, D.T.; Park, J.S.; Almer, J.; Zhang, G.M.; et al. In situ synchrotron tensile investigations on 14YWT, MA957, and 9-Cr ODS alloys. *J. Nucl. Mater.* **2016**, *471*, 289–298. [CrossRef]
40. Phillion, A.B.; Cockcroft, S.L.; Lee, P.D. A new methodology for measurement of semi-solid constitutive behavior and its application to examination of as-cast porosity and hot tearing in aluminum alloys. *Mater. Sci. Eng. A* **2008**, *491*, 237–247. [CrossRef]
41. Gan, Y.Y.; Mo, K.; Yun, D.; Hoelzer, D.T.; Miao, Y.B.; Liu, X.; Lan, K.C.; Park, J.S.; Almer, J.; Chen, T.Y.; et al. Temperature effect of elastic anisotropy and internal strain development in advanced nanostructured alloys: An in-situ synchrotron X-ray investigation. *Mater. Sci. Eng. A* **2017**, *692*, 53–61. [CrossRef]
42. Wang, S.G.; Sun, M.Y.; Song, Z.Q.; Xu, J. Cast defects induced sample-size dependency on compressive strength and fracture toughness of Mg-Cu-Ag-Gd bulk metallic glass. *Intermetallics* **2012**, *29*, 123–132. [CrossRef]

43. Marcadon, V.; Davoine, C.; Passilly, B.; Boivin, D.; Popoff, F.; Rafray, A.; Kruch, S. Mechanical behaviour of hollow-tube stackings: Experimental characterization and modelling of the role of their constitutive material behaviour. *Acta Mater.* **2012**, *60*, 5626–5644. [CrossRef]
44. Siegkas, P.; Tagarielli, V.; Petrinic, N.; Lefebvre, L.P. Rate dependency of the response of Ti foams in compression. *Material* **2012**, *2*, 661–666.
45. Patterson, B.M.; Henderson, K.; Gilbertson, R.D.; Tornga, S.; Cordes, N.L.; Chavez, M.E.; Smith, Z. Morphological and performance measures of polyurethane foams using X-ray CT and mechanical testing. *Microsc. Microanal.* **2014**, *20*, 1284–1293. [CrossRef] [PubMed]
46. Ou, X.X.; Zhang, X.; Lowe, T.; Blanc, R.; Rad, M.N.; Wang, Y.; Batail, N.; Pham, C.; Shokri, N.; Garforth, A.A.; et al. X-ray micro computed tomography characterization of cellular SiC foams for their applications in chemical engineering. *Mater. Charact.* **2017**, *123*, 20–28. [CrossRef]
47. Hu, X.F.; Wang, L.B.; Xu, F.; Xiao, T.Q.; Zhang, Z. In situ observations of fractures in short carbon fiber/epoxy composites. *Carbon* **2014**, *67*, 368–376. [CrossRef]
48. Bayraktar, E.; Bessri, K.; Bathias, C. Deformation behaviour of elastomeric matrix composites under static loading conditions. *Eng. Fract. Mech.* **2008**, *75*, 2695–2706. [CrossRef]
49. Patterson, B.M.; Cordes, N.L.; Henderson, K.; Williams, J.J.; Stannard, T.; Singh, S.S.; Ovejero, A.R.; Xiao, X.H.; Robinson, M.; Chawla, N. In situ X-ray synchrotron tomographic imaging during the compression of hyper-elastic polymeric materials. *J. Mater. Sci.* **2016**, *51*, 171–187. [CrossRef]
50. Tourret, D.; Mertens, J.C.E.; Lieberman, E.; Imhoff, S.D.; Gibbs, J.W.; Henderson, K.; Fezzaa, K.; Deriy, A.L.; Sun, T.; Lebensohn, R.A.; et al. From solidification processing to microstructure to mechanical properties: A multi-scale X-ray study of an Al-Cu alloy sample. *Metall. Mater. Trans. A* **2017**, *48A*, 5529–5546. [CrossRef]
51. Daudin, R.; Terzi, S.; Lhuissier, P.; Salvo, L.; Boller, E. Remelting and solidification of a 6082 Al alloy containing submicron yttria particles: 4D experimental study by in situ X-ray microtomography. *Mater. Des.* **2015**, *87*, 313–317. [CrossRef]
52. Daudin, R.; Terzi, S.; Lhuissier, P.; Tamayo, J.; Scheel, M.; Babu, N.H.; Eskin, D.G.; Salvo, L. Particle-induced morphological modification of Al alloy equiaxed dendrites revealed by sub-second in situ microtomography. *Acta Mater.* **2017**, *125*, 303–310. [CrossRef]
53. Schleef, S.; Lowe, H.; Schneebeli, M. Hot-pressure sintering of low-density snow analyzed by X-ray microtomography and in situ microcompression. *Acta Mater.* **2014**, *71*, 185–194. [CrossRef]
54. Wu, Z.; Sun, L.C.; Wang, J.Y. Effects of sintering method and sintering temperature on the microstructure and properties of porous Y_2SiO_5. *J. Mater. Sci. Technol.* **2015**, *31*, 1237–1243. [CrossRef]
55. Cai, B.; Wang, J.; Kao, A.; Pericleous, K.; Phillion, A.B.; Atwood, R.C.; Lee, P.D. 4D synchrotron X-ray tomographic quantification of the transition from cellular to dendrite growth during directional solidification. *Acta Mater.* **2016**, *117*, 160–169. [CrossRef]
56. Alvarez-Murga, M.; Perrillat, J.P.; Le Godec, Y.; Bergame, F.; Philippe, J.; King, A.; Guignot, N.; Mezouar, M.; Hodeau, J.L. Development of synchrotron X-ray micro-tomography under extreme conditions of pressure and temperature. *J. Synchrotron Radiat.* **2017**, *24*, 240–247. [CrossRef] [PubMed]
57. Liu, D.; Gludovatz, B.; Barnard, H.S.; Kuball, M.; Ritchie, R.O. Damage tolerance of nuclear graphite at elevated temperatures. *Nat. Commun.* **2017**, *8*, 15942. [CrossRef] [PubMed]
58. Galli, S.; Hammel, J.U.; Herzen, J.; Damm, T.; Jimbo, R.; Beckmann, F.; Wennerberg, A.; Willumeit-Romer, R. Evaluation of the degradation behavior of resorbable metal implants for in vivo osteosynthesis by synchrotron radiation based X-ray tomography and histology. *Proc. SPIE* **2016**, *9967*, 996704.
59. Singh, S.S.; Stannard, T.J.; Xiao, X.H.; Chawla, N. In situ X-ray microtomography of stress corrosion cracking and corrosion fatigue in aluminum alloys. *JOM* **2017**, *69*, 1404–1414. [CrossRef]
60. Wang, S.G.; Wang, S.C.; Zhang, L. Application of high resolution transmission X-ray tomography in material science. *Acta Metall. Sin.* **2013**, *49*, 897–910. [CrossRef]
61. Brushett, F.R.; Trahey, L.; Xiao, X.H.; Vaughey, J.T. Full-field synchrotron tomography of nongraphitic foam and laminate anodes for lithium-ion batteries. *ACS Appl. Mater. Int.* **2014**, *6*, 4524–4534. [CrossRef] [PubMed]
62. Drozhzhin, O.A.; Sumanov, V.D.; Karakulina, O.M.; Abakumov, A.M.; Hadermann, J.; Baranov, A.N.; Stevenson, K.J.; Antipov, E.V. Switching between solid solution and two-phase regimes in the $Li_{1-x}Fe_{1-y}Mn_yPO_4$ cathode materials during lithium (de)insertion: Combined pitt, in situ XRPD and electron diffraction tomography study. *Electrochim. Acta* **2016**, *191*, 149–157. [CrossRef]

63. Harris, S.J.; Lu, P. Effects of inhomogeneities-nanoscale to mesoscale-on the durability of Li-ion batteries. *J. Phys. Chem. C* **2013**, *117*, 6481–6492. [CrossRef]
64. Michailidis, N. Strain rate dependent compression response of Ni-foam investigated by experimental and FEM simulation methods. *Mater. Sci. Eng. A* **2011**, *528*, 4204–4208. [CrossRef]
65. Michailidis, N.; Stergioudi, F.; Omar, H.; Papadopoulos, D.; Tsipas, D.N. Experimental and FEM analysis of the material response of porous metals imposed to mechanical loading. *Colloid Surf. A* **2011**, *382*, 124–131. [CrossRef]
66. Petit, C.; Maire, E.; Meille, S.; Adrien, J. Two-scale study of the fracture of an aluminum foam by X-ray tomography and finite element modeling. *Mater. Des.* **2017**, *120*, 117–127. [CrossRef]
67. Buffiere, J.Y.; Maire, E.; Adrien, J.; Masse, J.P.; Boller, E. In situ experiments with X-ray tomography: An attractive tool for experimental mechanics. *Exp. Mech.* **2010**, *50*, 289–305. [CrossRef]
68. Mayo, S.C.; Miller, P.R.; Wilkins, S.W.; Davis, T.J.; Gao, D.; Gureyev, T.E.; Paganin, D.; Parry, D.J.; Pogany, A.; Stevenson, A.W. Quantitative X-ray projection microscopy: Phase-contrast and multi-spectral imaging. *J. Microsc.* **2002**, *207*, 79–96. [CrossRef] [PubMed]
69. Wang, S.; Lau, S.H.; Tkachuk, A.; Druewer, F.; Chang, H.; Feser, M.; Yun, W.B. Non-destructive 3D imaging of nano-structures with multi-scale X-ray microscopy. *NSTI Nanotech.* **2008**, *1*, 822–825.
70. Als-Nielsen, J.; McMorrow, D. *Elements of Modern X-ray Physics*; Wiley: New York, NY, USA; Chichester, UK, 2001.
71. Villanova, J.; Daudin, R.; Lhuissier, P.; Jauffres, D.; Lou, S.Y.; Martin, C.L.; Laboure, S.; Tucoulou, R.; Martinez-Criado, G.; Salvo, L. Fast in situ 3D nanoimaging: A new tool for dynamic characterization in materials science. *Mater. Today* **2017**, *20*, 354–359. [CrossRef]
72. Lee, S.; Kwon, I.H.; Kim, J.Y.; Yang, S.S.; Kang, S.; Lim, J. Early commissioning results for spectroscopic X-ray nano-imaging beamline BL 7C SXNI at PLS-II. *J. Synchrotron Radiat.* **2017**, *24*, 1276–1282. [CrossRef] [PubMed]
73. Baruchel, J. *X-ray Tomography in Material Science*; Hermes Science: Paris, France, 2000; p. 204.
74. Vedula, V.S.V.M.; Munshi, P. An improved algorithm for beam-hardening corrections in experimental X-ray tomography. *Ndt&E Int.* **2008**, *41*, 25–31.
75. Rau, C.; Somogyi, A.; Simionovici, A. Microimaging and tomography with chemical speciation. *Nucl. Instrum. Meth. B* **2003**, *200*, 444–450. [CrossRef]
76. Meirer, F.; Cabana, J.; Liu, Y.J.; Mehta, A.; Andrews, J.C.; Pianetta, P. Three-dimensional imaging of chemical phase transformations at the nanoscale with full-field transmission X-ray microscopy. *J. Synchrotron Radiat.* **2011**, *18*, 773–781. [CrossRef] [PubMed]
77. Arhatari, B.D.; Gureyev, T.E.; Abbey, B. Elemental contrast X-ray tomography using ross filter pairs with a polychromatic laboratory source. *Sci. Rep. UK* **2017**, *7*, 218. [CrossRef] [PubMed]
78. Fullagar, W.; Uhlig, J.; Walczak, M.; Canton, S.; Sundstrom, V. The use and characterization of a backilluminated charge-coupled device in investigations of pulsed X-ray and radiation sources. *Rev. Sci. Instrum.* **2008**, *79*, 103302. [CrossRef] [PubMed]
79. Fullagar, W.K.; Paziresh, M.; Latham, S.J.; Myers, G.R.; Kingston, A.M. The index of dispersion as a metric of quanta—Unravelling the fano factor. *Acta Crystallogr. B* **2017**, *73*, 675–695. [CrossRef] [PubMed]
80. Jacques, S.D.M.; Egan, C.K.; Wilson, M.D.; Veale, M.C.; Seller, P.; Cernik, R.J. A laboratory system for element specific hyperspectral X-ray imaging. *Analyst* **2013**, *138*, 755–759. [CrossRef] [PubMed]
81. Egan, C.K.; Jacques, S.D.M.; Wilson, M.D.; Veale, M.C.; Seller, P.; Beale, A.M.; Pattrick, R.A.D.; Withers, P.J.; Cernik, R.J. 3D chemical imaging in the laboratory by hyperspectral X-ray computed tomography. *Sci. Rep. UK* **2015**, *5*, 15979. [CrossRef] [PubMed]
82. Withers, P.J. X-ray nanotomography. *Mater. Today* **2007**, *10*, 26–34. [CrossRef]
83. Wang, Y.X.; Duewer, F.; Kamath, S.; Scott, D.; Yun, W.B. A novel X-ray microtomography system with high resolution and throughput. *NSTI Nanotech.* **2004**, *3*, 503–507.
84. Bleuet, P.; Laloum, D.; Audoit, G.; Torrecillas, R.; Gaillard, F.X. SEM-based system for 100nm X-ray tomography for the analysis of porous silicon. In *Developments in X-ray Tomography IX*; SPIE: Bellingham, WA, USA, 2014; Volume 9212, p. 92120Z.
85. Perini, L.A.G.; Bleuet, P.; Filevich, J.; Parker, W.; Buijsse, B.; Kwakman, L.F.T. Developments on a SEM-based X-ray tomography system: Stabilization scheme and performance evaluation. *Rev. Sci. Instrum.* **2017**, *88*, 063706.

86. Yin, G.C.; Song, Y.F.; Tang, M.T.; Chen, F.R.; Liang, K.S.; Duewer, F.W.; Feser, M.; Yun, W.B.; Shieh, H.P.D. 30 nm resolution X-ray imaging at 8 keV using third order diffraction of a zone plate lens objective in a transmission microscope. *Appl. Phys. Lett.* **2006**, *89*, 221122. [CrossRef]
87. Tkachuk, A.; Duewer, F.; Cui, H.T.; Feser, M.; Wang, S.; Yun, W.B. X-ray computed tomography in zernike phase contrast mode at 8 keV with 50-nm resolution using Cu rotating anode X-ray source. *Z. Kristallogr.* **2007**, *222*, 650–655. [CrossRef]
88. Harding, G.; David, B.; Schlomka, J.P. Liquid metal anode X-ray sources and their potential applications. *Nucl. Instrum. Methods B* **2004**, *213*, 189–196. [CrossRef]
89. Snigireva, I.; Vaughan, G.B.M.; Snigirev, A. High-energy nanoscale-resolution X-ray microscopy based on refractive optics on a long beamline. *AIP Conf. Proc.* **2011**, *1365*, 188–191.
90. Rau, C.; Crecea, V.; Richter, C.P.; Peterson, K.M.; Jemian, P.R.; Neuhausler, U.; Schneider, G.; Yu, X.; Braun, P.V.; Chiang, T.C.; et al. A hard X-ray KB-FZP microscope for tomography with sub-100-nm resolution. In Proceedings of the Developments in X-Ray Tomography V, San Diego, CA, USA, 15–17 August 2006; Volume 6318, p. 63181G.
91. Matsuyama, S.; Yasuda, S.; Yamada, J.; Okada, H.; Kohmura, Y.; Yabashi, M.; Ishikawa, T.; Yamauchi, K. 50-nm-resolution full-field X-ray microscope without chromatic aberration using total-reflection imaging mirrors. *Sci. Rep. UK* **2017**, *7*, 46358. [CrossRef] [PubMed]
92. Feng, B.G.; Deng, B.; Ren, Y.Q.; Wang, Y.D.; Du, G.H.; Tan, H.; Xue, Y.L.; Xiao, T.Q. Full-field X ray nano-imaging system designed and constructed at ssrf. *Chin. Opt. Lett.* **2016**, *14*, 093401. [CrossRef]
93. Bradley, R.S.; Liu, Y.; Burnett, T.L.; Zhou, X.; Lyon, S.B.; Withers, P.J.; Gholinia, A.; Hashimoto, T.; Graham, D.; Gibbon, S.R.; et al. Time-lapse lab-based X-ray nano-CT study of corrosion damage. *J. Microsc. Oxf.* **2017**, *267*, 98–106. [CrossRef] [PubMed]
94. Patterson, B.M.; Henderson, K.C.; Gibbs, P.J.; Imhoff, S.D.; Clarke, A.J. Laboratory micro- and nanoscale X-ray tomographic investigation of Al-7 at%Cu solidification structures. *Mater. Charact.* **2014**, *95*, 18–26. [CrossRef]
95. Wu, X.Z.; Liu, H.; Yan, A. Phase-contrast X-ray tomography: Contrast mechanism and roles of phase retrieval. *Eur. J. Radiol.* **2008**, *68*, S8–S12. [CrossRef] [PubMed]
96. Bronnikov, A.V. Phase-contrast CT: Fundamental theorem and fast image reconstruction algorithms. *Proc. SPIE* **2006**, *6318*, Q3180.
97. McCann, M.T.; Nilchian, M.; Stampanoni, M.; Unser, M. Fast 3D reconstruction method for differential phase contrast X-ray CT. *Opt. Express* **2016**, *24*, 14564–14581. [CrossRef] [PubMed]
98. Salvo, L.; Suery, M.; Marmottant, A.; Limodin, N.; Bernard, D. 3D imaging in material science: Application of X-ray tomography. *C. R. Phys.* **2010**, *11*, 641–649. [CrossRef]
99. Wu, S.C.; Xiao, T.Q.; Withers, P.J. The imaging of failure in structural materials by synchrotron radiation X-ray microtomography. *Eng. Fract. Mech.* **2017**, *182*, 127–156. [CrossRef]
100. Rowenhorst, D.J.; Voorhees, P.W. Measurement of interfacial evolution in three dimensions. *Annu. Rev. Mater. Res.* **2012**, *42*, 105–124. [CrossRef]
101. Zhao, S.; Li, S.J.; Wang, S.G.; Hou, W.T.; Li, Y.; Zhang, L.C.; Hao, Y.L.; Yang, R.; Misra, R.D.K.; Murr, L.E. Compressive and fatigue behavior of functionally graded Ti-6Al-4V meshes fabricated by electron beam melting. *Acta Mater.* **2018**, *150*, 1–15. [CrossRef]
102. Choy, S.Y.; Sun, C.N.; Leong, K.F.; Wei, J. Compressive properties of functionally graded lattice structures manufactured by selective laser melting. *Mater. Des.* **2017**, *131*, 112–120. [CrossRef]
103. Limmahakhun, S.; Oloyede, A.; Sitthiseripratip, K.; Xiao, Y.; Yan, C. Stiffness and strength tailoring of cobalt chromium graded cellular structures for stress-shielding reduction. *Mater. Des.* **2017**, *114*, 633–641. [CrossRef]
104. Li, S.J.; Murr, L.E.; Cheng, X.Y.; Zhang, Z.B.; Hao, Y.L.; Yang, R.; Medina, F.; Wicker, R.B. Compression fatigue behavior of Ti-6Al-4V mesh arrays fabricated by electron beam melting. *Acta Mater.* **2012**, *60*, 793–802. [CrossRef]
105. Sugimura, Y.; Rabiei, A.; Evans, A.G.; Harte, A.M.; Fleck, N.A. Compression fatigue of a cellular Al alloy. *Mater. Sci. Eng. A* **1999**, *269*, 38–48. [CrossRef]
106. Li, S.J.; Zhao, S.; Hou, W.T.; Teng, C.Y.; Hao, Y.L.; Li, Y.; Yang, R.; Misra, R.D.K. Functionally graded Ti-6Al-4V meshes with high strength and energy absorption. *Adv. Eng. Mater.* **2016**, *18*, 34–38. [CrossRef]

107. Qu, R.T.; Wang, S.G.; Wang, X.D.; Liu, Z.Q.; Zhang, Z.F. Revealing the shear band cracking mechanism in metallic glass by X-ray tomography. *Scr. Mater.* **2017**, *133*, 24–28. [CrossRef]
108. Luo, J.; Shi, Y.F. Tensile fracture of metallic glasses via shear band cavitation. *Acta Mater.* **2015**, *82*, 483–490. [CrossRef]
109. Flores, K.M.; Sherer, E.; Bharathula, A.; Chen, H.; Jean, Y.C. Sub-nanometer open volume regions in a bulk metallic glass investigated by positron annihilation. *Acta Mater.* **2007**, *55*, 3403–3411. [CrossRef]
110. Shao, Y.; Yang, G.N.; Yao, K.F.; Liu, X. Direct experimental evidence of nano-voids formation and coalescence within shear bands. *Appl. Phys. Lett.* **2014**, *105*, 181909. [CrossRef]
111. Maass, R.; Birckigt, P.; Borchers, C.; Samwer, K.; Volkert, C.A. Long range stress fields and cavitation along a shear band in a metallic glass: The local origin of fracture. *Acta Mater.* **2015**, *98*, 94–102. [CrossRef]
112. Zhao, Y.Y.; Zhang, G.L.; Estevez, D.A.; Chang, C.T.; Wang, X.M.; Li, R.W. Evolution of shear bands into cracks in metallic glasses. *J. Alloys Compd.* **2015**, *621*, 238–243. [CrossRef]
113. Zhang, Z.F.; He, G.; Eckert, J.; Schultz, L. Fracture mechanisms in bulk metallic glassy materials. *Phys. Rev. Lett.* **2003**, *91*, 045505. [CrossRef] [PubMed]
114. Wang, S.G.; Xu, J. Strengthening and toughening of Mg-based bulk metallic glass via in-situ formed B2-type AgMg phase. *J. Non-Cryst. Solids* **2013**, *379*, 40–47. [CrossRef]
115. Wu, S.J.; Wang, X.D.; Qu, R.T.; Zhu, Z.W.; Liu, Z.Q.; Zhang, H.F.; Zhang, Z.F. Gradual shear band cracking and apparent softening of metallic glass under low temperature compression. *Intermetallics* **2017**, *87*, 45–54. [CrossRef]
116. Anton, D.L.; Giamei, A.F. Porosity distribution and growth during homogenization in single-crystals of a nickel-base superalloy. *Mater. Sci. Eng.* **1985**, *76*, 173–180. [CrossRef]
117. Link, T.; Zabler, S.; Epishin, A.; Haibel, A.; Bansal, A.; Thibault, X. Synchrotron tomography of porosity in single-crystal nickel-base superalloys. *Mater. Sci. Eng. A* **2006**, *425*, 47–54. [CrossRef]
118. Li, X.W.; Wang, L.; Dong, J.S.; Lou, L.H.; Zhang, J. Evolution of micro-pores in a single-crystal nickel-based superalloy during solution heat treatment. *Metall. Mater. Trans. A* **2017**, *48A*, 2682–2686. [CrossRef]
119. Bokstein, B.S.; Epishin, A.I.; Link, T.; Esin, V.A.; Rodin, A.O.; Svedov, I.L. Model for the porosity growth in single-crystal nickel-base superalloys during homogenization. *Scr. Mater.* **2007**, *57*, 801–804. [CrossRef]
120. Yang, C.L.; Yang, H.J.; Zhang, Z.J.; Zhang, Z.F. Recovery of tensile properties of twinning-induced plasticity steel via electropulsing induced void healing. *Scr. Mater.* **2018**, *147*, 88–92. [CrossRef]
121. Zhang, W.; Sui, M.L.; Zhou, Y.Z.; Li, D.X. Evolution of microstructures in materials induced by electropulsing. *Micron* **2003**, *34*, 189–198. [CrossRef]
122. Qin, R.S.; Su, S.X. Thermodynamics of crack healing under electropulsing. *J. Mater. Res.* **2002**, *17*, 2048–2052. [CrossRef]
123. Guo, J.D.; Wang, X.L.; Dai, W.B. Microstructure evolution in metals induced by high density electric current pulses. *Mater. Sci. Technol. Lond.* **2015**, *31*, 1545–1554. [CrossRef]
124. Zhang, S.D.; Zhang, W.L.; Wang, S.G.; Gu, X.J.; Wang, J.Q. Characterisation of three-dimensional porosity in an Fe-based amorphous coating and its correlation with corrosion behaviour. *Corros. Sci.* **2015**, *93*, 211–221. [CrossRef]
125. Suegama, P.H.; Fugivara, C.S.; Benedetti, A.V.; Fernandez, J.; Delgado, J.; Guilemany, J.M. Electrochemical behavior of thermally sprayed stainless steel coatings in 3.4% NaCl solution. *Corros. Sci.* **2005**, *47*, 605–620. [CrossRef]
126. Suegama, P.H.; Fugivara, C.S.; Benedetti, A.V.; Fernandez, J.; Delgado, J.; Guilemany, J.M. Electrochemical behaviour of thermally sprayed Cr3C2-NiCr coatings in 0.5 m H_2SO_4 media. *J. Appl. Electrochem.* **2002**, *32*, 1287–1295. [CrossRef]
127. Gu, X.N.; Li, N.; Zhou, W.R.; Zheng, Y.F.; Zhao, X.; Cai, Q.Z.; Ruan, L.Q. Corrosion resistance and surface biocompatibility of a microarc oxidation coating on a Mg-Ca alloy. *Acta Biomater.* **2011**, *7*, 1880–1889. [CrossRef] [PubMed]
128. Wang, W.; Wan, P.; Liu, C.; Tan, L.; Li, W.; Li, L.; Yang, K. Degradation and biological properties of Ca-P contained micro-arc oxidation self-sealing coating on pure magnesium for bone fixation. *Regen. Biomater.* **2015**, *2*, 107–118. [CrossRef] [PubMed]
129. Han, J.J.; Wan, P.; Sun, Y.; Liu, Z.Y.; Fan, X.M.; Tan, L.L.; Yang, K. Fabrication and evaluation of a bioactive Sr-Ca-P contained micro-arc oxidation coating on magnesium strontium alloy for bone repair application. *J. Mater. Sci. Technol.* **2016**, *32*, 233–244. [CrossRef]

130. Zhou, G.M.; Li, L.; Ma, C.Q.; Wang, S.G.; Shi, Y.; Koratkar, N.; Ren, W.C.; Li, F.; Cheng, H.M. A graphene foam electrode with high sulfur loading for flexible and high energy Li-S batteries. *Nano Energy* **2015**, *11*, 356–365. [CrossRef]
131. Zhou, G.M.; Pei, S.F.; Li, L.; Wang, D.W.; Wang, S.G.; Huang, K.; Yin, L.C.; Li, F.; Cheng, H.M. A graphene-pure-sulfur sandwich structure for ultrafast, long-life lithium-sulfur batteries. *Adv. Mater.* **2014**, *26*, 625–631. [CrossRef] [PubMed]
132. Fang, R.P.; Zhao, S.Y.; Hou, P.X.; Cheng, M.; Wang, S.G.; Cheng, H.M.; Liu, C.; Li, F. 3D interconnected electrode materials with ultrahigh areal sulfur loading for Li-S batteries. *Adv. Mater.* **2016**, *28*, 3374–3382. [CrossRef] [PubMed]
133. Hu, G.J.; Xu, C.; Sun, Z.H.; Wang, S.G.; Cheng, H.M.; Li, F.; Ren, W.C. 3D graphene-foam-reduced-graphene-oxide hybrid nested hierarchical networks for high-performance Li-S batteries. *Adv. Mater.* **2016**, *28*, 1603–1609. [CrossRef] [PubMed]
134. Zhou, G.M.; Li, L.; Wang, D.W.; Shan, X.Y.; Pei, S.F.; Li, F.; Cheng, H.M. A flexible sulfur-graphene-polypropylene separator integrated electrode for advanced Li-S batteries. *Adv. Mater.* **2015**, *27*, 641–647. [CrossRef] [PubMed]
135. Nelson, J.; Misra, S.; Yang, Y.; Jackson, A.; Liu, Y.J.; Wang, H.L.; Dai, H.J.; Andrews, J.C.; Cui, Y.; Toney, M.F. In operando X-ray diffraction and transmission X-ray microscopy of lithium sulfur batteries. *J. Am. Chem. Soc.* **2012**, *134*, 6337–6343. [CrossRef] [PubMed]
136. Zielke, L.; Barchasz, C.; Walus, S.; Alloin, F.; Lepretre, J.C.; Spettl, A.; Schmidt, V.; Hilger, A.; Manke, I.; Banhart, J.; et al. Degradation of Li/S battery electrodes on 3D current collectors studied using X-ray phase contrast tomography. *Sci. Rep. UK* **2015**, *5*, 10921. [CrossRef] [PubMed]
137. McDonald, S.A.; Holzner, C.; Lauridsen, E.M.; Reischig, P.; Merkle, A.P.; Withers, P.J. Microstructural evolution during sintering of copper particles studied by laboratory diffraction contrast tomography (LabDCT). *Sci. Rep. UK* **2017**, *7*, 5251. [CrossRef] [PubMed]
138. Sun, J.; Lyckegaard, A.; Zhang, Y.B.; Catherine, S.A.; Patterson, B.R.; Bachmann, F.; Gueninchault, N.; Bale, H.; Holzner, C.; Lauridsen, E.; et al. 4D study of grain growth in armco iron using laboratory X-ray diffraction contrast tomography. In Proceedings of the 38th Riso International Symposium on Materials Science, San Diego, CA, USA, 4–8 September 2017; IOP Science: San Diego, CA, USA, 2017.
139. King, A.; Reischig, P.; Adrien, J.; Ludwig, W. First laboratory X-ray diffraction contrast tomography for grain mapping of polycrystals. *J. Appl. Crystallogr.* **2013**, *46*, 1734–1740. [CrossRef]
140. Wang, B.; Pan, B.; Lubineau, G. Morphological evolution and internal strain mapping of pomelo peel using X-ray computed tomography and digital volume correlation. *Mater. Des.* **2018**, *137*, 305–315. [CrossRef]
141. Lachambre, J.; Rethore, J.; Weck, A.; Buffiere, J.Y. Extraction of stress intensity factors for 3D small fatigue cracks using digital volume correlation and X-ray tomography. *Int. J. Fatigue* **2015**, *71*, 3–10. [CrossRef]
142. Khoshkhou, D.; Mostafavi, M.; Reinhard, C.; Taylor, M.P.; Rickerby, D.S.; Edmonds, I.M.; Evans, H.E.; Marrow, T.J.; Connolly, B.J. Three-dimensional displacement mapping of diffused Pt thermal barrier coatings via synchrotron X-ray computed tomography and digital volume correlation. *Scr. Mater.* **2016**, *115*, 100–103. [CrossRef]
143. Germaneau, A.; Doumalin, P.; Dupre, J.C. Comparison between X-ray micro-computed tomography and optical scanning tomography for full 3D strain measurement by digital volume correlation. *Ndt&E Int.* **2008**, *41*, 407–415.
144. Leclerc, H.; Roux, S.; Hild, F. Projection savings in CT-based digital volume correlation. *Exp. Mech.* **2015**, *55*, 275–287. [CrossRef]
145. Taillandier-Thomas, T.; Roux, S.; Hild, F. Soft route to 4D tomography. *Phys. Rev. Lett.* **2016**, *117*, 025501. [CrossRef] [PubMed]

© 2018 by the authors. Licensee MDPI, Basel, Switzerland. This article is an open access article distributed under the terms and conditions of the Creative Commons Attribution (CC BY) license (http://creativecommons.org/licenses/by/4.0/).